Systems & Control: Foundations & Applications

Founding Editor

Christopher I. Byrnes, Washington University

The International Institute for Applied Systems Analysis

is an interdisciplinary, nongovernmental research institution founded in 1972 by leading scientific organizations in 12 countries. Situated near Vienna, in the center of Europe, IIASA has been for more than two decades producing valuable scientific research on economic, technological, and environmental issues.

IIASA was one of the first international institutes to systematically study global issues of environment, technology, and development. IIASA's Governing Council states that the Institute's goal is: *to conduct international and interdisciplinary scientific studies to provide timely and relevant information and options, addressing critical issues of global environmental, economic, and social change, for the benefit of the public, the scientific community, and national and international institutions.* Research is organized around three central themes:

- Global Environmental Change;
- Global Economic and Technological Change;
- Systems Methods for the Analysis of Global Issues.

The Institute now has national member organizations in the following countries:

Austria
The Austrian Academy of Sciences

Bulgaria
The National Committee for Applied Systems Analysis and Management

Canada
The Canadian Committee for IIASA

Czech Republic
The Czech Committee for IIASA

Finland
The Finnish Committee for IIASA

Germany
The Association for the Advancement of IIASA

Hungary
The Hungarian Committee for Applied Systems Analysis

Italy
The Italian Committee for IIASA

Japan
The Japan Committee for IIASA

Kazakstan
The National Academy of Sciences

Netherlands
The Netherlands Organization for Scientific Research (NWO)

Poland
The Polish Academy of Sciences

Russian Federation
The Russian Academy of Sciences

Slovak Republic
The Slovak Committee for IIASA

Sweden
The Swedish Council for Planning and Coordination of Research (FRN)

Ukraine
The Ukrainian Academy of Sciences

United States of America
The American Academy of Arts and Sciences

Alexander Kurzhanski
István Vályi

Ellipsoidal Calculus for Estimation and Control

IIASA
International Institute for
Applied Systems Analysis
A-2361 Laxenburg/Austria

Springer Science+Business
Media, LLC

Alexander B. Kurzhanski
Moscow State University
Faculty of Computational
Mathematics & Cybernetics
Moscow 119899
Russia

István Vályi
Hungarian National Bank
MNB
Budapest H-1850
Hungary

Library of Congress Cataloging-in-Publication Data

Kurzhanskii, A. B.
Ellipsoidal calculus for estimation and control / Alexander Kurzhanski and István Vályi.
p. cm. -- (Systems & control)
Includes bibliographical references.
ISBN 978-1-4612-6690-7 ISBN 978-1-4612-0277-6 (eBook)
DOI 10.1007/978-1-4612-0277-6
1. Control theory--Data processing. 2. Elliptic functions.
I. Vályi, István, 1950- . II. Title. III. Series.
QA402.3.K773 1996
629.8'312--dc20 96-5738
CIP

Printed on acid-free paper

Originally published by Birkhäuser Boston in 1997
Softcover reprint of the hardcover 1st edition 1997

ISBN 978-1-4612-6690-7
Typeset by the Authors in LaTeX.

9 8 7 6 5 4 3 2 1

Contents

Preface

It is well known that the emphasis of mathematical modelling on the basis of available observations is first – to *use the data to specify or refine the mathematical model*, then – *to analyze the model* through available or new mathematical tools, and further on – to use this analysis in order *to predict or prescribe (control) the future course of the modelled process.* This is particularly done by specifying *feedback control strategies (policies)* that realize the desired goals. An important component of the overall process is *to verify the model* and its performance over the actual course of events.

The given principles are also among the objectives of modern control theory, whether directed at traditional (aerospace, mechanics, regulation, technology) or relatively new applications (environment, population, finances and economics, biomedical issues, communication, and transport).

Among the specific features of the controlled processes in the mentioned areas are usually their *dynamic nature* and the *uncertainty* in their description.

Following this reasoning, one may claim that control theory is a science of assigning *feedback control or regulation laws* for dynamic processes on the basis of available information on the system model and its performance goals (given both through on-line observations and through data known in advance). It is on how to construct, in some appropriate sense, *the best or the better* control laws. It is also used to indicate how the level of uncertainty and the amount of information used for designing the feedback control laws affects the result of the controlled process, particularly, the values of the cost functions or the aspired guaranteed performance levels.

However, it is not any type of theory that is desired. Not the least objective is to develop, among the possible approaches, a solution theory that allows analytical designs that are relatively simple for practical implementations or lead, at least, to effective numerical algorithms. These, desirably, should match the abilities of modern computer technology, e.g., allow parallel calculations and graphic animation.

The present book is devoted to an array of selected key problems in dynamic modelling, state estimation, viability, and feedback control under uncertainty. Its aim is to present a unified framework for effectively solving these problems and their generalizations *to the end* through modern computer tools.

The model of uncertainty considered here is deterministic, with set-membership description of the uncertain items. These are taken to be *unknown but bounded* with preassigned bounds and no statistical information whatever. The set-membership model of uncertainty reflects many actual information situations in applied problems. Particularly, it appears relevant in estimating nonrepetitive processes, processes with limited numbers of observations, incomplete knowledge of the problem data and no available statistics. It is a common approach in pursuit-evasion differential games, in robust stabilization, control and disturbance attenuation, particularly, under unmodelled dynamics. Needless to say, it also reflects the research preferences, interests and experiences of the authors.[1]

The problems treated here are described through *set-valued functions* and are thus to be treated through *set-valued analysis*. However, the aim of this book is not to produce any type of set-valued technique, but a calculus that allows effective solutions of the selected problems and their generalizations with fairly simple control designs and the possibility of graphic animation. The attempt is based on introducing an *"ellipsoidal*

[1]The references and some historical comments are given in the introductions to each part and throughout the text. The authors apologize that among the enormous literature on the subject they were able to mention only a very limited, representative, rather than exhaustive number of publications available to them, with an emphasis on those directly related to the topics of this publication and those that would allow, as we hope, to pursue the further directions indicated here.

calculus" that allows us to represent the exact set-valued solutions of the respective problems through ellipsoidal-valued functions. The solutions are thus constructed of elements that involve only ellipsoidal sets or ellipsoidal-valued functions and operations over such sets or functions. This further allows us to parallelize the calculations and to animate the solutions through computer graphic tools.

It is necessary to indicate that the ellipsoidal techniques of this particular book are not confined to approximation of convex sets by *one or several ellipsoids only* as done in other publications, but indicate *ellipsoidal representations of the exact solutions.* Namely, each convex set or convex set-valued function considered here is represented by a parametrized variety of ellipsoids or ellipsoidal-valued functions, which, while their number increases, jointly allow (through their sums, unions, or intersections), a more and more accurate representation with exact one in the limit. The scheme includes approximations by single ellipsoids as a particular element of the overall approach.

A particular emphasis of this book is on the possibility of computer-graphic representations. The animation of the problems in estimation, feedback control, and game-type dynamics not only allows us to present the rather sophisticated mathematical solutions in visible forms (and literally, to peer into the multidimensional spaces through computer windows or more sophisticated computer tools). The authors believe that it may also give new insights into the mathematical structure of the solutions. (Thus, some assertions of a general nature proved in this book were first noticed during the animation experiments.) The authors also hope that though applying their techniques to a specially selected array of problems, they also demonstrate an approach applicable to many other situations that spread quite beyond the topics addressed here. These are certainly not confined only to control applications, but cover a broad variety of problems in systems modelling.

The book is divided into four parts designed along the following lines. *The first part* gives exact solutions to the problems of evolution – attainability (reachability) and solvability, as well as of estimation, viability, and feedback terminal target control. The exact theory (both known and new) is rewritten within a unified framework that involves trajectory tubes and their set-valued descriptions either through evolution equations of the "funnel type", or through the evolution of support functions or through level sets of appropriate H-J-B (Hamilton-Jacobi-Bellman) equations. The feedback control designs are based on using set-valued solvability tubes which may be interpreted as "bridges" introduced by N.N. Krasovski as a basis for further "aiming rules", as well as on Dynamic Programming considerations. The principal schemes for this framework are specifically directed towards the desired transition, through ellipsoidal-valued representations, to parallelizable computation schemes.

The second part of the book describes the ellipsoidal calculus itself. It covers external and internal ellipsoidal representations for basic set-valued operations-geometrical ("Minkowski") sums and differences as well as intersections of ellipsoids and integrals of ellipsoidal-valued functions. Though written for applications to Problems of Part I, the text of this chapter may be also considered as a separate theory with motivations and applications coming from topics other than discussed here. Particularly, from optimization under uncertainty and multiobjective optimization, experiment planning, problems in probability and statistics, interval analysis and its generalizations, adaptive systems and robotics, image processing, mathematical morphology, and related areas of theoretical and applied research.

The third and fourth parts indicate the applications of ellipsoidal calculus to problems of Part I. Thus, *the third part* describes (both in forward and backward time) the internal and external ellipsoidal representations of attainability (reachability) tubes for systems without and with uncertainty. In the latter case these are more complicated, of course, being related to reachability or solvability *under uncertainty or counteraction* and allowing, particularly, a direct interpretation in terms of the above mentioned bridges – the key elements of game-theoretic feedback control. The third part also deals with feedback control. The respective control designs are based on applying ellipsoidal versions of the exact solutions. This leads to a nonlinear control synthesis in the form of *analytical designs*, except for a scalar parameter whose dependence on the state space vector may be calculated in advance, through the solution of a simple algebraic equation. These analytical designs are possible due to the fact that the internal ellipsoidal tubes that approximate the solvability domains under uncertainty are precisely such, that they possess the property of being an "ellipsoidal-valued bridge". The latter property arrives due to two basic features : the fact that the respective ellipsoidal-valued mappings possess a *semigroup property* and the fact that the internal tubes are *inclusion-maximal* among all other internal ellipsoidal tubes.

The fourth part deals with state-estimation under unknown but bounded errors, with attainability under state constraints and viability problems. It also indicates the applicability of the suggested schemes to problems posed within the so-called H_∞ approaches, when the value of the error bound is not specified. This is due to the involvement of Dynamic Programming techniques, particularly, of one and the same H-J-B equation for the treatment of uncertain dynamics in both of the settings investigated here. Other topics include new types of dynamic relations for the treatment of information sets and vector-valued guaranteed estimators as well as an interpretation of the state estimation problem as one of *tracking an unknown motion under unspecified but bounded errors*. This links the problem with those of viability under counteraction. The final Section 4.6 deals with problems of state estimation under "bad noise", which are approached through the incorporation of *singular perturbation techniques*. This approach allows us to treat discontinuous observations described by measurable functions and also to deal with viability problems under measurable constraints. Numerical examples complement the theoretical parts.

The narrative stops just short of control under measurement feedback and adaptive control. These are areas which require separate serious consideration and explanation. However, the application of ellipsoidal techniques would be especially useful in these areas, as we believe. The respective challenges are beyond the material of the present book.

As already mentioned, this book indicates a unified concise framework for problems of state estimation, viability and feedback control under set-membership uncertainty for systems with linear dynamics. It introduces an ellipsoidal calculus to develop the solutions *from theory to algorithms and computer animation*, and thus to solve the problem to the end.

This book is not a collection of numerous facts or artifacts in set-valued analysis or control theory. It is rather a book on *basic problems and principles for calculating their solutions* through set-valued models. Whether reached or not, our aim is also to stimulate and encourage further investigation in the spirit of the present approach as well as implementations in real-life modelling. (The latter issue could be the topic of a separate monograph.)

In this text we are confined to linear-convex systems and problems. However, the control synthesis given here is nonlinear and the synthesized systems are nonlinear systems. Moreover, the Dynamic Programming approaches applied here open the routes to further penetration into generically nonlinear classes of systems. The algorithmization and animation in these cases is certainly a worthy challenge.[2]

Another important aspect hardly discussed here is the accuracy and *computational complexity* of the underlying algorithms.

The conceptual approaches to controlled dynamics that served as a background for this work were influenced by the research of N.N. Krasovski and his associates at Yekaterinburg (Sverdlovsk), where the first of the authors had earlier worked for many years. In the necessity of studying and applying set-valued analysis we share the views of J-P. Aubin and his colleagues.[3]

The principal parts of this book and the underlying ellipsoidal representation approach for the set-valued functions were worked out throughout the authors participation at the SDS (Systems and Decision Sciences) Program of IIASA – the International Institute of Applies Systems Analysis at Laxenburg, Austria. The serious but friendly atmosphere, pleasant working conditions, and possibility of regular contacts with a broad spectrum of researchers certainly stimulated our work at the Institute and the direction of our efforts. The authors are grateful to the Directors of IIASA – Thomas Lee, Robert Pry, and Peter De Jánosi and to the Chairman of the IIASA Council in 1987–1992, the late Vladimir Sergeevich Michalevich, for their support of methodological research at IIASA, particularly of our own investigations.

We wish to thank our colleagues at IIASA and its Advisory Committee on Methodology – J.P. Aubin, H. Frankowska, A. Gaivoronski,

[2]The nondifferentiable version of Dynamic Programming has been substantially developed in the recent years (see references [82], [83], [290]), becoming an effective tool in nonlinear control theory, particularly. For covering the needs of this book we use its simple versions that do not extend beyond the use of subdifferential calculus.

[3]A set-valued approach to state estimation and uncertain dynamics was emphasized in [181].

P. Kenderov, G. Pflug, R.T. Rockafellar, W. Runggaldier, K. Sigmund, M. Thoma, V. Veliov, R. Wets, A. Wierzbicki for their stimulating discussions and support, K. Fedra and M. Makowski for their help in computer graphics, the SDS secretaries E. Gruber and C. Enzlberger-Vaughan for preparing papers and manuscripts used in this book. We thank T. Filippova, O. Nikonov, M. Tanaka, K. Sugimoto who have coauthored some of our IIASA Working Papers used here and also E.K. Kostousova and O.A. Schepunova for their help in arranging the final version of the manuscript.

Throughout the last years we had the pleasant opportunity to discuss the topics treated in this book with Z. Artstein, J. Baras, F. Chernousko, M. Gusev, A. Isidori, P. Kall, R. Kalman, H. Knobloch, A. Krener, G. Leitmann, C. Martin, M. Milanese, E. Mischenko, S. Mitter, J. Norton, Yu. Osipov, B. Pschenichnyi, S. Robinson, A. Rusczynski, P. Saint-Pierre, A. Subbotin, T. Tanino, V. Tikhomirov, P. Varaiya, S. Veres, and J. Willems. Their valuable comments certainly helped to shape the contents.

Our special thanks go to C. Byrnes – Editor of the Birkhäuser Series on Systems and Control: Foundations and Applications and to the Birkhäuser staff for their support, patience, and understanding of the problems faced by the authors in preparing the manuscript.

Part I. EVOLUTION and CONTROL: The EXACT THEORY

Introduction

The present first part of the book is a narrative on constructive techniques for modelling and analyzing an array of key problems in uncertain dynamics, estimation, and control. It presents a unified approach to these topics based on descriptions involving the notions of *trajectory tubes* and *evolution equations* for these tubes. The class of systems treated here are *linear time-variant systems*

$$\dot{x} = A(t)x + u + f(t), \quad x(t_0) = x_0 \ ,$$

with *magnitude bounds* on the controls u and the uncertain items $f(t), x_0$. The target control processes and the estimation problems are considered within *finite time intervals*: $t \in [t_0, t_1]$. This requires a rather detailed investigation of the system dynamics. The present topic thus differs from problems in which the objective lies only in the achievement of an appropriate asymptotic behaviour of the trajectories with perhaps some desired quality of the transient process.

The first step is the description of *attainability (reachability) domains* $\mathcal{X}[t]$ for systems without uncertainty. The evolution of these in time is naturally described by set-valued ("Aumann") integrals with variable upper time limit. This immediately leads to set-valued "attainability tubes" whose crossections are the attainability domains. However, it is not unimportant to introduce some sort of *evolution equation with set-valued state-space variable* that would describe the dynamics of sets $\mathcal{X}[t]$ in time. The tubes $\mathcal{X}[t]$ could then be interpreted as trajectories of some generalized dynamic systems. (Among the first investigations with this emphasis are the works of E. Barbashin and E. Roxin [34], [270].) The serious obstacle for deriving such an equation in a differential form is the difficulty in defining an appropriate derivative for set-valued functions. The objective is nevertheless reached through *evolution equations of the funnel type* that do not involve such derivatives. Though somewhat cumbersome at first glance, these equations indicate set-valued discrete-time schemes important for calculations.[4]

[4] Among the recent investigations on evolution equations for trajectory tubes are papers [246], [299], [193], [17].

The attainability tubes may be also constructed in backward time in which case they are referred to as *solvability tubes.* The solvability tubes are used here in synthesizing *feedback control strategies* for problems of *terminal target control.* Namely, if the solvability tube ends at the target set $\mathcal{M}$, then the synthesized control strategy should be designed to keep the trajectory within this tube (or bridge) throughout the process. This idea is the essence of the "extremal aiming rule" introduced by N.N. Krasovski [168], [169] and used in Section 1.4. A key element that allows us to use the solvability tubes for the control synthesis problem is that the respective multivalued maps satisfy a *semigroup property* and therefore generate a generalized dynamic system with set-valued trajectories.

A similar type of strategy may be derived through a Dynamic Programming technique with cost function being the square of the Euclid distance $d^2(x[t_1], \mathcal{M})$ from endpoint $x[t_1]$ to the target set $\mathcal{M}$, for example (see Section 1.5). Selecting a starting position $\{t, x\}, x = x(t)$, we may minimize the cost function by selecting an appropriate optimal control (in the class of either open-loop or closed-loop controls). Finding the optimal value of the cost function for any position $\{t, x\}$ we come to *the value function* $V(t, x)$. One should note that *in the absence of uncertainty the value function* $V(t, x)$ *is the same both for open-loop (programmed) control and for closed-loop (positional) feedback control.* For the linear-convex problems of this Section 1.5 the function $V(t, x)$ may be therefore calculated through standard methods of convex analysis used traditionally for solving related problems of open-loop control [167], [266], [181]. The function $V(t, x)$ then satisfies a corresponding generalized H-J-B (Hamilton-Jacobi-Bellman) equation.

The next stage is the treatment of systems with input uncertainty $f(t)$ (unknown but bounded, with magnitude bounds). In this case the *attainability set under counteraction* (in forward time) and *solvability set under uncertainty* (in backward time) are in general far more complicated than in the absence of uncertainty (see Sections 1.6–1.8). *One should now distinguish,* for example, *the open-loop solvability tubes from the closed-loop solvability tubes* (Section 1.6). Under some nondegeneracy conditions, the latter ones may be again interpreted as Krasovski's bridges (now for uncertain systems) and may be used for designing feedback strategies through the extremal aiming rules (Sections 1.7, 1.8). The backward procedure for solvability tubes is also similar in nature to the schemes introduced by P. Varaiya et al [308] and B. Pschenichnyi [259]. In the linear-convex case considered here the constructive description of solvability tubes may be given by a set-valued integral known as L.S. Pontryagin's *alternated integral* [257] (Section 1.7). It is indicated here that they also satisfy some special *evolution equations of the funnel type.* There is a particular case, however, when the open-loop and closed-loop solvability tubes *coincide.* This is when the system satisfies the so-called *matching condition* which means that the bounds on the controls u and the disturbances f are similar in some sense (Section 1.6). The calculation of the solvability tubes is then as simple as in the absence of uncertainty.

One may also apply Dynamic Programming to the mentioned uncertain systems. Taking the cost function $d^2(x(t_1), \mathcal{M})$, for example, we note that now it should be *minimaximized* over the control u and the input disturbance f respectively. But the value of this minmax, when calculated over *closed loop* controls is *different*, in general, from its value calculated for *open loop controls.* It is the former value that may be described through a respective H-J-B-I (Hamilton-Jacobi-Bellman-Isaacs) equation (see [109], [171], [219], [290]). There is an exception again, nevertheless. Namely, *if the matching conditions are satisfied*, then the minmax of the cost function or, in other words, *the value function* $V(t,x)$ *is the same, whether calculated over open-loop or closed-loop controls.*

Having in mind the previous remarks, one may observe that the value functions $V(t,x)$ used in this book play the role of Liapunov functions used in respective approaches to the design of feedback controllers for uncertain systems (see [214], [215]). We should also emphasize that the problem treated here is to reach the goal in finite time attenuating the unknown disturbances. Namely, it is to ensure that the system is steered by control u to the target set $\mathcal{M}$ (at given time t_1) under persistent disturbances f rather than to figure out the saddle points of a positional (feedback) dynamic game between two equal players u and f which is the emphasis of the theory of differential games (see [37], [171], [50], [119]).

The further problems are similar to the previous ones but complicated by state constraints (viability restrictions). Evolution funnel equations are introduced for the dynamics of attainability sets under state constraints (in forward time) and respective solvability sets (in backward time). The latter ones are similar to *viability kernels* introduced by J.P. Aubin [15], within the framework of *viability theory* (Section 1.9). The control synthesis problem is now to ensure *viability* (Section 1.10) or *viability under counteraction* (or under persistent disturbances, in another interpretation), while also reaching the terminal set (Section 1.11).

The last problem of the first part is the one of *state estimation* under unknown but bounded errors and disturbances (Section 1.12).[5]

The main objects of investigation here are *the information sets* consistent with the system dynamics, the available measurement and the constraints on the uncertain items. The information sets are actually the attainability domains under a state constraint that is induced by the measurement equation and therefore arrives *on-line*, together with the result of the measurement. The evolution equation for the information set acts as a guaranteed filtering equation and the guaranteed state estimate is then the "Chebyshev center" of this set (namely, the center of the smallest ball that includes the information set). This first part of the book gives but a general introduction to the problem, while constructive techniques are introduced in Parts III and IV, where one may also find some connections with other approaches to deterministic filtering (particularly, the H_∞ approach [94], in the interpretation of J. Baras and M. James [30]).

A synopsis of the results and some suggestions on why ellipsoids were undertaken to be studied finish this part.

We now proceed with the main text, commencing with the basic notions.

1.1 The System

In this book we consider dynamic models described in general by a *linear time-variant system*

$$\dot{x}(t) = A(t)x(t) + u + f(t) \qquad (1.1.1)$$

[5]The first investigations of state estimation problems under unknown but bounded inputs date to papers [166], [318], [178]. A systematic investigation of the set-valued approach in continuous time seems to have started with [54], [277], [179], [181].

with finite-dimensional state space vector $x(t) \in \mathbb{R}^n$ and inputs u (*the control*) and $f(t)$ (*the disturbance* or external forcing term). The $n \times n$ matrix function $A(t)$ is taken to be continuous on a preassigned interval of time $T = \{t \in \mathbb{R} : t_0 \leq t \leq t_1\}$ within which we consider the forthcoming problems, with $u(t), f(t)$ assumed Lebesgue-measurable in $t \in T$.

The values u of the controls are assumed to be restricted for almost all t by a *magnitude or geometrical constraint*

$$u \in \mathcal{P}(t) \ , \tag{1.1.2}$$

where $\mathcal{P}(t)$ is a multivalued function $\mathcal{P} : T \to \text{conv}\mathbb{R}^n$, continuous in t.

Here and further on symbol $\text{conv}\mathbb{R}^n$ stands for the variety of *closed convex sets* in finite-dimensional space $\mathbb{R}^n$ while $\text{comp}\mathbb{R}^n$ stands for the variety of *convex compact sets* in $\mathbb{R}^n$.

We shall further consider two types of controls which are:

- *open loop*, when $u = u(t)$ is a function of time t, measurable on T (a measurable control), and
- *closed loop*, when $u = \mathcal{U}(t, x(t))$ is a *multivalued map*, namely,

$$\mathcal{U} : T \times \mathbb{R}^n \longrightarrow \text{conv}\mathbb{R}^n$$

  measurable in t and *upper semicontinuous in* x, being a function of the *position* $\{t, x\}$ of the system.

(The definition of upper semicontinuity is standard, it may be found in [176], [68], [20], [22] and other related publications.)

In the first case we come to a *linear differential equation*

$$\dot{x}(t) = A(t)x(t) + u(t) + f(t) \tag{1.1.3}$$

with $u(t) \in \mathcal{P}(t), t \in T$, being an *open-loop control.* The class of functions $u(\cdot) = u(t), t \in T$, measurable in $t \in T$ and restricted as in (1.1.2) is further denoted as $U^O_{\mathcal{P}}$.

In the second case we come to a *nonlinear differential inclusion*

$$\dot{x}(t) \in A(t)x(t) + \mathcal{U}(t, x(t)) + f(t) \quad , \tag{1.1.4}$$

where

$$\mathcal{U}(t, x) \subseteq \mathcal{P}(t), \quad t \in T \quad , \tag{1.1.5}$$

is a *feedback (closed-loop) control strategy.*

The class $U_{\mathcal{P}}^{c} = \{\mathcal{U}(t, x)\}$ of *feasible control strategies* consists of all convex compact-valued multifunctions that are measurable in t, upper semicontinuous in x, being restricted by (1.1.5) and such that *equation (1.1.4) does have a solution* extensible to any finite time interval T for any $x^0 = x(t_0) \in \mathbb{R}^n$. The latter means that there exists an absolutely continuous function $x(t), t \in T$, that yields the inclusion

$$\dot{x}(t) \in A(t)x(t) + \mathcal{U}(t, x(t)) + f(t)$$

for almost all $t \in T$.

The existence of solution for system (1.1.3) is a standard property of linear differential equations [61], [142], [167], [248].

Systems (1.1.3), (1.1.4) may be transformed into simpler relations.

Let $S(t, \tau)$ stand for the matrix solution to the equation

$$\frac{\partial}{\partial t} S(t, \tau) = -S(t, \tau)A(t), \quad S(\tau, \tau) = I \quad , \tag{1.1.6}$$

which also satisfies the equation

$$\frac{\partial}{\partial \tau} S(t, \tau) = A(\tau)S(t, \tau), \quad S(t, t) = I \quad .$$

As it is well known, the solution to (1.1.3) with initial value

$$x(t_0) = x^0 \tag{1.1.7}$$

is given by the formula

$$x(t) = S(t_0, t)x^0 + \int_{t_0}^{t} S(\tau, t)(u(\tau) + f(\tau))d\tau \quad .$$

Taking the transformation

$$z(t) = S(t, t_1)x(t) \tag{1.1.8}$$

and substituting x for z in (1.1.3) we come to the equation

$$\dot{z}(t) = S(t, t_1)u(t) + S(t, t_1)f(t) \tag{1.1.9}$$

$$z^0 = z(t_0) = S(t_0, t_1)x^0 \ . \tag{1.1.10}$$

Clearly, there is a one-to-one correspondence of type (1.1.6) between the solutions $x(t)$ and $z(t)$ to equations (1.1.3) and (1.1.9), respectively. The initial values for these are related through (1.1.10). Therefore, instead of the systems (1.1.3), (1.1.4), constraint (1.1.2) and initial condition (1.1.7), we come to

$$\dot{z}(t) = w(t) + g(t), \quad z(t_0) = S(t_0, t_1)x^0 \ , \tag{1.1.11}$$

$$\dot{z}(t) \in \mathcal{W}(t, z) + g(t) \tag{1.1.12}$$

with constraint

$$w(t) \in \mathcal{P}_0(t) \ . \tag{1.1.13}$$

Here obviously

$$\begin{aligned} w(t) &= S(t, t_1)u(t) \\ g(t) &= S(t, t_1)f(t) \\ \mathcal{W}(t, z) &= S(t, t_1)\mathcal{U}(t, S^{-1}(t_0, t)z) \\ \mathcal{P}_0(t) &= S(t, t_1)\mathcal{P}(t) \end{aligned}$$

and the set-valued function $\mathcal{P}_0(t)$ remains continuous. The new feedback strategies $\mathcal{W}(t, z)$ belong to the class defined by the constraint $\mathcal{P}_0(t)$, but otherwise the same as before:

$$\mathcal{W}(t, z) \in U^{c}_{\mathcal{P}_0} \ .$$

Without loss of generality we may therefore further treat systems (1.1.11)–(1.1.13) rather than (1.1.2)–(1.1.4). It is compulsory however that the constraint function $\mathcal{P}_0(t)$ would be time-variant. In other terms, *without loss of generality* we may further follow the notations of (1.1.2)–(1.1.4) with $A(t) \equiv 0$.

One should realize, however, that the described substitution (1.1.8) allows us to consider the forthcoming problems for $A(t) \equiv 0$ within the time range $\{t \leq t_1\}$. A similar result may be also obtained by substitution

(1.1.14) $$z(t) = S(t, t_0)\, x(t) \ .$$

Then the original system may be again, without loss of generality, taken with $A(t) \equiv 0$, but the time range for which the respective substitution is true will be $\{t \geq t_0\}$.

We shall often make use of the indicated facts in the sequel in the hope that this will enable us to demonstrate the basic techniques without overloading the text with unessential but cumbersome procedures. The reader will always be able to return to $A(t) \not\equiv 0$ as a healthy exercise.

The first issue to discuss is the description of the set of states that can be reached in finite time due to systems (1.1.3), (1.1.4) under restriction (1.1.2) and (1.1.5).

1.2 Attainability and the Solution Tubes

Taking system (1.1.3), (1.1.2) for $A(t) \equiv 0$, we have

(1.2.1) $$\dot{x}(t) = u(t) + f(t), \quad t \in T,$$

with constraint (1.1.2) $u(t) \in \mathcal{P}(t)$. We also presume that the initial state $x^0 = x(t_0)$ is restricted by the inclusion

(1.2.2) $$x^0 \in \mathcal{X}^0, \quad \mathcal{X}^0 \in \text{comp}\, \mathbb{R}^n.$$

One of the first questions that arise in control theory is to describe the variety of all states $x = x(t)$ that can be reached by the system trajectories that start at a prescribed set $\mathcal{X}^0$.

Let $x[t] = x(t, t_0, x^0)$ denote an *isolated trajectory* of system (1.2.1) that starts at instant t_0 from state x_0, being driven by a certain control $u(t)$. We will be further interested in the *union* of all such isolated trajectories over all possible initial states $x^0 \in X^0$ and measurable controls $u(t) \in \mathcal{P}(t)$. Therefore we denote

$$\mathcal{X}[t] = \mathcal{X}(t, t_0, \mathcal{X}^0) = \bigcup \{x(t, t_0, x^0) : x^0 \in \mathcal{X}^0, u(t) \in \mathcal{P}(t), t \in T\}.$$

For the mapping $\mathcal{X}(t, t_0, \cdot) : \text{comp}\mathbb{R}^n \rightarrow \text{comp}\mathbb{R}^n$ it is not difficult to check that it satisfies the following *semigroup property*:

$$\mathcal{X}(t, t_0, \mathcal{X}^0) = \mathcal{X}(t, \tau, \mathcal{X}(\tau, t_0, \mathcal{X}^0)),$$

whatever are the values t, τ with $t_1 \geq t \geq \tau \geq t_0$.

Definition 1.2.1 *The set* $\mathcal{X}[t] = \mathcal{X}(t, t_0, \mathcal{X}^0)$ *is referred to as the* **attainability domain** *for system (1.1.3) or (1.2.1), (1.1.2.)* **at time** t, **from set** $\mathcal{X}^0$.

The attainability domain $\mathcal{X}[t]$ is often said to be *the reachability domain.*

The set-valued map

$$\mathcal{X}[t] = \mathcal{X}(t, t_0, \mathcal{X}^0) \ , \ t \in T,$$

defines a *solution tube* to the differential inclusion

$$\dot{x}(t) \in \mathcal{P}(t) + f(t), \tag{1.2.3}$$

that starts from set $\mathcal{X}^0$. In other words, the set $\mathcal{X}[t] = \{x^*\}$ consists of all those vectors x^* for each of which there exists an isolated trajectory $x[\tau] = x(\tau, t_0, x^0), t_0 \leq \tau \leq t$, of (1.2.3) that satisfies the boundary conditions $x[t_0] \in \mathcal{X}^0, x[t] = x^*$.

It is clear that the control

$$u(t) = \dot{x}[t] - f(t)$$

is the one that corresponds to $x[t]$, so that we could also indicate

$$\mathcal{X}[t] = \{x[t] : \ \dot{x}[\tau] - f(\tau) \in \mathcal{P}(\tau), \ t_0 \leq \tau \leq t, \ x[t_0] \in \mathcal{X}^0\}. \tag{1.2.4}$$

The multivalued function $\mathcal{X}[t]$, $t \in T$, $\mathcal{X}[t_0] = \mathcal{X}^0$ is also known as the *solution tube* to system (1.2.1), under restriction (1.1.2), from set $\mathcal{X}^0$, for the interval $t \in [t_0, t_1] = T$.[6]

As a preliminary exercise it is not difficult to prove the following

[6] Other terminology says that $\mathcal{X}[t]$ is the *trajectory assembly* generated by system (1.2.3) and set $\mathcal{X}[t_0] = \mathcal{X}^0$ [181].

Lemma 1.2.1 *The multifunction $\mathcal{X}[t]$ is convex compact-valued ($\mathcal{X}[t] \in$ conv$\mathbb{R}^n$) and continuous on the interval T.*

Remark 1.2.1 *One of the popular problems studied on the subject of attainability is the following: given $\mathcal{X}^0 = \{0\}$, will the set*

$$\mathcal{X} = \cup\{\mathcal{X}[t], t \in [t_0, \infty)\},$$

coincide with the whole space $\mathbb{R}^n$?

An affirmative answer will indicate that any point in $\mathbb{R}^n$ may be reached in finite time through a bounded control $u(\cdot) \in U_{\mathcal{P}}^{O}$. Otherwise one is to specify $\mathcal{X}$ as a subset *of $\mathbb{R}^n$.*

Exercise 1.2.1. Investigate the problem of Remark 1.2.1.

Passing to the differential inclusion

$$\dot{x}(t) \in \mathcal{U}(t, x) + f(t), \tag{1.2.5}$$

$$\mathcal{U}(\cdot, \cdot) \in U_{\mathcal{P}}^{c}, \tag{1.2.6}$$

where the class of feasible feedback strategies $U_{\mathcal{P}}^{c}$ is as defined in Section 1.1, we come to the following questions.

Let $\mathcal{X}_{\mathcal{U}}[t] = \mathcal{X}_{\mathcal{U}}(t, t_0, x^0)$ be the crossection of the set of all isolated trajectories $x[t]$ that satisfy the relation

$$\dot{x}[t] \in \mathcal{U}(t, x[t]) + f(t), \quad x[t_0] = x^0,$$

for a given multivalued map $\mathcal{U}(\cdot, \cdot) \in U_{\mathcal{P}}^{c}$. A particular element of $U_{\mathcal{P}}^{c}$ is the set-valued map $\mathcal{P}(t)$ itself, so that (1.2.3) could be viewed as a particular case of equation (1.2.5), when $\mathcal{U}(t, x) \equiv \mathcal{P}(t)$.

Denote for a fixed $\mathcal{U}$ of (1.2.6):

$$\mathcal{X}_{\mathcal{U}}[t] = \mathcal{X}_{\mathcal{U}}(t, t_0, \mathcal{X}^0) = \bigcup\{\mathcal{X}_{\mathcal{U}}(t, t_0, x^0) : x^0 \in \mathcal{X}^0\},$$

and further

$$\mathcal{X}^*[t] = \mathcal{X}^*(t, t_0, \mathcal{X}^0) = \bigcup\{\mathcal{X}_{\mathcal{U}}(t, t_0, \mathcal{X}^0) : \mathcal{U}(\cdot, \cdot) \in U_{\mathcal{P}}^{c}\}.$$

Then one may want to know what is the relation between the tubes $\mathcal{X}^*[t]$ obtained for the closed-loop system (1.2.5), (1.2.6) and $\mathcal{X}[t]$ obtained for the open-loop system (1.2.3)?

Theorem 1.2.1 *With $\mathcal{X}^*[t_0] = \mathcal{X}[t_0] = \mathcal{X}^0$ the following relation is true:*

$$\mathcal{X}[t] \equiv \mathcal{X}^*[t], \quad t \in T.$$

To prove this assertion we observe that every single-valued function $u(t)$ can be treated as an element of $U_{\mathcal{P}}^{c}$. Therefore $\mathcal{X}[t] \subseteq \mathcal{X}^*[t], t \in T$. To show the opposite assume there exists a trajectory $x^*[t] = x^*(t, t_0, x^0), x^0 \in \mathcal{X}$, which satisfies the inclusion $x^*[t] \in \mathcal{X}^*[t], t \in T$, namely,

$$\dot{x}^*[\tau] \in \mathcal{U}^*(\tau, x^*[\tau]) + f(\tau),\ \tau \leq t,.$$

for some $\mathcal{U}(\cdot,\cdot) = \mathcal{U}^*(\cdot,\cdot) \in U_{\mathcal{P}}^{c}$. Then obviously

$$\dot{x}^*[\tau] \in \mathcal{U}^*(\tau, x^*[\tau]) + f(\tau) \subseteq P(\tau) + f(\tau),\ \tau \leq t,$$

and due to (1.2.4) this yields $x^*[t] \in \mathcal{X}[t], t \in T$.

The main conclusion given by Theorem 1.2.1 is such that with function $f(t)$ given (there is *no uncertainty* in system (1.1.1), (1.1.2)), the solvability tube $\mathcal{X}[t]$ for system (1.1.1), (1.1.2) taken in the class $U_{\mathcal{P}}^{0}$ of *open-loop controls* $u[t]$ is the same as the solvability tube $\mathcal{X}^*[t]$ taken in the class $U_{\mathcal{P}}^{c}$ of *closed-loop controls* $\mathcal{U}(t,x)$. This conclusion is also true when the closed loop controls are selected among appropriate classes of single-valued functions $u = u(t,x) \in \mathcal{P}(t)$ that allow the existence and prolongation of solutions of (1.1.1) with $u = u(t,x), t \in T$.

The next question is whether it would be possible to describe the evolution of sets $\mathcal{X}[t]$ in time t through some type of *evolution equation with set-valued states* $\mathcal{X} = \mathcal{X}[t]$.

1.3 The Evolution Equation

We shall now introduce an *evolution equation* with state space variable $\mathcal{X} \in \mathrm{conv}\mathbb{R}^n$, whose solution will be precisely the tube $\mathcal{X}[t]$ of Section 1.1.2.

Obviously

$$\mathcal{X}[t] = \mathcal{X}^0 + \int_{t_0}^{t} \mathcal{P}(\tau)d\tau + \int_{t_0}^{t} f(\tau)d\tau \ , \tag{1.3.1}$$

where the second term in the right hand is the set-valued Lebesgue integral ("the Aumann integral" [25]) for the function $\mathcal{P}$. The question is therefore whether one could construct an evolution equation for describing $\mathcal{X}[t]$.

Denote $\mathcal{S} = \{x : (x,x) \leq 1\}$ to be *the unit ball* in $\mathbb{R}^n$.

Definition 1.3.1 *The Hausdorff semidistance $h_+(\mathcal{X},\mathcal{Y})$ between sets $\mathcal{X},\mathcal{Y} \in conv\ \mathbb{R}^n$ is introduced as*

$$h_+(\mathcal{X},\mathcal{Y}) = \min\{\gamma \geq 0 : \mathcal{X} \subseteq \mathcal{Y} + \gamma\mathcal{S}\}$$

or equivalently

$$h_+(\mathcal{X},\mathcal{Y}) = \max_x \min_y \{(x-y, x-y)^{\frac{1}{2}} | x \in \mathcal{X}, y \in \mathcal{Y}\} \ .$$

Similarly

$$h_-(\mathcal{X},\mathcal{Y}) = h_+(\mathcal{Y},\mathcal{X}) \ .$$

The following properties are true for $\mathcal{X},\mathcal{Y},\mathcal{Z} \in \mathrm{conv}\mathbb{R}^n$:

(i) $h_+(\mathcal{X},\mathcal{Y}) = 0$ implies $\mathcal{X} \subseteq \mathcal{Y}$
(and $h_-(\mathcal{X},\mathcal{Y}) = 0$ implies $\mathcal{Y} \subseteq \mathcal{X}$).

(ii) $h_+(\mathcal{X},\mathcal{Z}) + h_+(\mathcal{Z},\mathcal{Y}) \geq h_+(\mathcal{X},\mathcal{Y})$.

Definition 1.3.2 *The Hausdorff distance $h(\mathcal{X},\mathcal{Y})$ between sets $\mathcal{X},\mathcal{Y} \in$ conv$\mathbb{R}^n$ is introduced as*

$$h(\mathcal{X},\mathcal{Y}) = \max\{h_+(\mathcal{X},\mathcal{Y}), h_-(\mathcal{X},\mathcal{Y})\} \ .$$

Obviously

(iii) $\quad h(\mathcal{X},\mathcal{Y}) = 0$ implies $\mathcal{X} = \mathcal{Y}$ for $\mathcal{X},\mathcal{Y} \in \mathrm{conv}\mathbb{R}^n$

As it is well known, a closed convex set $\mathcal{X} \in \mathrm{conv}\mathbb{R}^n$ may be described by its support function

$$\rho(l|\mathcal{X}) = \sup\{(l,x) \mid x \in \mathcal{X}\}$$

which is a positively homogeneous convex function of l, namely,

$$\rho(\alpha|\mathcal{X}) \;=\; \alpha\rho(l|\mathcal{X}) \;\text{ for }\; \alpha \geq 0$$

and

$$\rho(\alpha_1 l_1 + \alpha_2 l_2 \,|\mathcal{X}) \;\leq \alpha_1\rho(l_1|\mathcal{X}) \;+\; \alpha_2\rho(l_2|\mathcal{X}) \;\;,$$

where $\alpha_1 > 0$, $\alpha_2 \geq 0$, $\alpha_1 + \alpha_2 = 1$.

For $\mathcal{X} \in \text{comp}\mathbb{R}^n$ we have $\rho(l|\mathcal{X}) < \infty, \forall l \in \mathbb{R}^n$. A well-known property is given by

Lemma 1.3.1 *The inclusion $x \in \mathcal{X}$, $\mathcal{X} \in \text{conv}\mathbb{R}^n$, is equivalent to the inequality*

$$(l, x) \;\leq \rho(l|\mathcal{X}), \quad \forall l \in \mathbb{R}^n \;\;.$$

Direct calculation gives us the following formulae:

$$h_+(\mathcal{X}, \mathcal{Y}) \;= \max\,\{\, \rho(l|\mathcal{X}) - \rho(l|\mathcal{Y}) \;:\; \|l\| \;\leq\; 1\} \;\;,$$

$$(1.3.2) \qquad h(\mathcal{X}, \mathcal{Y}) \;=\; \max\,\{\, |\rho(l|\mathcal{X}) - \rho(l|\mathcal{Y})| \;:\; \|l\| \;\leq\; 1\} \;\;.$$

Definition 1.3.3 *A function $\mathcal{X} : T \to \text{conv}\mathbb{R}^n$ is said to be absolutely h-continuous on T if for any $\varepsilon > 0$ there exists a $\delta > 0$ such that condition*

$$\sum_i (t_i'' - t_i') \;<\; \delta, \; t_i'' \geq t_i'$$

yields

$$\sum_i h(\mathcal{X}[t_i'], \; \mathcal{X}[t_i'']) \;\leq \varepsilon \;\;.$$

The definition of *absolute h_+-continuity* is given by mere substitution of h by h_+ in Definition 1.3.3.

Lemma 1.3.2 *A function $\mathcal{X} : T \to \mathbb{R}^n$ is absolutely h-continuous if the support function $\rho(l|\mathcal{X}[t]) = f(l, t)$ is absolutely continuous in $t \in T$ uniformly in $l \in \mathcal{S}$.*

Now we may consider the equation

$$\lim_{\sigma\to 0} \sigma^{-1} h(\mathcal{X}[t+\sigma],\ \mathcal{X}[t] + \sigma\mathcal{P}(t) + \sigma f(t)) \;=\; 0 \tag{1.3.3}$$

with initial value

$$\mathcal{X}[t_0] \;=\; \mathcal{X}^0\ . \tag{1.3.4}$$

Definition 1.3.4 *A multivalued function $\mathcal{Z} : T \to \mathrm{conv}\mathbb{R}^n$ is said to be a solution of (1.3.3), (1.3.4) if it is absolutely h-continuous and satisfies (1.3.3) for almost all $t \in T$, together with (1.3.4).*

Let us see whether $\mathcal{X}[t]$ is a solution to (1.3.3) in the sense of the last definition.

Rewriting (1.3.1) in terms of support functions, we come to

$$\rho(l|\mathcal{X}[t]) \;=\; \rho(l|\mathcal{X}_0) + \int_{t_0}^{t} \rho(l|\mathcal{P}(\tau))d\tau + \int_{t_0}^{t} (l, f(\tau))d\tau\ . \tag{1.3.5}$$

Here we made use of the fact that for a continuous map $\mathcal{P} : T \to \mathrm{conv}\mathbb{R}^n$, the following is true

$$\rho(l|\int_{t_0}^{t} \mathcal{P}(\tau)d\tau) \;=\; \int_{t_0}^{t} \rho(l|\mathcal{P}(\tau))d\tau\ .$$

To calculate

$$h(\mathcal{X}[t+\sigma], \mathcal{X}[t] + \sigma\mathcal{P}(t) \;+\; \sigma f(t)) \;=\; H(\sigma, t)\ ,$$

due to (1.3.3), (1.3.5) at first we have

$$R(l,\sigma,t) \;=\; \rho(l|\mathcal{X}[t+\sigma]) \;-\; \rho(l|\mathcal{X}[t]) \;-\; \sigma\rho(l|\mathcal{P}(t)) - \sigma(l, f(t))\ ,$$

$$R(l,\sigma,t) \;=\; \int_{t}^{t+\sigma} [(\rho(l|\mathcal{P}(\tau)) \;+\; (l, f(\tau))]d\tau - \sigma\rho(l|\mathcal{P}(t)) \;-\; \sigma(l, f(t))$$

In case of continuous $f(t)$ and $\mathcal{P}(t)$ we further have

$$\sigma^{-1} \int_{t}^{t+\sigma} f(\tau)d\tau \;\to\; f(t), \quad \sigma \to 0 \tag{1.3.6}$$

for all t and

$$(1.3.7) \qquad \sigma^{-1} \int_{t}^{t+\sigma} \rho(l|\mathcal{P}(\tau))d\tau \rightarrow \rho(l|\mathcal{P}(t)), \quad \sigma \rightarrow 0 \ .$$

If $f(t)$ is not continuous, being only measurable in t, relation (1.3.6) is still true, but now only for almost all of the values of t, which are the points of density of $f(t)$ [232]. A similar remark is true for set-valued function $\mathcal{P}(t)$ and thus for (1.3.7) with $\rho(l|\mathcal{P}(t))$ measurable in t [68], [21].

Taking into account the equality

$$H(\sigma, t) \ = \ \max\{|R(l, \sigma, t)| \ : \ \|l\| = 1\}$$

and the relation

$$\lim_{\sigma \rightarrow 0} \sigma^{-1} R(l, \sigma, t) \ = \ 0 \ ,$$

that follows from (1.3.6), (1.3.7), and is uniform in $l \in \mathcal{S}$, being true for almost all $t \in T$, we observe

$$\lim_{\sigma \rightarrow 0} \sigma^{-1} H(\sigma, t) \ = \ 0$$

for almost all $t \in T$. This proves the following assertion:

Theorem 1.3.1 *The map $\mathcal{X} : T \rightarrow \text{conv}\mathbb{R}^n$, is a solution to the evolution equation (1.3.3).*

Theorem 1.2.1 implies the following

Corollary 1.3.1 *The map $\mathcal{X}^*[t]$ of Section 1.2.1 is a solution to the evolution equation (1.3.3).*

It is not uninteresting to write down a formal analogy of equation (1.3.3) when $A(t) \not\equiv 0$.

This is as follows:

$$(1.3.8) \qquad \lim_{\sigma \rightarrow 0} \sigma^{-1} h(\mathcal{X}[t + \sigma], \ (I + \sigma A(t))\mathcal{X}[t] \ + \ \sigma \mathcal{P}(t) \ + \ \sigma f(t)) \ = \ 0 \ .$$

A solution $\mathcal{X}[t]$ to (1.3.7) with given initial state $\mathcal{X}[t_0] = \mathcal{X}^0, \mathcal{X}^0 \in \operatorname{conv}\mathbb{R}^n$ is one that satisfies Definition 1.3.4, but with equation (1.3.3) substituted by (1.3.8).

Let us now have a look at what would equation (1.3.8) be when $\mathcal{X}[t] = \{x[t]\}$ and $\mathcal{P}(t) = \{p(t)\}$ are *single-valued.* Then, clearly,

$$h(x', x'') = d(x', x'') = (x' - x'', x' - x'')^{1/2}$$

and for almost all $t \in T$

$$(1.3.9) \quad x[t+\sigma] = (I + \sigma A(t))x[t] + \sigma p(t) + \sigma f(t) + o(t, \sigma) \ ,$$

where $\sigma^{-1} o(t,\sigma) \to 0$ with $\sigma \to 0$. This yields

$$\sigma^{-1}(x[t+\sigma] - x[t]) = A(t)x[t] + p(t) + f(t) + \sigma^{-1} o(t,\sigma)$$

for almost all $t \in T$ or after a limit transition $\sigma \to 0$:

$$(1.3.10) \qquad \dot{x}[t] = A(t)x[t] + p(t) + f(t), \quad x[t_0] = x^0$$

for almost all $t \in T$.

Thus, equation (1.3.8) is clearly a *set-valued analogy of the ordinary differential equation* (ODE) (1.3.10) that may also be presented in a form similar to (1.3.8), which is (1.3.9).

There is no special point, however, in presenting an ODE in the form (1.3.9). It is not so for set-valued maps, where equation (1.3.8) may be quite convenient, particularly because here we avoid the unpleasant operation of subtraction of sets or set-valued functions.

Equation (1.3.3) *may be integrated.* The integral form for its solution $\mathcal{X}[t]$ is given by (1.3.1) and is described by a *multivalued Lebesgue integral* [25].

The support function for a solution $\mathcal{X}[t]$ of (1.3.3), (1.3.4) satisfies, as one may directly conclude from these relations, the partial differential equation

$$(1.3.11) \qquad \frac{\partial \rho}{\partial t}(l|\mathcal{X}[t]) = \rho(l|\mathcal{P}(t)) + (l, f(t)) \ ,$$

$$(1.3.12) \qquad \rho(l|\mathcal{X}[t_0]) = \rho(l|\mathcal{X}^0), \quad t \in T, \ l \in \mathbb{R}^n$$

for almost all $t \in T$ and all $l \in \mathbb{R}^n$, that follows due to (1.3.2). From (1.3.11), (1.3.12) it is then not difficult to observe that $\mathcal{X}[t]$ is *the only solution* to (1.3.3), (1.3.4):

Lemma 1.3.3 *The solution $\mathcal{X}[t]$ to equation (1.3.3), (1.3.4) is unique.*

To conclude this paragraph we shall introduce another version of the evolution equation (1.3.3), namely, by substituting the Hausdorff distance $h()$ for a *semidistance* $h_+()$. This gives

$$(1.3.13)\quad \lim_{\sigma\to+0} \sigma^{-1}h_+(\mathcal{Z}[t+\sigma], \mathcal{Z}[t] + \sigma\mathcal{P}(t) + \sigma f(t)) = 0 \ ,$$

$$(1.3.14)\quad \mathcal{Z}[t_0] \subseteq \mathcal{X}^0 \ .$$

A solution $\mathcal{Z}[t]$ to (1.3.12), is specified as in Definition 1.3.4 but with equation (1.3.3) substituted by (1.3.13).[7] Here a solution $\mathcal{Z}[t]$ to (1.3.12) satisfies an inclusion

$$\mathcal{Z}[t+\sigma] \subseteq \mathcal{Z}[t] + \sigma\mathcal{P}(t) + \sigma f(t) + o(t,\sigma)\mathcal{S} \ ,$$

rather than an equality, ($\subseteq$ instead of $=$,) which would be the case for (1.3.3). This directly yields a *partial differential inequality* (true for all $l \in \mathbb{R}^n$) and almost all $t \in T$)

$$(1.3.15)\quad \frac{\partial\rho}{\partial t}(l|\mathcal{Z}[t]) \le \rho(l|\mathcal{P}(t)) + (l, f(t))$$

for the solution $\mathcal{Z}[t]$. The initial condition also satisfies an inequality

$$(1.3.16)\quad \rho(l|\mathcal{Z}[t_0]) \le \rho(l|\mathcal{X}^0) \ .$$

It is not difficult to observe that (1.3.13) has a *nonunique* solution. Particularly, any single valued trajectory $x(t)$ driven by a control $u(t) \in \mathcal{P}(t)$ with $x^0 \in \mathcal{X}^0$ will be one of these.

Integrating (1.3.13), we come to

$$\rho(l|\mathcal{Z}[t]) \le \rho(l|\mathcal{X}[t_0]) + \int_{t_0}^{t} (\rho(l|\mathcal{P}(\tau)) + (l, f(\tau)))d\tau = \rho(l|\mathcal{X}[t]) \ ,$$

in view of (1.3.15) and (1.3.16). This leads us to the assertion

[7] **In the sequel in all the equations of the funnel type that involve Hausdorff *semidistance* h_+ we shall presume that $\sigma \to +0$ without additional indication.**

Lemma 1.3.4 *The solutions $\mathcal{Z}[t]$, $\mathcal{X}[t]$ to the evolution equations (1.3.3) (1.3.4) and (1.3.13), (1.3.4), respectively, satisfy the inclusion*

$$\mathcal{Z}(t) \subseteq \mathcal{X}[t] \ ,$$

for all $t \in T$.

We emphasize again that (1.3.3), (1.3.4) has a unique solution, while, in general, the solution to (1.3.13), (1.3.4) is nonunique.

Definition 1.3.5 *A solution $\mathcal{Z}^0[t]$ to (1.3.13) is maximal if*

$$\mathcal{Z}[t] \subseteq \mathcal{Z}^0[t], \quad \forall t \in T \ ,$$

for any solution $\mathcal{Z}[t]$ to (1.3.13) with the same initial condition (1.3.4).

As an exercise the reader may prove the following

Lemma 1.3.5 *The maximal solution $\mathcal{Z}^0[t]$ to (1.3.13), (1.3.14) exists and coincides with the unique solution $\mathcal{X}[t]$ to (1.3.3), (1.3.4).*

We will further use the evolution equations (1.3.3), (1.3.8), (1.3.18) and their generalizations as an essential tool for describing the topics of this book. Among the first of these is the problem of Control Synthesis.

We shall first present a constructive technique for Control Synthesis based on set-valued calculus and further used here in the sections devoted to ellipsoidal-valued dynamics. A still further Section 1.1.5 is intended to indicate that the technique of the next Section 1.1.4 is not an isolated approach, but allows an equivalent representation in conventional terms of Dynamic Programming as applied in either a standard or a nondifferentiable version.

1.4 The Problem of Control Synthesis: A Solution Through Set-Valued Techniques

Consider system (1.1.1)–(1.1.2) and a terminal set $\mathcal{M} \in \text{conv}\mathbb{R}^n$.

Definition 1.4.1 *The problem of* **control synthesis** *consists in specifying a solvability set* $\mathcal{W}^*(\tau, t_1, \mathcal{M})$ *and a* **feedback control strategy** $u = \mathcal{U}(t, x)$, $\mathcal{U}(\cdot, \cdot) \in U_{\mathcal{P}}^c$ *such that all the solutions to the differential inclusion*

$$\dot{x}(t) \in \mathcal{U}(t, x) + f(t) \tag{1.4.1}$$

that start from any given position $\{\tau, x_\tau\}$, $x_\tau = x[\tau]$, $x_\tau \in \mathcal{W}^*(\tau, t_1, \mathcal{M})$, $\tau \in [t_0, t_1)$, *would reach the terminal set* $\mathcal{M}$ *at time* t_1*:* $x[t_1] \in \mathcal{M}$.

The definition is nonredundant provided $x_\tau \in \mathcal{W}^*(\tau, t_1, \mathcal{M}) \neq \emptyset$, where the *solvability* set $\mathcal{W}^*(\tau, t_1, \mathcal{M}) = \mathcal{W}^*[\tau]$ is the largest set of states from which the solution to the problem of control synthesis does exist at all. (More precisely this will be specified below).

Taking $\mathcal{W}^*(t, t_1, \mathcal{M})$ for any instant $t \in [t_0, t_1]$, we come to a set-valued map $\mathcal{W}^*[t] = \mathcal{W}^*(t, t_1, \mathcal{M})$, $t \in T$, (the *solvability tube*) where $\mathcal{W}^*[t_1] = \mathcal{M}$.

To describe the tube $\mathcal{W}^*[t]$ we first start from the following

Definition 1.4.2 *The open-loop solvability set* $\mathcal{W}(\tau, t_1, \mathcal{M})$ *is the set of all states* $x_\tau \in \mathbb{R}^n$ *such that there exists a control* $u(t) \in \mathcal{P}(t)$, $\tau \leq t \leq t_1$ *that steers the system from* x_τ *to* $\mathcal{M}$ *due to a respective trajectory* $x[t]$, $\tau \leq t \leq t_1$, *so that* $x[\tau] = x_\tau$, *and* $x[t_1] \in \mathcal{M}$.

The set $\mathcal{W}[\tau] = \mathcal{W}(\tau, t_1, \mathcal{M})$ is nothing more than the *attainability domain* at instant τ for system (1.1.1), (1.1.2), from set $\mathcal{M}$, but calculated in *backward time*, namely, from t_1 to τ. The respective map $\mathcal{W}[t]$, $t \in T$, $\mathcal{W}(t_1) = \mathcal{M}$ is defined as the *open-loop solvability tube* for set $\mathcal{M}$, on the interval T.

A direct consequence of Theorem 1.3.1 and the definition of $\mathcal{W}[t]$ is the following

Theorem 1.4.1 *The set-valued function $\mathcal{W}[t]$ satisfies the evolution equation*

$$(1.4.2) \qquad \lim_{\sigma\to 0} \sigma^{-1} h(\mathcal{W}[t-\sigma], \mathcal{W}[t] - \sigma\mathcal{P}(t) - \sigma f(t)) = 0$$

$$(1.4.3) \qquad \mathcal{W}[t_1] = \mathcal{M}$$

and the semi-group property

$$\mathcal{W}(\tau, t_1, \mathcal{M}) = \mathcal{W}(\tau, t, \mathcal{W}(t, t_1, \mathcal{M}))$$

for all $t_0 \leq \tau \leq t \leq t_1$.

Its solution is obviously

$$(1.4.4) \qquad \mathcal{W}[t] = \mathcal{M} - \int_t^{t_1} \mathcal{P}(\tau)d\tau - \int_t^{t_1} f(\tau)d\tau \ .$$

Equation (1.4.2) is the same as (1.3.3), but is treated *in backward time.* The definition of the solution is, naturally, also the same.

Definition 1.4.3 *The closed-loop solvability set $\mathcal{W}^*(\tau, t_1, \mathcal{M})$ is the set of all states $x_\tau \in \mathbb{R}^n$ such that there exists a control strategy $u = \mathcal{U}(t,x)$, $\mathcal{U}(\cdot,\cdot) \in U^c_\mathcal{P}$ that ensures every trajectory $x[t]$ of the differential inclusion (1.4.1) that starts at τ, $x[\tau] = x_\tau$, to end in set $\mathcal{M} : x[t_1] \in \mathcal{M}$.*

The respective map $\mathcal{W}^*[t] = \mathcal{W}^*(t, t_1, \mathcal{M})$, $t \in T$, $\mathcal{W}^*[t_1] = \mathcal{M}$ defines the *closed-loop solvability tube* $\mathcal{W}[\cdot]$ for set $\mathcal{M}$.

From Theorem 1.2.1 we come to

Lemma 1.4.1 *With $\mathcal{W}[t_1] = \mathcal{W}^*[t_1] = \mathcal{M}$ the open-loop and closed-loop solvability tubes, which are $\mathcal{W}[t]$ and $\mathcal{W}^*[t]$, do coincide, namely,*

$$\mathcal{W}[t] \equiv \mathcal{W}^*[t],;\ t \in T \ .$$

Tube $\mathcal{W}^*[t]$ therefore satisfies the evolution equation (1.4.2), (1.4.3). With $A(t) \not\equiv 0$ we have

(1.4.5)
$$\lim_{\sigma \to 0} \sigma^{-1} h\left(\mathcal{W}[t-\sigma], (I - \sigma A(t))\, \mathcal{W}[t] - \sigma \mathcal{P}(t) - \sigma f(t)\right) \equiv 0 \ .$$

The solutions to (1.4.2), (1.4.3), or (1.4.5), (1.4.3) are unique and are given by convex compact-valued functions.

Substituting Hausdorff distance $h(\cdot)$ for semidistance $h_+(\cdot)$, we come to the equation

(1.4.6) $$\lim_{\sigma \to 0} \sigma^{-1} h_+\left(\mathcal{Z}[t-\sigma], \mathcal{Z}[t] - \sigma \mathcal{P}(t) - \sigma f(t)\right) = 0$$

(1.4.7) $$\mathcal{Z}[t_1] \subseteq \mathcal{M}$$

which is the same as (1.3.13), (1.3.14) but taken in backward time. The definition of its solution and maximal solution are analogies of those given in the previous section for direct (forward) time. By analogy with Lemma 1.3.5 we also come to

Lemma 1.4.2 *With $\mathcal{W}[t_1] = \mathcal{M}$, the map $\mathcal{W}[t]$, $t \in T$, is the maximal solution to (1.4.6), (1.4.7).*

This is a consequence of the definition of the solvability sets. It is important to emphasize that the condition

$$\mathcal{W}(\tau, t_1, \mathcal{M}) \neq \emptyset \ \ , \forall \tau \in T \ \ ,$$

is *necessary and sufficient* for the solvability of the control synthesis problem of Definition 1.4.1. An essential element in constructing the respective strategy $\mathcal{U} = \mathcal{U}(t,x)$ is the tube $\mathcal{W}[t]$, $\tau \leq t \leq t_1$.

Assume $x \in \mathbb{R}^n$ and set $\mathcal{W}[\tau]$ to be given. Let us introduce a synthesizing function $V(\tau, x) = d^2[\tau, x]$, where

$$d[\tau, x] = h_+(x, \mathcal{W}[\tau]) \ ,$$

$$h_+(x, \mathcal{W}[\tau]) = \min\{\|x - w\| \mid w \in \mathcal{W}[\tau]\} \ .$$

Clearly,

$$V[\tau, x] = 0 \ \text{ implies } \ x \in \mathcal{W}[\tau] \ ,$$

and

$$V[\tau, x] > 0, \quad \text{implies} \quad x \notin \mathcal{W}[\tau] \ .$$

(One may observe that

$$\mathcal{W}[\tau] = \{x : V(\tau, x) \leq 0\}$$

is the *level set* for $V(\tau, x)$.)

We may now investigate the derivative

$$\frac{d}{dt}V(t, x)\Big|_{(1.2.1)}$$

along the trajectories of system (1.2.1). The control set $\mathcal{U}^0(t, x)$ will then consist, as we shall see, of all the values $u(t) \in \mathcal{P}(t)$ that minimize this derivative, namely,

$$\mathcal{U}^0(t, x) = \arg\min\{\frac{d}{dt}V(t, x)\Big|_{(1.2.1)} : u \in \mathcal{P}(t)\} \ .$$

Let us specify this in detail. A direct differentiation yields

$$\text{(1.4.8)} \qquad \frac{d}{dt}V(t, x)\Big|_{(1.2.1)} = 2d[t, x](\frac{d}{dt}d[t, x]\Big|_{(1.2.1)}) \ ,$$

where

$$\text{(1.4.9)} \qquad d[t, x] = \max\{(l, x) - \rho(l|\mathcal{W}[t]) \ : \ \|l\| \leq 1\} =$$
$$= (l^0, x) - \rho(l^0|\mathcal{W}[t]) \ ,$$

and where $l^0 \neq 0$, $\|l^0\| = 1$, is a *unique* maximizer for $d[t, x] > 0$. We will always choose the maximizer to be $l^0 = \{0\}$ if $d[t, x] = 0$.

Since $\mathcal{W}[t]$ is absolutely continuous, it is not difficult to prove the following property that justifies the differentiation.

Lemma 1.4.3 *Let $x^*[t]$ be an absolutely continuous function on an interval e where $d^*(t) = h_+(x^*[t], \mathcal{W}[t]) > 0$. Then the function $d^*(t)$ is absolutely continuous on the same interval.*

We further need the derivative $d(d[t,x])/dt$ when $d[t,x] > 0$, due to the system (1.2.1), which is

$$\dot{x}(t) = u(t) + f(t) .$$

For this we obtain:

(1.4.10)
$$\left.\frac{d}{dt}d[t,x]\right|_{1.2.1} = \frac{\partial}{\partial t}\, d[t,x] + \left(\frac{\partial}{\partial x}d[t,x], \dot{x}(t)\right)$$
$$= (l^0, \dot{x}(t)) - \frac{\partial}{\partial t}\, \rho(l^0|\mathcal{W}[t])$$
$$= (l^0, u(t) + f(t)) + \rho(-l^0|\mathcal{P}(t)) - (l^0, f(t)),$$

and therefore

$$\frac{d}{dt}d[t,x(t)] = (l^0, u(t)) + \rho(-l^0|\mathcal{P}(t)) . \tag{1.4.11}$$

Here we have used the formula

$$\rho(l|\mathcal{W}[t]) = \rho(l|\mathcal{M}) + \int_t^{t_1} \rho(-l|\mathcal{P}(\tau))d\tau - \int_t^{t_1} (l, f(\tau))d\tau$$

that follows from (1.4.4) and the fact that in calculating the derivative (1.4.10) for $d[t,x]$ which is the maximum over l in (1.4.9), we should avoid differentiation in l^0.

Indeed, following [86], [265], [261], we observe that for a differentiable function of type

$$h(t,x(t)) = \max\{H(t,x(t),l) : \|l\| = 1\} ,$$

with unique maximizer l^0 we have

$$\frac{dh(t,x)}{dt} = \frac{\partial H(t,x,l^0)}{\partial t} + \left(\frac{\partial H(t,x,l^0)}{\partial x}, \dot{x}\right) ,$$

where $l^0 = \arg\max\, \{H(t,x,l) : \|l\| = 1\}$.

Remark 1.4.1 *The direct calculation of $\partial\rho(l|\mathcal{W}[t])/\partial t$ introduced in (1.4.10) also indicates that with $\mathcal{Z}(t) = \mathcal{W}[t]$ the inequality (1.3.15) turns out to be an equality.*

We now proceed with specifying the feedback strategy $\mathcal{U}^0(t,x)$. Since l^0 depends on t and x, we further use the notation $l^0 = l^0(t,x)$.

The strategy $\mathcal{U}^0(t,x)$ has to be specified both in the domain $\{x \notin \mathcal{W}[t]\}$ (or $V(t,x) > 0$) and in $\{x \in \mathcal{W}[t]\}$ (or $V(t,x) = 0$).

Assume $V(t,x) > 0$. Then $\mathcal{U}^0(t,x)$ is defined as

$$\mathcal{U}^0(t,x) = \arg\min\left\{\frac{d}{dt}\, d[t,x] \,\middle|\, u \in \mathcal{P}(t)\right\} =$$

$$= \arg\min\left\{\frac{d}{dt}\, V(t,x) \,\middle|\, u \in \mathcal{P}(t)\right\} .$$

(We further omit the index (1.2.1) that indicates the system along whose solutions we calculate the derivative.)

Due to (1.4.11), this turns into

$$\mathcal{U}^0(t,x) = \arg\min\{(l^0(t,x),u) \,|\, u \in \mathcal{P}(t)\}$$

or, what is the same,

$$\mathcal{U}^0(t,x) = \arg\max\{(-l^0(t,x),u) \,|\, u \in \mathcal{P}(t)\} . \tag{1.4.12}$$

(One should observe that with $V(t,x) = 0$ we have $l^0 = 0$ and therefore $\mathcal{U}^0(t,x) = \mathcal{P}(t)$.)

Relations (1.4.9), (1.4.10) yield the following assertion

Lemma 1.4.4 *With $d[t,x] > 0$ the derivative*

$$\frac{d}{dt}d[t,x] \geq 0 \quad \textit{for any} \quad u \in \mathcal{P}, \tag{1.4.13}$$

$$\frac{d}{dt}d[t,x] = 0 \quad \textit{for} \quad u(t) \in \mathcal{U}^0(t,x), \tag{1.4.14}$$

where $\mathcal{U}^0(t,x)$ is defined by relation (1.4.11).

Combining this with (1.4.8), (1.4.11), (1.4.12), we come to

Lemma 1.4.5 *For any position $\{t, x\}$the derivative*

$$\frac{d}{dt}V(t,x)\bigg|_{(1.2.1)} \leq 0 \ , \tag{1.4.15}$$

provided $u \in \mathcal{U}^0(t, x)$

The latter relations allow us to prove

Theorem 1.4.2 *The strategy $u(t) = \mathcal{U}^0(t, x)$ defined by equation (1.4.11), does solve the problem of control synthesis specified in Definition 1.4.1.*

Assume that $x^0 \in \mathcal{W}[t_0]$ and that the inclusion (1.4.1) is run by strategy $\mathcal{U} = \mathcal{U}^0(t, x)$, which, in general, generates a tube

$$\mathcal{X}[t, \mathcal{U}^0] = \mathcal{X}(t, t_0, x^0 | \mathcal{U}^0) = \{x^0(t, t_0, x^0)\}, \quad t \in T,$$

of isolated trajectories $x^0[t] = x^0(t, t_0, x^0)$ to (1.4.1), $(\mathcal{U} = \mathcal{U}^0)$.

The proof of Theorem 1.4.2 is based on the following

Lemma 1.4.6 *The tube $\mathcal{X}[t|\mathcal{U}^0], t \in T$; $\mathcal{X}[t_0|\mathcal{U}^0] = x^0$, $x^0 \in \mathcal{W}[t_0]$, satisfies the inclusion*

$$\mathcal{X}[t|\mathcal{U}^0] \subseteq \mathcal{W}[t], \quad t \in T,$$

therefore

$$\mathcal{X}[t_1|\mathcal{U}^0] \subseteq \mathcal{M}.$$

Proof. Assume $x[t] = x(t, t_0, x^0)$ is a trajectory of inclusion (1.4.1), $\mathcal{U} = \mathcal{U}^0$, with $x^0 \in \mathcal{W}[t_0]$ or equivalently, with $V(t_0, x_0) \leq 0$ and $x[t] \in \mathcal{X}[t|\mathcal{U}^0]$, $t \in T$. We shall prove that $x[t] \in \mathcal{W}[t]$, or equivalently, that $V(t, x[t]) \leq 0$, for all $t \in T$. Then, for any value of $t \in (t_0, t_1]$, we observe that the integral

$$\int_{t_0}^{t} \frac{dV(\tau, x[\tau])}{d\tau} d\tau = V(t, x[t]) - V(t_0, x[t_0]) \leq 0$$

due to (1.4.15). Since $x[t_0] \in \mathcal{W}[t_0]$, this yields

$$V(t, x[t]) \leq V(t_0, x[t_0]) \leq 0$$

for any $t \in (t_0, t_1]$ and thus proves Lemma 1.4.6 from which Theorem 1.4.1 follows directly.

The same property is true if x^0 is substituted by a set $\mathcal{X}^0 \subseteq \mathcal{W}[t^0]$.

Corollary 1.4.1 *With $\mathcal{X}^0 = \mathcal{X}(t^0|\mathcal{U}^0) \subseteq \mathcal{W}[t_0]$ the respective tube of solutions $\mathcal{X}[t|\mathcal{U}^0] = \mathcal{X}(t, t_0, \mathcal{X}^0|\mathcal{U}^0)$ to the differential inclusion (1.4.1) generated by strategy $\mathcal{U}(t,x) = \mathcal{U}^0(t,x)$, satisfies the relation*

$$\mathcal{X}[t|\mathcal{U}^0] \subseteq \mathcal{W}[t], \quad t \in T,$$

and therefore $\mathcal{X}[t_1|\mathcal{U}^0] \subseteq \mathcal{M}$.

We thus observe that if for an instant $\tau \in T$ the inclusion $\mathcal{X}_\tau = \mathcal{X}[\tau|\mathcal{U}^0] \subseteq \mathcal{W}[\tau]$ is true, then

$$\mathcal{X}[t|\mathcal{U}^0] \subseteq \mathcal{W}[t] \tag{1.4.16}$$

will hold for all $t \geq \tau$, i.e., for the whole trajectory tube $\mathcal{X}[t|\mathcal{U}^0], t \in T_\tau, T_\tau = [\tau, t_1]$ generated by system $(1.4.1), \mathcal{U} = \mathcal{U}^0$.

The tube $\mathcal{X}[t|\mathcal{U}^0]$ therefore satisfies (1.4.16) as a *state constraint.* According to the terminology of [17] $\mathcal{X}_\tau$ is *strongly invariant* relative to tube $\mathcal{W}[t], t \in T_\tau$, for time τ. The latter means that every trajectory of the inclusion $(1.4.1), \mathcal{U} = \mathcal{U}^0$ that evolves from set $\mathcal{X}_\tau$ remains within $\mathcal{W}[t]$. It is now obvious that the largest strongly invariant set for time τ, relative to $\mathcal{W}[t], t \in T_\tau$, is $\mathcal{W}[\tau]$ itself.

The feedback strategy $\mathcal{U}^0(t,x)$ may be rewritten in terms of the notion of *subdifferential* [265], [267], [261]. We recall that a subdifferential $\partial_l f(t, l^0)$ (in the variable l, at point $l = l^0$) of a function $f(t,l)$ convex in l, is the set of all vectors q such that

$$f(t,l) - f(t,l^0) \geq (q, l - l^0), \quad \forall l \in \mathbb{R}^n. \tag{1.4.17}$$

Assume $f(t,l) = \rho(l|\mathcal{P}(t))$. Then, due to the definition, a vector $q \in \partial_l f(t, l^0)$ if and only if

$$\rho(l|\mathcal{P}(t)) - \rho(l^0|\mathcal{P}(t)) \geq (q, l - l^0), \quad \forall l \in \mathbb{R}^n.$$

From here it follows (taking $l = 2l^0$), that

$$\rho(l^0|\mathcal{P}(t)) - (l^0, q) \geq 0 \ , \tag{1.4.18}$$

and therefore,

$$\rho(l|\mathcal{P}(t)) - (q, l) \geq (l^0|\mathcal{P}(t)) - (l^0, q) \geq 0, \quad \forall l \in \mathbb{R}^n \ ,$$

whence $q \in \mathcal{P}(t)$.

On the other hand, with $l = 0$, we come to

$$(l^0, q) \geq \rho(l^0|\mathcal{P}(t)) \ . \tag{1.4.19}$$

A comparison of (1.4.18), (1.4.19) and a substitution of q for u yields

Lemma 1.4.7 *With $f(t, l) = \rho(l|\mathcal{P}(t))$, the respective subdifferential $\partial_l f(t, l)$ is given by:*

$$\partial_l f(t, l^0) = \{u \in \mathbb{R}^n \mid (l^0, u) = \rho(l^0|\mathcal{P}(t))\} \ .$$

Clearly, for $l^0 = 0$ we have $\partial_l f(t, l^0) = \mathcal{P}(t)$, and therefore

$$\mathcal{U}^0(t, x) = \partial_l f(t, -l^0(t, x)) \ ,$$

where $l^0(t, x)$ is the maximizer for (1.4.9).

Summarizing the reasoning of the above, we conclude the following.

Theorem 1.4.3 *The feedback strategy $\mathcal{U}^0(t, x)$ that solves the problem of control synthesis may be defined as*

$$\mathcal{U}^0(t, x) = \partial_l f(t, -l^0(t, x)) \ , \tag{1.4.20}$$

where $f(t, l) = \rho(l|\mathcal{P}(t))$ and $l^0(t, x)$ is the maximizer for problem (1.4.9).

One may check, as an exercise, that the strategy $\mathcal{U}^0(t, x) \in U^c_{\mathcal{P}}$ belongs to the class of feasible strategies introduced in Section 1.2.

1.5 Control Synthesis Through Dynamic Programming Techniques

For a control-theorist experienced in the methods of this theory the geometrical techniques of set-valued calculus, as introduced in the above and further used in the sequel, may seem, at first glance, to be somewhat unusual. It may be demonstrated, however, that they are quite in line with the well-known fundamentals of control theory. We therefore feel obliged to indicate, in a very concise form, a conventional way of looking at the problems under discussion.

Assume a position $\{\tau, x\}$ due to system (1.1.1) to be given together with a terminal set $\mathcal{M} \in conv\mathbb{R}^n$. (Although matrix $A(t) \not\equiv 0$ is present in the first part of this section, one may always take $A(t) \equiv 0$ as shown in Section 1.2). Let us indicate a *cost criterion* $\mathcal{I}(\tau, x)$ for the problem of control synthesis, assuming that our objective will be to find an optimal control strategy $u = u(t, x)$ that minimizes this criterion. We shall further look for the solution in the class $u = \mathcal{U}(t, x) \in U_{\mathcal{P}}^{c}$.[8]

More specifically, let us assume

$$\mathcal{I}(\tau, x^*) = h_+^2(x[t_1], \mathcal{M}]) , \tag{1.5.1}$$

where $x[t_1] = x(t_1, \tau, x^*)$.

The optimal value

$$\mathcal{I}^0(\tau, x) = \min \{\mathcal{I}(\tau, x) | \, u(\cdot, \cdot) \in U_{\mathcal{P}}^{c}\}$$

when taken for any position $\{\tau, x\}$ will be further referred to as the *value function*

$$\mathcal{V}(\tau, x) = \mathcal{I}^0(\tau, x) .$$

It is obvious that $\mathcal{V}(\tau, x) = 0$, if $x[t_1] \in \mathcal{M}$ and $\mathcal{V}(\tau, x) > 0$ if $x[t_1] \notin \mathcal{M}$.

[8]The standard schemes of Dynamic Programming presume that the strategy $u = u(t, x)$ is single-valued. Following [168], we shall however allow them to be multivalued, observing that for the classes of problems considered here the multivalued functions appear naturally, from the same technique.

Therefore, the *solvability domain* $W[\tau]$ of Section 1.4 is actually *the level set*

$$W[\tau] = \{x : \mathcal{V}(\tau, x) \leq 0\} .$$

Let us now calculate function $\mathcal{V}(\tau, x)$ by formally writing down the H-J-B (Hamilton-Jacobi-Bellman) equation for the problem of minimizing cost criterion (1.5.1) along the trajectories of system (1.1.1) with $u = u(t,x) \in U^c_{\mathcal{P}}$. (The respective theory may be found in [109], [53].) This is

(1.5.2)

$$\frac{\partial \mathcal{V}(\tau,x)}{\partial \tau} + \min\left\{\left(\frac{\partial \mathcal{V}(\tau,x)}{\partial x}, A(\tau)x + u + f(\tau)\right)\Big| u \in \mathcal{P}(\tau)\right\} = 0$$

with boundary condition

$$\mathcal{V}(t_1, x) = h_+^2(x, \mathcal{M}) \qquad (1.5.3)$$

or, more precisely,

(1.5.4)

$$\frac{\partial \mathcal{V}(\tau,x)}{\partial \tau} + \left(A'(\tau)\frac{\partial \mathcal{V}(\tau,x)}{\partial x}, x\right) - \rho\left(\frac{\partial(-\mathcal{V}(\tau,x))}{\partial x}\Big| \mathcal{P}(\tau) + f(\tau)\right) = 0$$

with same boundary condition (1.5.3).

We have to check, however, whether these formal operations are justified. We shall do that by calculating the value $\mathcal{I}(\tau, x)$ directly, through the technique of convex analysis.

Obviously, the function

$$\phi(x) = h_+^2(x, \mathcal{M}) = \min\{(x-q, x-q) | q \in \mathcal{M}\}$$

has a conjugate

$$\phi^*(l) = \max\{(l,x) - \phi(x) | x \in \mathbb{R}^n\} =$$

$$= \max\{\max\{(l,x) - (x-q, x-q) | x \in \mathbb{R}^n\} | q \in \mathcal{M}\} =$$

$$= \max\{(l,q) + \frac{1}{4}(l,l) | q \in \mathcal{M}\}$$

which is

$$\phi^*(l) = \rho(l|\mathcal{M}) + \frac{1}{4}(l,l) . \qquad (1.5.5)$$

Our problem is to find

$$\mathcal{I}^0(\tau, x) = \min\{\phi(x(t_1))|u(t) \in \mathcal{P}(t), \tau \leq t \leq t_1\}$$

over the trajectories of system (1.2.1), (1.1.2) with given initial position $\{\tau, x\}$. We have

$$\min_{u(\cdot)} \phi(x(t_1)) = \min_{u(\cdot)} \max_{l}\{(l, x(t_1)) - \phi^*(l)\} =$$

$$= \max_{l}\{\min_{u(\cdot)}\{(l, x(t_1)) - \phi^*(l)\}\} .$$

The function in the brackets in the right-hand side is linear in $u(\cdot)$ and concave in l, with $\phi(l) \to \infty$ as $\|l\| \to \infty$. This indicates that the operations of min and max are interchangeable [101].

Denoting $s(t, t_1, l)$ to be the solution of the adjoint equation

$$\dot{s} = -sA(t) , \; s(t_1) = l, \; t \leq t_1 ,$$

(s is a *vector-row*) and using the notations of (1.1.6), we may rewrite

$$(l, x(t_1)) = (l, S(\tau, t_1)x) + \int_{\tau}^{t_1} (l, S(t, t_1)u(t) + f(t)) \, dt =$$

$$= s(\tau, t_1, l)x + \int_{\tau}^{t_1} s(t, t_1, l)(u(t) + f(t)) \, dt .$$

Hence

(1.5.6) $$\mathcal{I}^0(\tau, x) = \max\{\Phi(\tau, x, l)|l \in \mathbb{R}^n\},$$

where

$$\Phi(\tau, x, l) = s(\tau, t_1, l)x -$$

$$- \int_{\tau}^{t_1} \rho(-s(t, t_1, l)|\mathcal{P}(t) + f(t)) \, dt - \phi^*(l) .$$

The function $\Phi(\tau, x, l)$ is concave in l (moreover, even strictly concave, due to the quadratic term). The maximum in (1.5.6) is therefore attained at a *single vector* $l^0 = l^0(\tau, x)$, whatever is the position $\{\tau, x\}$.

Lemma 1.5.1 *The maximizer $l^0(\tau, x)$ of (1.5.6) is continuous in $\{\tau, x\}$.*

This is a well-known property in convex analysis (see, for example, [261], Chapter II.3).

Denote $\mathcal{X}^0[\tau] = \{x : \mathcal{I}^0(\tau, x) \leq 0\}$.

Lemma 1.5.2 *For any $x \in \mathcal{X}^0[\tau]$ we have*

$$l^0(\tau, x) = \arg\max\{\Phi(\tau, x, l) | l \in \mathbb{R}^n\} = \{0\} \quad .$$

This follows from the explicit expression for $\Phi(\tau, x, l)$.

A direct differentiation of $\mathcal{I}^0(\tau, x)$ in x now gives

$$\text{(1.5.7)} \qquad \frac{\partial \mathcal{I}^0(\tau, x)}{\partial x} = s(\tau, t_1, l^0(\tau, x)).$$

(Recall that since $l^0(\tau, x)$ is unique, the respective formula is as follows

$$\frac{\partial}{\partial x}\left(\max\{\Phi(\tau, x, l) | l\}\right) = \frac{\partial}{\partial x}\Phi(\tau, x, l)\bigg|_{l = l^0(\tau.x)} \quad) \; .$$

Similarly, with $l = l^0(t, x)$,

$$\text{(1.5.8)} \quad \frac{\partial \mathcal{I}^0(\tau, x)}{\partial \tau} = \rho(-s(\tau, t_1, l) | \mathcal{P}(\tau) + f(\tau)) - s(\tau, t_1, l)A(\tau)x.$$

Taking $\mathcal{V}(t, x) = \mathcal{I}^0(t, x)$ and substituting into (1.5.2), we have, in view of (1.5.7) and (1.5.8),

$$\rho(-s(\tau, t_1, l^0(\tau, x)) | \mathcal{P}(\tau)) - s(\tau, t_1, l^0(\tau, x))(A(\tau)x + f(\tau)) +$$

$$\text{(1.5.9)} \; \min\{s(\tau, t_1, l^0(\tau, x))(A(\tau)x + u + f(\tau)) | u \in \mathcal{P}(\tau)\} = 0 \; .$$

In order to check the boundary condition (1.5.3), we may formally observe from formula (1.5.6)

$$\text{(1.5.10)} \; \mathcal{I}^0(t_1, x) = \max\{(l, x) - \rho(l | \mathcal{M}) - \frac{1}{4}(l, l) | l \in \mathbb{R}^n\} = \psi^*(x) \quad ,$$

where

$$\psi(l) = \rho(l | \mathcal{M}) + \frac{1}{4}(l, l) = \phi^*(l)$$

due to (1.5.5.). Hence $\psi^*(x) = (\phi^*)^*(x) = \phi(x) = h_+^2(x, \mathcal{M})$. This is a consequence of the obvious relation

$$\mathcal{I}^0(t_1, x) = h_+^2(x, \mathcal{M}) \quad .$$

Therefore, the following assertion turns out to be true

Theorem 1.5.1 *The value function $\mathcal{V}(\tau, x) = \mathcal{I}^0(\tau, x)$ given by formula (1.5.6) satisfies the Dynamic Programming (H-J-B) equation (1.5.2) ((1.5.4)) with boundary condition (1.5.3).*

To compare this section with the previous one we further change the variable τ to t in the relations for the value function $V(\tau, x)$. The respective control $u = u(t, x)$ is then formally determined from (1.5.2) and (1.5.9) as

$$(1.5.11) \quad u^0(t,x) = \arg\max\left\{ \left(\frac{\partial(-\mathcal{V}(t,x))}{\partial x}, u \right) \Big| u \in \mathcal{P}(t) \right\}.$$

Particularly, with $\mathcal{V}(t, x) = 0$, this gives

$$(1.5.12) \qquad u^0(t,x) \equiv \mathcal{P}(t) \ .$$

(This reflects that $\partial\mathcal{V}(t, x)/\partial x = 0$ if $\mathcal{V}(t, x) = 0$, in view of Lemma 1.5.2.)

The control $u^0(t, x)$ is thus similar to $\mathcal{U}^0(t, x)$ defined in Section 1.4, while $(-\partial\mathcal{V}(t, x)/\partial x)$ plays the role of vector $l^0(t, x)$ in (1.4.12) and (1.4.20).

We therefore come to an equivalent of Theorem 1.4.2.

Theorem 1.5.2 *The solution strategy $u^0(t, x)$ is given by by relations (1.5.11), (1.5.12), where $\mathcal{V}(t, x)$ is the value function $\mathcal{I}^0(t, x)$.*

Let us now calculate the value

$$r[t,x] = \min\{h_+(x[t_1], \mathcal{M}) | u(s) \in P(s), t \le s \le t_1\} \ ,$$

assuming $A(t) \equiv 0$. Following the scheme for calculating (1.5.6), we have

$$r[t,x] = \max\{\Psi(t,x,l) | \|l\| \le 1\} \ ,$$

where

$$\Psi(t,x,l) = (l,x) - \int_t^{t_1} \rho(-l|\mathcal{P}(s) + f(s))ds - \rho(l|\mathcal{M}) =$$

$$= (l,x) - \rho(l|\mathcal{W}[t])$$

and $\mathcal{W}[t]$ is defined by (1.4.4). This yields $r[t, x] = d[t, x]$ and therefore leads us to

Lemma 1.5.3 *With $A(t) \equiv 0$ the value function*

$$\mathcal{I}^0(t,x) = \mathcal{V}(t,x) = d^2[t,x] \ .$$

Thus, under condition $A(t) \equiv 0$ (which does not imply any loss of generality, as we have seen), *the solution given in Section 1.4 through set-valued techniques is precisely the one derived in this section through Dynamic Programming.*

We shall further continue to indicate the Dynamic Programming interpretations of the outcoming relations which, of course, shall be somewhat more complicated in the case of uncertain systems and state constraints. Nevertheless, in the problems of this book, aimed particularly at the applicability of *ellipsoidal calculus* , the value functions will turn out to be convex in x. They will be directionally differentiable and therefore allow a more or less clear propagation of the notions of Dynamic Programming (DP).

In the more general case of nonlinear systems and an arbitrary terminal cost $\phi(x)$ the main inconvenience is that *there may be nondifferentiable function* $\mathcal{V}(t,x)$ that solves the DP (H-J-B) equation (a nonlinear analogy of equation (1.5.2)), whereas if we look for nondifferentiable functions, then the partial derivatives of $\mathcal{V}$ may not be continuous or may not even exist at all. The solution to the DP equation should then be interpreted in some *generalized sense.* Particularly, it may be interpreted as a *viscosity solution* [82], [109], or its equivalent – *the minmax solution* [290].

Looking at the solution (1.5.11) and (1.5.12), one may observe that for defining $u^0(t,x)$ through DP techniques, one needs to know the following elements:

- the level sets
$$W^0[t] \ = \ \{x \ : \mathcal{V}(t,x) \ \leq \ 0\}$$
- the partial derivatives $\partial\mathcal{V}/\partial x$ in the domain $\{x \ : \ \mathcal{V}(t,x) \ > \ 0\}$.

For the problems treated in this book these elements may be determined *without integrating* equation (1.5.4) but through direct constructive techniques which, particularly, are those formulated in Section 1.4. One just has to recognise that $d^2[t,x] \ = \ \mathcal{V}(t,x)$ and therefore that the level set

$W^0[t]$ is the *solvability set* $\mathcal{W}[t]$, ($W^0[t] \equiv \mathcal{W}[t](!)$), while the antigradient $(-\partial\mathcal{V}/\partial x)$ is collinear with $l^0(t,x)$ in (1.4.12).

Needless to say, the elements $\mathcal{V}(t,x)$, $\partial\mathcal{V}(t,x)/\partial x$, may be, of course, calculated by integrating equation (1.5.4) or its analogies (in a generalized sense, perhaps). This integration will be an essential tool for the treatment of those nonlinear problems for which the techniques of this book cease to be effective.

Example 1.5.1

Let us write down equation (1.5.2), with boundary condition (1.5.3) for the particular case when the system is autonomous, $A \equiv 0$, and $\mathcal{M}, \mathcal{P}(t)$ are *nondegenerate ellipsoids*, namely,

$$\mathcal{M} = \{x : (x-m, M^{-1}(x-m)) \leq 1\} = \mathcal{E}(m, M) \ ,$$

$$\mathcal{P}(t) = \{u : (u-p(t), P^{-1}(t)(u-p(t))) \leq 1\} = \mathcal{E}(p(t), P(t)) \ ,$$

where

$$M, \ P(t) > 0$$

are positive definite and $P(t), p(t)$ are continuous. We have

$$\frac{\partial\mathcal{V}(t,x)}{\partial t} + \left(\frac{\partial\mathcal{V}(t,x)}{\partial x}, \ f(t)+p\right) + \tag{1.5.13}$$

$$+\left(\frac{\partial\mathcal{V}(t,x)}{\partial x}, P^{-1}\frac{\partial\mathcal{V}(t,x)}{\partial x}\right)^{\frac{1}{2}} = 0$$

with

(1.5.14)

$$\mathcal{V}(t_1,x) = (x-m, M(x-m))(1-(x-m, M(x-m))^{-\frac{1}{2}})^2, \ \ x \notin \mathcal{E}(m,M) \ ,$$

$$\mathcal{V}(t_1,x) = 0 \ , \ x \in \mathcal{E}(m,M) \ .$$

Relation (1.5.14) follows from (1.5.10) by direct calculation.

Exercise 1.5.1. With $A(t) \not\equiv 0$ indicate the cost criterion $\mathcal{I}^*$ for which the value function $\mathcal{V}^*$ would be

$$\mathcal{V}^*(t,x) = h_+^2(x, \mathcal{W}[t]) \ ,$$

where $\mathcal{W}[t]$ is the *solvability set* of Section 1.4.

Let us now indicate another relation for the solvability set $W[\tau]$ under the conditions of Example 1.5.1. Taking system (1.2.1) and position $\{\tau, x\}, \tau \in [t_0, t_1], x = x(\tau)$, solve the following problem: minimize the functional

(1.5.15) $$\Psi(\tau, x, u(\cdot)) = \max\{I_T, I_1\},$$

where

$$I_T^2 = (x(t_1) - m, M(x(t_1) - m)) \ ,$$

$$I_1^2 = \text{esssup}_t(u(t) - p(t), P(t)(u(t) - p(t))), \ \ t \in [\tau, t_1] \ .$$

Introduce value function

$$V(\tau, x) = \min\{\Psi(\tau, x, u(\cdot)) | u(\cdot)\} \ .$$

Then, clearly

(1.5.16) $$W[\tau] = \{x : V(\tau, x) \le 1\} \ .$$

We shall now indicate an explicit relation for $V(\tau, x)$.

First, consider set

$$W_\mu[\tau] = m + \mu\mathcal{E}(0, M) - \int_\tau^{t_1} (p(t) + f(t) + \mu\mathcal{E}(0, P(t)))dt \ .$$

This set is similar to set $\mathcal{W}[\tau]$ of (1.4.4) with $\mathcal{M} = m + \mu\mathcal{E}(0, M), \ \ \mathcal{P}(t) = p(t) + \mu\mathcal{E}(0, P(t))$.

Its support function

$$\rho(l | W_\mu[\tau]) = (l, x^*(\tau)) + \mu\Big((l, Ml)^{1/2} + \int_\tau^{t_1} (l, P(t)l)^{1/2} dt\Big) \ ,$$

where $x^*(t)$ is the solution to system

$$\dot{x}^* = p(t) + f(t), \ \ x^*(t_1) = m \ .$$

Second, for a given position $\{\tau, x\}$ find smallest μ for which $x \in W_\mu[\tau]$. We have $x \in W_\mu[\tau]$ if

$$(l, x) \le \rho(l | W_\mu[\tau]), \ \ \forall l \in \mathbb{R}^n \ ,$$

or otherwise

$$(l, x - x^*(\tau))(H(\tau, l))^{-1} \le \mu, \ \ \forall l \ ,$$

where

$$H(\tau, l) = (l, Ml)^{1/2} + \int_\tau^{t_1} (l, P(t)l)^{1/2} dt \ .$$

This immediately yields

Lemma 1.5.4 *The value function*

$$V(\tau,x) = \max\{(l, x - x^*(\tau))(H(\tau,l))^{-1} | l \in \mathbb{R}^n\} \quad . \tag{1.5.17}$$

(Check that here, with $M > 0$, the maximum is attained.)

Exercise 1.5.2. Try to write a formal H-J-B equation for cost criterion $\Psi(\tau, x, u(\cdot))$. Check whether this equation does have a classical solution. In what sense could function $V(\tau, x)$ be considered a solution to this equation? Would it be a viscosity solution [109]? (See [82].)

Later, in Part IV, Sections 4.1–4.3, we shall indicate an approach for approximating the solution of the H-J-B equation of Exercise 1.5.2, rather than solving it explicitly.

Naturally, the description of attainability domains also allows an application of DP. Indeed, since $\mathcal{W}[t]$ is similar to the attainability domain $\mathcal{X}[t]$, if the latter is calculated in backward time, it is possible to formulate an optimization problem, such that $\mathcal{X}[t]$ would be the level set for the respective value function. (We ask the reader to specify the formulation of such a problem.) Later, in Sections 4.1–4.4, we shall discuss this issue in conjunction with ellipsoidal techniques.

Our next subject will be the issue of *uncertainty* in the knowledge of the system inputs.

1.6 Uncertain Systems: Attainability Under Uncertainty

We are returning to systems (1.1.1), (1.1.2) and (1.1.4), but now the disturbance (or forcing term) $f(t)$ will be taken to be *unknown but bounded,* namely, the information on $f(t)$ will be restricted to the inclusion

$$f(t) \in \mathcal{Q}(t) \quad , \tag{1.6.1}$$

where $\mathcal{Q}(t)$ is a given multi-valued map $\mathcal{Q} : T \to \text{conv}\mathbb{R}^n$, continuous in t.

We therefore come to the following systems:

(i) the linear differential equation

(1.6.2) $$\dot{x} = u(t) + f(t), \; x(t_0) \in \mathcal{X}^0, \; t \in T \; ,$$

that reflects the availability of only *open-loop controls* $u(\cdot) \in U_{\mathcal{P}}^{0}$ and also has an unknown disturbance $f(t)$ subject to a given constraint (1.6.1),

(ii) the nonlinear differential inclusion

(1.6.3) $$\dot{x} \in \mathcal{U}(t,x) + f(t), \quad x^0 \in \mathcal{X}^0 \; ,$$

where $\mathcal{U}(\cdot,\cdot) \in U_{\mathcal{P}}^{c}$ and $f(t)$ is unknown, but bounded by constraint (1.6.1). This reflects the availability of *closed-loop (feedback) controls.*

What would be the notion of *attainability* now that the input $f(t)$ is unknown? It is quite obvious that the respective definitions for both open-loop and closed-loop controls could be presented in several ways. We shall start with the following *open-loop* construction that will be used in the sequel.

Definition 1.6.1 *An open-loop domain of attainability under counteraction for system (1.6.2), (1.6.1) from set $\mathcal{X}^0 = \mathcal{X}[t_0]$ at time t_1 is defined as the set $\mathcal{X}(t_1, t_0, \mathcal{X}^0) = \mathcal{X}[t_1]$ of all states x^*, such that for any $f^*(\cdot) \in Q(\cdot)$ there exists a pair $\{x^{0*}, u^*(\cdot)\}$, $x^{0*} \in \mathcal{X}^0, u^*(\cdot) \in U_{\mathcal{P}}^{0}$, that generates a trajectory $x[t]$ of system*

$$\dot{x} = u^* + f^*, \quad x[t_0] = x^{0*}, \quad t \in T \; ,$$

that satisfies the boundary conditions

$$x^{0*} \in \mathcal{X}^0, \quad x[t_1] = x^* \; .$$

Let us further add the symbol $f(\cdot)$, to the notation of the attainability domain $\mathcal{X}[t] = \mathcal{X}(t, t_0, \mathcal{X}^0)$ of Section 1.2 emphasizing its dependence on a given input $f(\cdot)$, namely,

$$\mathcal{X}_f[t] = \mathcal{X}(t, t_0, \mathcal{X}^0, f) \; .$$

In other terms $\mathcal{X}[t] = \mathcal{X}(t, t_0, \mathcal{X}^0, f)$ is the cross-section at instant t of the solution tube to the linear-convex differential inclusion

$$\dot{x} \in \mathcal{P}(t) + f(t), \quad \mathcal{X}(t_0) = \mathcal{X}^0 \ .$$

The set $\mathcal{X}[t_1]$ of Definition 1.6.1 may then be presented as

$$\mathcal{X}[t_1] = \mathcal{X}(t_1, t_0, \mathcal{X}^0) = \tag{1.6.4}$$

$$= \bigcap_f \bigcup_{x_0} \left\{ \mathcal{X}(t_1, t_0, \{x^0\}, f) \,\middle|\, x^0 \in \mathcal{X}^0, f(\cdot) \in Q(\cdot) \right\}$$

or

$$\mathcal{X}[t_1] = \bigcap \left\{ \mathcal{X}(t_1, t_0, \mathcal{X}^0, f) \,\middle|\, f(\cdot) \in Q(\cdot) \right\}$$

Remark 1.6.1 *Other types of attainability domains than those in Definition 1.6.1 may be defined by introducing operations of either intersection* $\cap$ *or union* $\cup$ *over* $x^0, f(\cdot)$*, in an order other than (1.6.4). We invite the reader to investigate this issue.*

Returning to (1.6.4) and taking $\mathcal{X}[t]$, for any $t \in T$, we come to the *open-loop solution tube* (under counteraction).

Let us see, whether it is possible to derive an *evolution equation* for the tube $\mathcal{X}[t]$. Obviously $x^* \in \mathcal{X}[t]$ if and only if

$$(l, x^*) \leq \rho(l | \mathcal{X}(t, t_0, \mathcal{X}^0, f)) \quad \forall l \in \mathbb{R}^n, \quad \forall f(\cdot) \in \mathcal{Q}(\cdot)$$

or

$$(l, x^*) \leq h(t, l) \ ,$$

with

$$h(t, l) = \inf \left\{ \rho(l | \mathcal{X}(t, t_0, \mathcal{X}^0, f) \mid f(\cdot) \in \mathcal{Q}(\cdot) \right\} \ .$$

Here direct calculation gives

$$h(t, l) = \rho(l | \mathcal{X}^0) + \int_{t_0}^{t} \rho(l | \mathcal{P}(\tau)) d\tau + \inf \left\{ \int_{t_0}^{t} (l, f(\tau)) d\tau \mid f(\cdot) \in \mathcal{Q}(\cdot) \right\}$$

$$h(t, l) = \rho(l | \mathcal{X}^0) + \int_{t_0}^{t} (\rho(l | \mathcal{P}(\tau)) - \rho(-l | \mathcal{Q}(\tau))) d\tau \ . \tag{1.6.5}$$

The function $h(t, l)$ is positively homogeneous in l, namely, $h(t, \alpha l) = \alpha h(t, l)$, for all $\alpha > 0$.

Assumption 1.6.1 *The function*

$$g(\tau, l) = \rho(l|\mathcal{P}(\tau)) - \rho(-l|\mathcal{Q}(\tau))$$

is convex in l (and finite valued: $g(\tau, l) > -\infty$, $\forall l, \forall \tau \in T$).

Under Assumption 1.6.1 the function $g(\tau, l)$ is convex in l. It is also positively homogeneous in l and therefore turns out to be *the support function* for a certain set $\mathcal{R}(\tau)$ that evolves continuously in time. A standard result in convex analysis consists in the following [265],

Lemma 1.6.1 *Under Assumption 1.6.1 the function*

$$g(\tau, l) = \rho(l|\mathcal{P}(\tau)) - \rho(-l|\mathcal{Q}(\tau))$$

is the support function for set

$$\mathcal{R}(\tau) = \mathcal{P}(\tau) \dot{-} (-\mathcal{Q}(\tau)) \ ,$$

namely,

$$\rho(l|\mathcal{R}(\tau)) = \rho(l|\mathcal{P}(\tau) \dot{-} (-\mathcal{Q}(\tau))) = \rho(l|\mathcal{P}(\tau)) \ - \ \rho(l| - \mathcal{Q}(\tau))$$

and $\mathcal{R}(\tau) \neq \emptyset$. The set-valued map $\mathcal{R}(\tau)$ is continuous.

Here $\mathcal{P} \dot{-} \mathcal{Q}$ stands for the *geometrical ("Minkowski") difference* between $\mathcal{P}$ and $\mathcal{Q}$:

$$\mathcal{P} \dot{-} \mathcal{Q} = \{x : x + \mathcal{Q} \subseteq \mathcal{P}\} \ .$$

Following the proof of Theorem 1.4.1 we come to

Theorem 1.6.1 *Under Assumption 1.6.1 the tube $\mathcal{X}[t]$ satisfies the following evolution equation*

$$\text{(1.6.6)} \quad \lim_{\sigma \to 0} \sigma^{-1} h(\mathcal{X}[t+\sigma], \ \mathcal{X}[t] + \sigma\mathcal{R}(t)) = 0, \quad \mathcal{X}[t_0] = \mathcal{X}^0 \ .$$

The solution $\mathcal{X}[t] = \mathcal{X}(t, t_0, \mathcal{X}^0)$ to equation (1.6.6) satisfies the semi-group property

$$\mathcal{X}(t, \tau, \mathcal{X}(\tau, t_0, \mathcal{X}^0)) = \mathcal{X}(t, t_0, \mathcal{X}^0) \ ,$$

for $t_0 \leq \tau \leq t \leq t_1$.

Remark 1.6.2 *A typical example for Assumption 1.6.1 is when* $0 \in \mathcal{P}$, $-\mathcal{Q} = \alpha\mathcal{P} + c$, $0 < \alpha < 1$, $c \in \mathbb{R}^n$, *so that*

$$\mathcal{P} \dot{-} (-\mathcal{Q}) = (1-\alpha)\mathcal{P} - c \ .$$

This case is known as the *matching condition* for the constraints on u, f in equation (1.2.1).

It is not difficult to formulate the necessary and sufficient condition for $\mathcal{X}[t]$ to be nonvoid. This is given by

Lemma 1.6.2 *The set* $\mathcal{X}[t]$ *is nonvoid* ($\mathcal{X}[t] \neq \emptyset$) *if and only if there exists a vector* $c \in \mathbb{R}^n$ *such that the function* $h(t,l) - (l,c) \geq 0$, $\forall l \in \mathbb{R}^n$.

The proof of this assertion is a standard exercise in convex analysis.

Once $\mathcal{X}^0 = \{x^0\}$ is a singleton, the function $h(t,l)$ satisfies the condition of Lemma 1.6.2 if the following assumption is fulfilled.

Assumption 1.6.2 *The geometrical difference of the following two integrals is not empty:*

$$\int_{t_0}^{t} \mathcal{P}(\tau)d\tau \dot{-} \int_{t_0}^{t} (-\mathcal{Q}(\tau))d\tau \neq \emptyset \ .$$

Thus the set $\mathcal{X}[t] \neq \emptyset$ if and only if $(\mathrm{co}_l h)(t,l) \neq -\infty$, for all l, where $(\mathrm{co}_l h)\ (t,l)$ is the *closed convex hull* of $h(t,l)$ in the variable l [265], [100].

Then $x^* \in \mathcal{X}[t]$ if and only if

$$(l, x^*) \leq (\mathrm{co}_l h)(t,l), \quad \forall l \in \mathbb{R}^n$$

or

(1.6.7) $$(\mathrm{co}_l h)(t,l) = \rho(l \mid \mathcal{X}[t]) \ .$$

Therefore

$$(1.6.8) \qquad \mathcal{X}[t] = (\mathcal{X}[t_0] + \int_{t_0}^{t} \mathcal{P}(\tau)d\tau) \dot{-} \int_{t_C}^{t} \mathcal{Q}(\tau)d\tau \ .$$

It follows under Assumption 1.6.1, that the set $\mathcal{X}[t] \neq \emptyset$ *for any* convex compact set $\mathcal{X}[t_0]$ and any $t \in T$, since

$$\int_{t_0}^{t} \mathcal{P}(\tau)d\tau \dot{-} \int_{t_0}^{t} \mathcal{Q}(\tau)d\tau \supseteq \int_{t_0}^{t} (\mathcal{P}(\tau) \dot{-} \mathcal{Q}(\tau))\, d\tau$$

(prove this inclusion).

Remark 1.6.3 *The results of convex analysis imply a formal calculation for determining the closed convex hull* $(co_l h)(t,l)$. *This is given by the relation*

$$(co_l h)(t,l) = h_l^{**}(t,l) \ ,$$

where $h_l^{**}(t,l)$ *is the second conjugate of* $h(t,l)$ *in the variable* l. *Recall that*

$$k^{**}(l) = (k^*)^*(l) \ ,$$

where

$$k^*(p) = \sup\{(l,p) - k(l) | l \in \mathbb{R}^n\} \ .$$

Remark 1.6.4 *It is not difficult to check that the tube* $\mathcal{X}[t]$ *of (1.6.8) may not satisfy the semigroup property.*

Exercise. Construct an example for this remark.

Let us now define an attainability domain under counteraction in the class of feedback (closed-loop) control strategies.

Given a strategy $u = \mathcal{U}^*(t,x)$, $\mathcal{U}^*(t,x) \in U_{\mathcal{P}}^c$, and $f(\cdot) \in \mathcal{Q}(t)$, we shall define the respective solution tube to system

$$(1.6.9) \qquad \dot{x} \in \mathcal{U}^*(t,x) + f(t) \ ,$$

$$x(t_0) = x^0$$

as $\mathcal{X}(t, t_0, x^0, f|\mathcal{U}^*)$, so that

$$\mathcal{X}(t, t_0, \mathcal{X}^0, f|\mathcal{U}^*) = \bigcup\left\{\mathcal{X}(t, t_0, x^0, f|\mathcal{U}^*) \ : \ x^0 \in \mathcal{X}^0\right\} .$$

The union of such tubes will be

$$\mathcal{X}(t, t_0, \mathcal{X}^0, f) = \bigcup\left\{\mathcal{X}(t, t_0, \mathcal{X}^0, f|\mathcal{U}) \ : \ \mathcal{U} \in U_{\mathcal{P}}^{c}\right\} .$$

Definition 1.6.2 *A closed-loop domain of attainability under counteraction for system (1.6.3), (1.6.1) from set $\mathcal{X}^0 = X[t_0]$ at time t_1 is defined as the set $X[t_1] = X(t_1, t_0, \mathcal{X}^0)$ of all states x^*, such that for any $f(\cdot) \in \mathcal{Q}(\cdot)$ there exists a vector $x^{0*} \in \mathcal{X}^0$, and a strategy $\mathcal{U}^* \in U_{\mathcal{P}}^{c}$, such that the pair $\{x^{0*}, \mathcal{U}^*\}$, generates a solution tube $\mathcal{X}(t_1, t_0, x^{0*}, f|\mathcal{U}^*)$ to system (1.6.9) that satisfies the boundary condition*

$$x^* \in \mathcal{X}(t_1, t_0, x^{0*}, f|\mathcal{U}^*) .$$

In other words, $X(t_1, t_0)$ can be described as

(1.6.10)
$$X(t_1, t_0, \mathcal{X}^0) =$$

$$= \bigcap_f \bigcup_{\mathcal{U}} \bigcup_{x_0} \left\{\mathcal{X}(t_1, t_0, x^0, f|\mathcal{U}) \ : \ x^0 \in \mathcal{X}^0, \mathcal{U} \in U_{\mathcal{P}}^{c}, \ f(\cdot) \in \mathcal{Q}(\cdot)\right\} =$$

$$= \bigcap\left\{\mathcal{X}(t_1, t_0, \mathcal{X}^0, f) \ : \ f(\cdot) \in \mathcal{Q}(\cdot)\right\} .$$

This also means that set $X[t] = X(t, t_0, \mathcal{X}^0)$ consists, for any fixed $t \in T$, of all those states x^* such that for any $f(\cdot) \in \mathcal{Q}(\cdot)$ there exists a solution $x[\tau], \tau \in [t_0, t]$ to (1.6.3) generated by some $x[t_0] \in \mathcal{X}^0$, $\mathcal{U} \in U_{\mathcal{P}}^{c}$, such that $x[t] = x^*$.

Other types of attainability domains under feedback and counteraction could be defined by introducing operations of intersection or union over $x^0, f, \mathcal{U}$ in an order other than in (1.6.10). This is left to the reader.

1.7 Uncertain Systems: The Solvability Tubes

An idea that is also important for control synthesis is that of the *solvability set.* We shall start with a respective definition for the case of open-loop controls.

Definition 1.7.1 *The open-loop solvability set under counteraction at time $t, t < t_1$, for terminal set $\mathcal{M}$ is the set $W[t] = W(t, t_1, \mathcal{M})$ of* **all states** $x^* \in \mathbb{R}^n$, *such that for every function $f(\tau) \in \mathcal{Q}(\tau)$, $t \le \tau \le t_1$, there exists an open-loop control $u = u(\tau)$, $u(\cdot) \in U_{\mathcal{P}}^{O}$ that steers system*

$$\dot{x} = u(\tau) + f(\tau) \tag{1.7.1}$$

from $x^* = x(t)$ to set $\mathcal{M}$, so that $x(t_1) \in \mathcal{M}$.

A direct calculation similar to that of Section 1.6 (see 1.6.5) gives:

$$x^* \in \mathcal{W}[t]$$

if and only if

$$(l, x^*) \le k(t, l), \quad \forall l \in \mathbb{R}^n \quad ,$$

where

$$k(t, l) = \rho(l|\mathcal{M}) + \int_t^{t_1} (\rho(l| - \mathcal{P}(\tau)) - \rho(l|\mathcal{Q}(\tau)))d\tau \quad .$$

In terms of set-valued maps this allows the relation

$$\mathcal{W}[t] = \left(\mathcal{M} + \int_t^{t_1} (-\mathcal{P}(\tau))d\tau\right) \dot{-} \int_t^{t_1} \mathcal{Q}(\tau)d\tau \quad . \tag{1.7.2}$$

If nonvoid, the function $\mathcal{W}[t] = \mathcal{W}(t, t_1, \mathcal{M})$ generates a multivalued map with convex compact values (provided $\mathcal{M} \in \text{conv } \mathbb{R}^n$). For the open-loop case considered here, it is clear that the *inverse problem* of finding $W[t]$ is precisely the one of constructing the *attainability domain* of Definition 1.6.1, but when the latter is taken *in backward time.* This does not mean, however, that $\mathcal{W}[t]$ would immediately satisfy a semigroup property and therefore, an evolution funnel equation, since an additional assumption is required here.

Lemma 1.7.1 *Under Assumption 1.6.1 the map $\mathcal{W}[t]$ satisfies the evolution equation*

$$\lim_{\sigma\to 0}\sigma^{-1}h(\mathcal{W}[t-\sigma],\ \mathcal{W}[t]-\sigma(\mathcal{P}(t)\dot{-}(-\mathcal{Q}(t)))=0 \tag{1.7.3}$$

$$\mathcal{W}[t_1]=\mathcal{M}\ .$$

This Lemma follows directly from Theorem 1.6.1. Particularly, under the conditions of Remark 1.6.2 (namely, $0\in\mathcal{P}(t)$, $-\mathcal{Q}(t)=\alpha\mathcal{P}(t)+c$, $\alpha\in(0,1)$), we have

$$\mathcal{P}(t)\dot{-}(-\mathcal{Q}(t))=(1-\alpha)\mathcal{P}(t)+c=\mathcal{R}(t)\ . \tag{1.7.4}$$

The evolution equation (1.7.3) in this case is precisely the equation (1.6.4), but its solution is evolving in backward time, starting at t_1 and moving towards given instant $t<t_1$.

Our aim, however, is to devise a feedback control strategy for an uncertain system that operates under *unknown but bounded input disturbances.* A precise definition of the problem as well as its solution will be given in the next section. This solution requires some preparatory work. Let us formally construct a set $\mathcal{W}^*(t,t_1,\mathcal{M})$ which shall be a certain superposition of the open-loop sets $\mathcal{W}(t,t_1,\mathcal{M})$ defined above.

Taking the interval $t\le\tau\le t_1$, introduce a subdivision $\Sigma=\{\sigma_1,\ldots,\sigma_k\}$,

$$t=t_1-\sum_{i=1}^{k}\sigma_i,\ldots,t_1-\sigma_1,t_1\ ,$$

where

$$\sigma_i>0,\quad \sum_{i=1}^{k}\sigma_i=t_1-t\ .$$

As a first step, starting at instant t_1, find the open-loop solvability set $\mathcal{W}[t_1-\sigma_1]=\mathcal{W}(t_1-\sigma_1,t_1,\mathcal{M})$. Due to (1.7.2) this gives

$$\mathcal{W}[t_1-\sigma_1]=\Big(\mathcal{M}+\int_{t_1-\sigma_1}^{t_1}(-\mathcal{P}(\tau))d\tau\Big)\dot{-}\int_{t_1-\sigma_1}^{t_1}\mathcal{Q}(\tau)d\tau\ . \tag{1.7.5}$$

Following the procedure, we come to

(1.7.6) $$\mathcal{W}(t-\sigma_1-\sigma_2, t_1-\sigma_1, \mathcal{W}[t_1-\sigma_1]) =$$

$$= \left(\mathcal{W}[t_1-\sigma_1] + \int\limits_{t_1-\sigma_1-\sigma_2}^{t_1-\sigma_1} (-\mathcal{P}(\tau))d\tau\right) \dot{-} \int\limits_{t_1-\sigma_1-\sigma_2}^{t_1-\sigma_1} \mathcal{Q}(\tau)d\tau$$

and may finally calculate the value

$$\mathcal{W}\Big(t, t+\sigma_k, \mathcal{W}(t+\sigma_k, t+\sigma_k+\sigma_{k-1}, \ldots, \mathcal{W}(t_1-\sigma_1, t_1, \mathcal{M})\ldots)\Big) =$$

(1.7.7) $$= \mathcal{J}(t, t_1, \mathcal{M}, \Sigma) \ .$$

The formal procedure described here presumes all the sets $\mathcal{W}(\cdot)$ of type (1.7.5)–(1.7.7) involved in the construction to be nonvoid.

Assumption 1.7.1 *There exists a continuous function $\beta(t) > 0, t \in T$, such that all the sets*

$$\mathcal{W}(t_1 - \sum_{i=1}^{j}\sigma_i,\ t_1 - \sum_{i=1}^{j-1}\sigma_i,\ \mathcal{W}(t_1 - \sum_{i=1}^{j-1}\sigma_1,\ t_1 - \sum_{i=1}^{j-2}\sigma_i, \ldots \ ,$$

$$\ldots, \mathcal{W}(t_1-\sigma_1, t_1, \mathcal{M}))\ldots) - \beta(t_1 - \sum_{i=1}^{j}\sigma_i)\mathcal{S}$$

are nonvoid with $j = 1, \ldots, k$, whatever is the subdivision

$$\Sigma = \{\sigma_1, \ldots, \sigma_k\}, \quad \sum_{i=1}^{k}\sigma_i = t_1 - t, \quad \sigma_i > 0 \ .$$

Assumption 1.7.1 clearly ensures $\mathcal{J}(t, t_1, \mathcal{M}, \Sigma) \neq \emptyset$ for any subdivision Σ.

Following (1.7.2), (1.7.5)–(1.7.7), we come to the analytical expression

$$\mathcal{J}(t, t_1, \mathcal{M}, \Sigma) =$$

$$= (\ldots((\mathcal{M} + \int\limits_{t_1-\sigma_1}^{t_1} (-\mathcal{P}(\tau)d\tau) \dot{-} \int\limits_{t_1-\sigma_1}^{t_1} \mathcal{Q}(\tau)d\tau) \ldots \dot{-} \int\limits_{t}^{t+\sigma_k} \mathcal{Q}(\tau)d\tau)) \ .$$

The set $\mathcal{J}(t, t_1, \mathcal{M}, \Sigma)$ is convex and compact for any subdivision Σ. We may consider the limit of these sets with $\max\{\sigma_i : i = 1, \ldots, k\} \to 0$.

Lemma 1.7.2 *Under Assumption 1.7.1 there exists a Hausdorff limit* $\mathcal{J}(t, t_1, \mathcal{M})$*:*

$$\lim h\Big(\mathcal{J}(t, t_1, \mathcal{M}, \Sigma)\,,\ \mathcal{J}(t, t_1, \mathcal{M})\Big) = 0$$

with

$$\max\{\sigma_i : i = 1, \ldots, k\} \to 0,\ k \to \infty, \sum_{i=1}^{k} \sigma_i = t_1 - t \ .$$

We shall refer to

$$\mathcal{J}(t, t_1, \mathcal{M}) = \mathcal{W}^*(t, t_1, \mathcal{M}) = \mathcal{W}^*[t] \tag{1.7.8}$$

as the *alternated solvability domain* and denote

$$\mathcal{J}(t, t_1, \mathcal{M}) = \int_{t_1, \mathcal{M}}^{t} ((-\mathcal{P}(\tau))d\tau \dot{-} \mathcal{Q}(\tau)d\tau) \ .$$

The set $\mathcal{J}(t, t_1, \mathcal{M})$ is actually the value of a certain type of *set-valued integral* that is known as the *"Alternated Integral of L.S. Pontryagin."* The integral was introduced and described in detail in papers [256], [257].

Definition 1.7.2 *With t varying, the set-valued function* $\mathcal{W}^*[t]$ *of (1.7.8) will be referred to as the "alternated solvability tube".*

Lemma 1.7.3 *Once* $\mathcal{W}^*[t] \neq \emptyset$, $t \in T$, *the set-valued function* $\mathcal{W}^*[t]$ *satisfies for all* $t \in T$ *the evolution equation*

$$\lim_{\sigma \to 0} \sigma^{-1} h_+(\mathcal{W}[t - \sigma] + \sigma \mathcal{Q}(t),\ \mathcal{W}[t] - \sigma \mathcal{P}(t)) = 0 \ , \tag{1.7.9}$$

$$\mathcal{W}[t_1] = \mathcal{M} \ . \tag{1.7.10}$$

Proof Obviously $\mathcal{W}^*[t_1] = \mathcal{M}$. Taking $\mathcal{W}^*$ at an arbitrary instant of time t and also $\mathcal{W}^*[t - \sigma]$, and following the definitions of these sets we observe that there exists a function $\gamma(\sigma)$, such that

$$\mathcal{W}^*[t - \sigma] \subseteq \mathcal{W}(t - \sigma, t, \mathcal{W}^*[t]) \ + \ \gamma(\sigma)\mathcal{S} \ , \tag{1.7.11}$$

where

$$\gamma(\sigma) > 0, \quad \sigma > 0; \quad \sigma^{-1}\gamma(\sigma) \to 0 \text{ with } \sigma \to 0, \quad (\text{see [257]}) .$$

Due to

(1.7.12)
$$\mathcal{W}(t-\sigma, t, \mathcal{W}^*[t]) = \\ = (\mathcal{W}^*[t] + \int_{t-\sigma}^{t} (-\mathcal{P}(\tau))d\tau) \dot{-} \int_{t-\sigma}^{t} \mathcal{Q}(\tau)d\tau \ ,$$

and to the definition of geometric difference, relation (1.7.11) yields the following:

(1.7.13)
$$\left(\mathcal{W}^*[t-\sigma] + \int_{t-\sigma}^{t} \mathcal{Q}(\tau)d\tau\right) \subseteq \\ \subseteq \mathcal{W}^*[t] + \int_{t-\sigma}^{t} (-\mathcal{P}(\tau))d\tau + \gamma(\sigma)\mathcal{S} \ .$$

The continuity of $\mathcal{P}(\tau)$, and $\mathcal{Q}(\tau)$ implies

$$\lim_{\sigma\to 0} \sigma^{-1} h\left(\int_{t-\sigma}^{t} \mathcal{P}(\tau)d\tau, \ \sigma\mathcal{P}(\tau)\right) = 0 \ ,$$

$$\lim_{\sigma\to 0} \sigma^{-1} h\left(\int_{t-\sigma}^{t} \mathcal{Q}(\tau)d\tau, \ \sigma\mathcal{Q}(t)\right) = 0 \ .$$

The latter relations, together with (1.7.1) give the inclusion

(1.7.14) $$\mathcal{W}^*[t-\sigma] + \sigma\mathcal{Q}(t) \subseteq \mathcal{W}^*[t] - \sigma\mathcal{P}(t) + \alpha(\sigma)\mathcal{S} \ ,$$

where $\sigma^{-1}\alpha(\sigma) \to 0$ with $\sigma \to 0$.

Q.E.D.

Relation (1.7.14) is equivalent to the existence of a solution to (1.7.9) at any given instant t, particularly, at $t = t_1$. The prolongability of the solution $\mathcal{W}^*[t]$ towards time t_0 follows from the condition $\mathcal{W}^*[t] \neq \emptyset$, $t \in T$ and from the boundedness of the tube $\mathcal{W}^*[t]$. This justifies the assertion of Lemma 1.7.3. Studying equation (1.7.9) it is possible to observe that its solution in nonunique (devise an example) and moreover, that $\mathcal{W}^*[t]$ satisfies the following properties.

Lemma 1.7.4 *The set-valued function $\mathcal{W}^*[t]$ is a maximal solution to equation (1.7.9).*

The proof of this assertion is left to the reader. It also follows from Lemma 1.8.3 of the next section.

Lemma 1.7.5 *The set-valued map $\mathcal{W}^*(t, t_1, \mathcal{M})$ satisfies the semigroup property (in backward time). Namely,*

$$\mathcal{W}^*(t, t_1, \mathcal{M}) = \mathcal{W}^*(t, \tau, \mathcal{W}^*(\tau, t_1, \mathcal{M})) \quad , \tag{1.7.15}$$

with $t \leq \tau \leq t_1$.

The proof of the relation (1.7.14) follows from the additivity properties of the alternated integral $\mathcal{J}(t, t_1, \mathcal{M})$ [257].

Remark 1.7.1 *The assertions of this section concerning* **the Alternated Solvability Tube** $\mathcal{W}^*[t]$ *have been all derived* **under Assumption 1.7.1.** *Hence, all the propositions that follow in the sequel and involve the tube $\mathcal{W}^*[t]$ are true only under this assumption.*

For future operation it may be sometimes more convenient to use an assumption of equivalent type

Assumption 1.7.2 *The alternated solvability tube $\mathcal{W}^*[t]$ is nondegenerate. Namely, there exists an absolutely continuous function $x(t)$ and a function $\beta(t) > 0$, $t_0 \leq t \leq t_1$, such that*

$$x(t) + \beta(t)\mathcal{S} \subset \mathcal{W}^*[t], \quad t_0 \leq t \leq t_1 \quad .$$

The given assumptions are also important in the sequel in Part III for the justification of ellipsoidal approximations of the present constructions.[9]

As we shall see in the next section, the tube $\mathcal{W}^*[t]$ coincides with the *solvability tube* for the problem of *control synthesis under uncertainty.*

[9]Once Assumptions 1.7.1 or 1.7.2 are not fulfilled, there is a degenerate situation which has to be approached separately, by means of a regularization procedure that allows us to keep up with the basic solution scheme. Such situations are not discussed in this book and are left for additional treatment.

1.8 Control Synthesis Under Uncertainty

Consider system (1.7.1) and terminal set $\mathcal{M}$.

Definition 1.8.1 *The problem of* **control synthesis under uncertainty** *consists in specifying a* **solvability set** $\mathcal{W}_*(\tau, t_1, \mathcal{M})$ *and a set-valued* **feedback control strategy** $u = \mathcal{U}(t,x)$, $\mathcal{U}(\cdot,\cdot) \in U_{\mathcal{P}}^c$ *such that all the solutions to the differential inclusion*

(1.8.1)
$$\dot{x} \in \mathcal{U}(t,x) + \mathcal{Q}(t)$$

that start from any given position $\{\tau, x_\tau\}$, $x_\tau = x[\tau] \in \mathcal{W}_*(\tau, t_1, \mathcal{M})$, $\tau \in [t_0, t_1)$, *would reach the terminal set* $\mathcal{M}$ *at time* $t_1 : x(t_1) \in \mathcal{M}$.

Set $\mathcal{W}_*[t] = \mathcal{W}_*(t, t_1, \mathcal{M})$ *is* **the set of all states** $x_\tau \in \mathbb{R}^n$ *that satisfy the above.*

Definition 1.8.1 is nonredundant if $\mathcal{W}_*(\tau, t_1, \mathcal{M}) \neq \emptyset$, where, as we have seen, $\mathcal{W}_*(\tau, t_1, \mathcal{M})$ is the *solvability* set, which is the largest set of states from which the solution to the problem does exist at all.

Taking $\mathcal{W}_*(t, t_1, \mathcal{M}) = \mathcal{W}_*[t]$, we come to a set-valued map (the solvability tube). We shall prove that the alternated solvability tube $\mathcal{W}^*[t]$ of Section 1.6 does coincide with $\mathcal{W}_*[t]$.

Let us first try to find a tube $\mathcal{Z}(t)$ that would also provide solvability of the problem of control synthesis, but would not necessarily be the largest solvability tube as required by Definition 1.8.1. Assume $\mathcal{Z}(t)$ to be a solution to the evolution equation (1.7.9) with boundary condition

$$\mathcal{Z}(t_1) \subseteq \mathcal{M}$$

and therefore an absolutely continuous set-valued map with convex compact values.

For every such solution $\mathcal{Z}(t)$ let us assign a feedback strategy $\mathcal{U}_{\mathcal{Z}}(t,x)$ constructed similar to the one in Section 1.4 (see (1.4.11), (1.4.14), and (1.4.15)). Thus

(1.8.2)
$$\mathcal{U}_{\mathcal{Z}}(t,x) = \partial_l f(t, -l^0_{\mathcal{Z}}(t,x)) \quad ,$$

where $f(t,l) = \rho(l|\mathcal{P}(t))$ and $l^0 = l^0_{\mathcal{Z}}(t,x)$ is the maximizer of the expression for calculating $d_{\mathcal{Z}}[t,x] = h_+(x,\mathcal{Z}(t)$, which is

$$d_{\mathcal{Z}}[t,x] = \max\{(l,x) - \rho(l|\mathcal{Z}(t)) \ : \ \|l\| \leq 1\} =$$

$$= (l^0,x) - \rho(l^0|\mathcal{Z}(t) \ ,$$

$$(l^0 = \{0\} \ \text{ for } \ d_{\mathcal{Z}}[t,x] = 0).$$

Relation (1.8.2) is formally similar to the definition of the extremal strategy (1.4.19).

Consider the derivative

$$\frac{d}{dt} d^2_{\mathcal{Z}}[t,x] \ = 2 \ d_{\mathcal{Z}}[t,x] \frac{d}{dt} d_{\mathcal{Z}}[t,x]$$

due to system (1.7.1). At a point $\{t,x\}$ that has $d_{\mathcal{Z}}[t,x] > 0$, a direct calculation yields

(1.8.3)
$$\frac{d}{dt} d_{\mathcal{Z}}[t,x] = (l^0,\dot{x}) - \frac{\partial \rho(l^0|\mathcal{Z}(t))}{\partial t} \ .$$

Lemma 1.8.1 *The following inequality is true*

(1.8.4)
$$\frac{\partial \rho(l|\mathcal{Z}(t))}{\partial t} \ \geq \ \rho(l|\mathcal{Q}(t)) - \rho(-l|\mathcal{P}(t)), \ l \in \ \mathbb{R}^n \ .$$

Proof. The evolution equation (1.7.9) leads to the inclusion

$$\mathcal{Z}(t-\sigma) + \sigma\mathcal{Q}(t) \subseteq \mathcal{Z}(t) - \sigma\mathcal{P}(t) + o(\sigma)\mathcal{S}$$

with

$$\sigma^{-1}o(\sigma) \to 0, \quad \sigma \to 0 \ ,$$

and further on,

$$\rho(l|\mathcal{Z}(t-\sigma)) + \sigma\rho(l|\mathcal{Q}(t)) \ \leq \rho(l|\mathcal{Z}(t)) + \sigma\rho(-l|\mathcal{P}(t)) + o(\sigma)(l,l)^{1/2}$$

or otherwise, the inequality

$$\sigma^{-1}(\rho(l|\mathcal{Z}(t) - \rho(l|\mathcal{Z}(t-\sigma)) \geq \rho(l|\mathcal{Q}(t)) - \rho(-l|\mathcal{P}(t)) + \sigma^{-1}o(\sigma)(l,l)^{1/2}$$

which gives, after a limit transition $\sigma \to 0$, the result (1.8.4) of the Lemma.

Q.E.D.

A consequence of Lemma 1.8.1 (see (1.8.2) and (1.8.3)) is that with $d_{\mathcal{Z}}[t,x] > 0$ for its derivative the following inequality holds:

$$(1.8.5) \quad \frac{d}{dt} d_{\mathcal{Z}}[t,x] \leq (l^0, u(t) + f(t)) - \rho(l^0 | \mathcal{Q}(t)) + \rho(-l^0 | \mathcal{P}(t)),$$

where

$$u(t) \in \mathcal{P}(t), \quad f(t) \in \mathcal{Q}(t) \quad .$$

Therefore, with $u = u^0$, where

$$-(l^0, u^0) = \rho(-l^0 | \mathcal{P}(t)) \quad ,$$

we will have

$$(1.8.6) \qquad \frac{d}{dt} d_{\mathcal{Z}}[t,x] \leq 0, \quad \forall f(t) \in \mathcal{Q}(t).$$

This leads us to

Lemma 1.8.2 *The derivative $\frac{d}{dt} d^2_{\mathcal{Z}}[t,x]$ calculated due to the system*

$$(1.8.7) \qquad \dot{x} \in \mathcal{U}_{\mathcal{Z}}(t,x) + f(t)$$

satisfies the inequality

$$\frac{d}{dt} d^2_{\mathcal{Z}}[t,x] \leq 0, \quad \forall f(t) \in \mathcal{Q}(t) \quad .$$

Some further reasoning yields the next assertion

Lemma 1.8.3 *With $x_\tau \in \mathcal{Z}[\tau]$, $\tau < t_1$, the tube $\mathcal{X}_{\mathcal{Z}}[t] = \mathcal{X}_{\mathcal{Z}}(t, \tau, x_\tau | f(\cdot))$ of all trajectories of system (1.8.7), $x[\tau] = x_\tau$, $\tau \leq t \leq t_1$, satisfies the inclusion*

$$(1.8.8) \qquad \mathcal{X}_{\mathcal{Z}}[t] \subseteq \mathcal{Z}[t], \quad \forall f(t) \in \mathcal{Q}(t), \quad \tau \leq t \leq t_1 \quad ,$$

and therefore, the boundary condition $Z(t_1) \subseteq \mathcal{M}$.

The proof of Lemma 1.8.3 is similar to that of Lemma 1.3.3 as the main relation used in the proof is the inequality (1.8.4).

Lemma 1.8.3 thus indicates that $\mathcal{U}_{\mathcal{Z}}(t,x)$ is a *synthesizing strategy* that solves the problem of control synthesis of Definition 1.8.1 with $\mathcal{Z}[\tau]$ being *the solvability domain* (but not necessarily the largest one). It is therefore possible, in principle, to solve the problem of control synthesis through any solution $\mathcal{Z}[t]$ of equation (1.7.9) with boundary condition

$$\mathcal{Z}[t_1] \subseteq \mathcal{M} \quad .$$

The set of states for which the problem of Definition 1.8.1 is solvable will then be restricted to $\mathcal{Z}[t]$. Our problem, however, is to find the *maximal solvability domain* $\mathcal{W}_*[t]$ for the problem of Definition 1.8.1 and the respective strategy $\mathcal{U}(t,x)$.

Referring to Lemmas 1.7.3, 1.7.4, we observe that tube $\mathcal{W}^*[t]$ is the maximal solution to equation (1.7.9) with an equality in the boundary condition ($\mathcal{W}^*[t_1] = \mathcal{M}$). The tube $\mathcal{W}^*[t]$ generates a strategy

$$\mathcal{U}^0(t,x) = \partial_l f(t, -l^0_{\mathcal{W}^*}(t,x)) \quad , \tag{1.8.9}$$

where $l^0 = l^0_{\mathcal{W}^*}(t,x)$ is the maximizer for the problem

$$d_{\mathcal{W}^*}[\tau, x] = \max\{(l,x) - \rho(l|\mathcal{W}^*[\tau])|\ \|l\| \leq 1\}$$

$$d_{\mathcal{W}^*}[\tau, x] = (l^0, x) - \rho(l^0|\mathcal{W}^*[\tau]) \quad . \tag{1.8.10}$$

The results of Lemmas 1.8.2, 1.8.3 imply

Lemma 1.8.4 *Strategy $\mathcal{U}^0(t,x)$ ensures the inclusion*

$$X_{\mathcal{W}^*}(t, \tau, x_\tau) \subseteq \mathcal{W}^*[t], \quad \tau \leq t \leq t_1 \quad , \tag{1.8.11}$$

provided $x_\tau \in \mathcal{W}^*[\tau]$.

Here $X_{\mathcal{W}^*}(t,\tau,x_\tau)$ is the solution tube for system (1.8.1), $x[\tau] = x_\tau$, with $\mathcal{U}(t,x) = \mathcal{U}^0(t,x)$. The results of the above may be summarized into

Theorem 1.8.1 *The synthesizing strategy $\mathcal{U}^0(t,x)$ of (1.8.9) resolves the problem of control synthesis under uncertainty of Definition 1.8.1.*

Remark 1.8.1 *It is necessary to emphasize that the last theorem is true in the absence of matching conditions of the Assumption 1.6.1 type. The result presumes however that the solution $\mathcal{W}^*[t]$ to the evolution equation (1.7.9) and (1.7.10) exists and that int$\mathcal{W}^*[t] \not\equiv \emptyset, \forall t \in T$. The latter is ensured by Assumption 1.7.1.*

Finally, to make the ends meet, we have to answer the following question: is the maximal solution $\mathcal{W}^*[t]$ to equation (1.7.9) and (1.7.10) also the maximal solution tube for the problem of control synthesis of Definition 1.8.1? As we shall see, the answer to the question is affirmative. This may be proved due to the inequality (1.8.4). Namely, once $x^* \notin \mathcal{W}^*[\tau]$, it is possible to select in the domain $d[x^*, \mathcal{W}^*[\tau]] > 0$ a strategy

(1.8.12) $$\mathcal{V}^0(t,x) \;=\; \{v \mid (l^0, v) \;=\; \rho(l^0|\mathcal{Q}(t))\}.$$

This strategy will affect the sign of the derivative $d(d_{\mathcal{W}^*}(t,x^*))/dt$ due to system (1.7.1).

Let us calculate this derivative solving the extremal problem

$$d_{\mathcal{W}^*}[t,x^*] = \max\{(l,x^*) - \rho(l|\mathcal{W}^*[t]) \;:\; \|l\| \leq 1\}$$

$$d_{\mathcal{W}^*}[t,x^*] = (l^0_*, x^*) - \rho(l^0_*|\mathcal{W}^*[t]) \;\;,$$

and

$$l^0_* = 0 \;\text{ for }\; d_{\mathcal{W}^*}[t,x^*] = 0 \;\;.$$

This gives

$$\frac{d}{dt} d_{\mathcal{W}^*}[t,x^*] = (l^0_*, \dot{x}^*) - \partial\rho(l^0_*|\mathcal{W}^*[t])/\partial t \;\;.$$

The calculation of the derivative $\partial\rho(l^0_*|\mathcal{W}^*[t])/\partial t$ can be done using the representations (1.7.12)–(1.7.14). Thus, in view of a relation of type (1.7.13) this further gives

$$\rho(l|\mathcal{W}^*[t{+}\sigma]){-}\rho(l|\mathcal{W}^*[t]) \geq \;-\;\rho(l|\int_t^{t+\sigma}(-\mathcal{P}(\tau))d\tau{+}\int_t^{t+\sigma}\mathcal{Q}(\tau)d\tau){+}o(t,\sigma)\;,$$

(1.8.13) $$\frac{\partial\rho(l|\mathcal{W}^*[t])}{\partial t} \geq \rho(l|\mathcal{Q}(t)) - \rho(l| - \mathcal{P}(t)) \;\;,$$

or, under Assumption 1.6.1,

$$\frac{\partial \rho(l|\mathcal{W}^*[t])}{\partial t} = \rho(l|\mathcal{Q}(t)) - \rho(l| - \mathcal{P}(t)) = \tag{1.8.14}$$

$$= -\rho(l|(-\mathcal{P}(t)) \dot{-} \mathcal{Q}(t)) \ .$$

We shall first continue under this assumption so that, with $Z(t) = \mathcal{W}^*(t)$, the inequality (1.8.5) turns to an equality. Differentiating $d_{\mathcal{W}^*}[t, x^*]$ for $x^* \notin \mathcal{W}^*[t]$, using (1.8.13) and also the respective rule indicated in Remark 1.4.1, we come to the following relation.

Lemma 1.8.5 *Under Assumption 1.6.1 the derivative $d_{\mathcal{W}^*}[t, x^*]/dt$ due to system (1.7.1) is given by the relation*

$$\frac{d}{dt} d_{\mathcal{W}^*}[t, x^*] = (l_*^0, u(t) + f(t)) + \rho(-l_*^0|\mathcal{P}(t)) - \rho(l_*^0|\mathcal{Q}(t)) \ . \tag{1.8.15}$$

Selecting $f \in \mathcal{V}^0(t, x)$, (1.8.12) and observing that

$$u \in \mathcal{P}(t) \ \text{implies} \ -(l_*^0, u) \leq \rho(l_*^0|\mathcal{P}(t)), \quad \forall u \in \mathcal{P}(t) \ ,$$

we arrive, due to (1.8.15), to the following relations

$$\left.\frac{d}{dt} d_{\mathcal{W}^*}[t, x^*]\right|_{f \in \mathcal{V}^0} \geq 0, \quad \forall u \in \mathcal{P}(t) \ ,$$

$$\left.\frac{d}{dt} d_{\mathcal{W}^*}[t, x^*]\right|_{f \in \mathcal{V}^0} = 0, \quad \forall u \in \mathcal{U}^0(t, x) \ .$$

This implies that any solution $x[t]$ to the differential inclusion

$$\frac{dx}{dt} \in \mathcal{U}(t, x) + f; \ f \in \mathcal{V}^0(t, x), \quad x[\tau] = x^* \ ,$$

that starts at a point $x^* \notin \mathcal{W}^*[t]$ or in other words, with $d[x^*, \mathcal{W}_*[\tau]] = r_\tau > 0$, does satisfy the inequality $d[x[t], \mathcal{W}_*[t]] \geq r_\tau, t \in [\tau, t_1]$ whatever is the strategy $\mathcal{U}(\cdot, \cdot) \in U_{\mathcal{P}}^c$. Under Assumption 1.6.1 we have therefore proved

Theorem 1.8.2 *(i) The alternated solvability tube $\mathcal{W}^*[t]$ coincides with the solvability tube $\mathcal{W}_*[t]$ of the problem of control synthesis under uncertainty, namely,*

$$\mathcal{W}_*[t] = \mathcal{W}^*[t], \quad t_0 \leq t \leq t_1 \ .$$

(ii) The set $\mathcal{W}^*[\tau]$, $\tau \in [t_0, t_1)$ is the largest solvability domain for this problem.

It should be emphasized that this theorem *remains true without the Assumption 1.6.1.* To prove (i) and (ii) of the last Theorem in the general case, one has to substitute strategy $\mathcal{V}^0(t, x)$ of (1.8.12) by another one $\mathcal{V}^*$ that would in some sense ensure a relation similar to the following type (for an appropriate vector l^0)

$$\frac{d}{dt} d_{W^*}[t, x] = (l^0, u + f(t)) \; + \; \rho(l^0 | (-\mathcal{P}(t)) \dot{-} \mathcal{Q}(t)) \geq 0$$

$$f \in \mathcal{V}^*(t, x), \quad \forall u \in \mathcal{P}(t) \; .$$

Since in general we have

$$\rho(l | (-\mathcal{P}) \dot{-} \mathcal{Q}) \; = \; co(f_1 - f_2)(l) \; , \tag{1.8.16}$$

$$f_1(l) = \rho(l | -\mathcal{P}), \;\; f_2(l) = \rho(l | \mathcal{Q}) \; ,$$

the desired strategy $\mathcal{V}^*$ may not exist in the explicit form of (1.8.12). It exists, however, in the class of *mixed strategies* (also known as *relaxed controls*), where $\mathcal{V}^*$ has to be specified as a probabilistic measure concentrated on $\mathcal{Q}$. Loosely speaking, the value v may be required to run around a variety $v^{(i)}$ of some extremal points of the set $\mathcal{Q}$, throughout any minor interval of time. We shall not specify the rigorous definition and precise construction of such strategies $\mathcal{V}^*$ as this would require us to discuss notions that are quite beyond the scope of this book, referring the reader to monographs [169], [170], [171], on *differential games*, where these topics are discussed in detail.

Let us now pass to the DP interpretation for this section. Consider equation (1.7.1) and target set $\mathcal{M}$. Introduce the value function

$$V_*(t, x) \; = \; \min_u \max_f \{\mathcal{I}(t, x) | \mathcal{U}(\cdot, \cdot) \in U_{\mathcal{P}}^c, f(\cdot) \in \mathcal{Q}(\cdot)\}$$

where $\mathcal{I}(t, x)$ is the same as in (1.5.1). Our aim is to minimaxize the cost $\mathcal{I}(t, x)$ over all the strategies $\mathcal{U}(\cdot, \cdot) \in U_{\mathcal{P}}^c$ and disturbances $f(\cdot) \in \mathcal{Q}(\cdot)$. Here the formal H-J-B equation for the value $V_*(t, x)$ looks as follows

$$\frac{\partial V}{\partial t} \; + \; \min_u \max_f \left(\frac{\partial V}{\partial x}, u + f \right) = 0 \tag{1.8.17}$$

with boundary condition

$$V(t_1, x) = h_+^2(x, \mathcal{M}) \ . \tag{1.8.18}$$

(When a minmax operation is involved, the latter H-J-B equation is often referred to as the *the H-J-B-I equation* with letter *I* being a reference to R. Isaacs and his contribution to differential games.)

Presuming $\mathcal{W}^*[t] \neq \emptyset$, consider the function $V(t,x) = d^2_{\mathcal{W}^*}[t,x]$. Obviously

$$V(t_1, x) = h_+^2(x, \mathcal{M}) \ . \tag{1.8.19}$$

Then in view of Lemma 1.8.2 one may observe that $V(t,x)$ satisfies the inequality

$$\frac{\partial V(t,x)}{\partial t} + \max_f \left(\frac{\partial V(t,x)}{\partial x}, \ u + f \right) \leq 0 \tag{1.8.20}$$

and boundary condition (1.8.18), provided

$$u \in \mathcal{U}_{\mathcal{W}^*}(t,x) = \partial_l f(t, -l^0_{\mathcal{W}^*}(t,x))$$

where $f(t,l) = \rho(l|\mathcal{P}(t))$ and $l^0_{\mathcal{W}^*}(t,x) = \partial d^2_{\mathcal{W}^*}[t,x]/\partial x$ with $d_{\mathcal{W}^*}[t,x] > 0$, $l^0_{\mathcal{W}^*}(t,x) = 0$ with $d_{\mathcal{W}^*}[t,x] = 0$.

Denoting $\mathcal{U}_{\mathcal{W}^*}(t,x) = \mathcal{U}^*(t,x)$, we may rewrite

$$\mathcal{U}^*(t,x) = \arg\min\{ \left(\frac{\partial V(t,x)}{\partial x}, u \right) \Big| u \in \mathcal{P}(t)\} = \tag{1.8.21}$$

$$= \partial_l f\left(t, \left(-\frac{\partial V(t,x)}{\partial x}\right)\right) \ .$$

Relations (1.8.20), (1.8.17) and (1.8.21), (1.8.9) then imply

Lemma 1.8.6 *Suppose $\mathcal{W}^*[t] \neq \emptyset$. Then the value function $V_*(t,x) \leq V(t,x)$ and the strategy $\mathcal{U}^*(t,x)$, (1.8.21), solves the problem of Control Synthesis of Definition 1.8.1.*

Indeed, inequality (1.8.20) ensures that once $V(t,x) \leq 0$, then $V(\tau, x[\tau]) \leq 0$ for any trajectory $x[\tau] = x(\tau, t, x), \tau \in [t, t_1]$, of the differential inclusion

$$\frac{dx(\tau)}{d\tau} \in \mathcal{U}^*(\tau, x) + f(\tau), \ x[t] = x \ , \tag{1.8.22}$$

whatever is the disturbance $f(\tau)$ that satisfies (1.6.1). The continuity of $\mathcal{W}^*[\tau]$ in τ implies the upper semicontinuity of $\mathcal{U}^*(\tau, x)$ in its variables and, therefore, the existence of solutions to the differential inclusion (1.8.22).

Under Assumption 1.6.1 relation (1.8.20) turns into an equality and $V_*(t, x) = V(t, x)$. In order to achieve this equality without such an assumption one has to allow the disturbance f to be selected as indicated in the comments after Theorem 1.8.2, namely, in the class of functions generated by a mixed strategy which may result in sliding modes or so-called chattering functions f. Then $V(t, x)$ will be the value of a respective *differential game* (see [171], [291]).

Our next issue is to deal with *state constraints.*

1.9 State Constraints and Viability

Let us return to system

(1.9.1) $$\dot{x}(t) \in \mathcal{P}(t) + f(t), \quad t_0 \leq t \leq t_1 \ ,$$

(1.9.2) $$x^0 \in \mathcal{X}^0 \ ,$$

with a fixed disturbance $f(t)$, taking it here to be *continuous.* We shall now introduce an additional *state constraint*

(1.9.3) $$Gx(t) \in \mathcal{Y}(t), \quad t_0 \leq t \leq t_1 \ ,$$

where G is a given matrix of dimensions $m \times n$ $(m \leq n)$ and $\mathcal{Y}(t)$ is a multivalued function continuous in t with convex compact values $(\mathcal{Y}(t) \in \text{comp}\mathbb{R}^n, \forall t)$.

We shall start from

Definition 1.9.1 *A trajectory $x[t] = x(t, t_0, x^0)$ of system (1.8.1) and (1.8.2) is said to be viable relative to constraint (1.9.3) if it satisfies the state constraint (1.9.3).*

Our interest is in describing the tube of such trajectories.[10] A detailed theory of viable trajectory tubes for differential inclusions may be found in [17],[193].

Definition 1.9.2 *A viability tube*

$$\mathcal{X}[t] = \mathcal{X}(t, t_0, \mathcal{X}^0)$$

is the union

$$\mathcal{X}[t] = \bigcup \{x(t, t_0, x^0) \ : \ x^0 \in \mathcal{X}^0\}$$

over $\mathcal{X}^0$ of all viable trajectories of system (1.9.1), (1.9.2) relative to constraint (1.9.3).

It is obvious that $\mathcal{X}[t_0] \in \mathcal{Y}(t_0)$.

Let us first calculate the support function $\rho(\ell|\mathcal{X}[t])$ of the crossection $\mathcal{X}[t]$ of the tube $\mathcal{X}[\cdot]$ at time t. This is also the *attainability domain* of system (1.9.1), (1.9.2) under *a state constraint* (1.9.3). The set $\mathcal{X}[t]$ is generated through relations (1.9.1)–(1.9.3). For a certain instant $t = \vartheta$ these relations yield

$$x(\vartheta) = x^0 + \int_{t_0}^{\vartheta} u(\tau)d\tau + \int_{t_0}^{\vartheta} f(\tau)d\tau, \tag{1.9.4}$$

$$Gx(t) = Gx^0 + \int_{t_0}^{t} Gu(\tau)d\tau + \int_{t_0}^{t} Gf(\tau)d\tau \ , \tag{1.9.5}$$

$$t_0 \leq t \leq \vartheta \ ,$$

with restrictions (1.9.3) and

$$x^0 \in \mathcal{X}^0, \quad u(t) \in \mathcal{P}(t), \quad t_0 \leq t \leq \vartheta \ . \tag{1.9.6}$$

It is not difficult to observe that (1.9.4) is equivalent to the equality

$$\ell' x(\theta) = \ell' x^0 + \int_{t_0}^{\vartheta} \ell' u(\tau)d\tau + \int_{t_0}^{\vartheta} \ell' f(\tau)d\tau \tag{1.9.7}$$

[10] This section gives a very concise description of the subject, being only an introduction to other parts of the book.

that should be true for any vector $\ell \in \mathbb{R}^n$, while (1.9.5) is equivalent to the equality

$$(1.9.8)\quad \int_{t_0}^{\vartheta} \lambda'(t)Gx(t)dt = \int_{t_0}^{\vartheta} \lambda'(t)Gx^0 dt + \int_{t_0}^{\vartheta}\left(\int_{\tau}^{\vartheta} \lambda'(t)Gdt\right) u(\tau)d\tau + \int_{t_0}^{\vartheta}\left(\int_{\tau}^{\vartheta} \lambda'(t)Gdt\right) f(\tau)d\tau$$

that should be true for any continuous m-dimensional vector-valued function $\lambda(t)$, ($\lambda(\cdot) \in C^m[t_0, \vartheta]$).

On the other hand, the inclusions (1.9.6) are equivalent to the following inequalities

$$(1.9.9)\qquad \ell' x^0 \le \rho(\ell|\mathcal{X}^0), \quad \forall \ell \in \mathbb{R}^n \ ,$$

$$(1.9.10)\qquad \ell' u(t) \le \rho(\ell|\mathcal{P}(t)), \quad \forall \ell \in \mathbb{R}^n, \quad t_0 \le t \le \vartheta \ ,$$

while (1.9.3) is equivalent to

$$(1.9.11)\qquad 0 \le -\int_{t_0}^{\vartheta} \lambda'(t)Gx(t)dt + \int_{t_0}^{\vartheta} \rho(\lambda(t)|\mathcal{Y}(t))dt \ ,$$

$$\forall \lambda(\cdot) \in C^n[t_0, \vartheta] \ .$$

The set $\mathcal{X}[\vartheta]$ will now consist of all those vectors $x[\theta]$ that satisfy (1.9.7), (1.9.8) under restrictions (1.9.9)–(1.9.11). In other terms, collecting relations (1.9.7)–(1.9.11), we observe that

$$(1.9.12)\qquad x(\theta) \in \mathcal{X}[\vartheta]$$

if and only if there exists a vector x^0 and a function $u(t)$ that respectively satisfy (1.9.9) and (1.9.10) and also the inequality

$$\ell' x(\theta) \le \left(\ell' - \int_{t_0}^{\vartheta} \lambda'(t)Gdt\right)x^0 + \int_{t_0}^{\vartheta}\left(\ell' - \int_{\tau}^{\vartheta} \lambda'(t)Gdt\right) u(\tau)d\tau + \int_{t_0}^{\vartheta}\left(\ell' - \int_{\tau}^{\vartheta} \lambda'(t)G\, dt\right) f(\tau)d\tau + \int_{t_0}^{\vartheta} \rho(\lambda(\tau)|\mathcal{Y}(\tau))d\tau \ ,$$

whatever are the elements $\ell \in \mathbb{R}^n$, $\lambda(\cdot) \in C^n[t_0, \vartheta]$.

Following the theory of convex analysis [100], it was proved in [181], that the latter requirement will be fulfilled if and only if

$$\ell' x(\theta) \le \Phi_{\vartheta}(\ell, \lambda(\cdot))$$

for any $\ell \in \mathbb{R}^n$, $\lambda(\cdot) \in C^m[t_0, \vartheta]$, where

$$\begin{aligned}\Phi_\vartheta(\ell, \lambda(\cdot)) = \quad & \rho(\ell' - \int_{t_0}^{\vartheta} \lambda'(t)G\, dt | \mathcal{X}^0) + \\ & \int_{t_0}^{\vartheta} \rho(\ell' - \int_{\tau}^{\vartheta} \lambda'(t)G\, dt | \mathcal{P}(\tau) + f(\tau))d\tau + \\ & \int_{t_0}^{\vartheta} \rho(\lambda(\tau) | \mathcal{Y}(\tau))d\tau \quad .\end{aligned}$$

This in its turn will be true if and only if

$$(1.9.13) \quad \ell' x(\theta) \leq \inf \{\Phi_\vartheta(\ell, \lambda(\cdot)) \mid \lambda(\cdot) \in \mathrm{C^m[t_0}, \vartheta]\} = \Phi_\vartheta[\ell] \quad .$$

Function $\Phi_\vartheta[\ell]$ happens to be convex and positively homogenous (this may be verified as an exercise) and, therefore, due to Lemma 1.3.1, is a support function of some set $\mathcal{X}_\vartheta$. Since (1.9.13) is necessary and sufficient for (1.9.12), we come to the equality $\mathcal{X}_\vartheta = \mathcal{X}[\vartheta]$, having proved

Theorem 1.9.1 *The support function for $\mathcal{X}[\vartheta]$ is given by*

$$(1.9.14) \qquad \rho(\ell | \mathcal{X}[\vartheta]) = \Phi_\vartheta[\ell] \quad .$$

A more detailed version of these calculations could be also found in paper [181]. Let us now introduce an *evolution equation* which will prove to be an appropriate description for $\mathcal{X}[t]$. This will be

$$(1.9.15) \quad \lim_{\sigma \to 0} \sigma^{-1} h_+(\mathcal{Z}[t+\sigma], \mathcal{Z}[t] \cap \mathcal{Y}(t) + \sigma \mathcal{P}(t) + \sigma f(t)) = 0$$

$$\mathcal{Z}[t_0] \subseteq \mathcal{X}^0 \quad .$$

A solution to (1.9.15) is a multifunction $\mathcal{Z}[t]$ that satisfies (1.9.15) almost everywhere and is also h_+ – absolutely continuous in the sense of Definition 1.3.2.

Let $\mathcal{X}[t] = \{x[t]\}$ be the union of all trajectories of (1.9.1), (1.9.2) viable relative to constraint (1.9.3). Then, obviously

$$(1.9.16) \qquad \mathcal{X}[t] \subset \mathcal{Y}[t], \quad t_0 \leq t \leq t_1 \quad ,$$

and

$$\mathcal{X}[t_0] = \mathcal{X}^0 \cap \mathcal{Y}[t_0] \quad .$$

At the same time $\mathcal{X}[t] = \{x[t]\}$ is a collection of all solutions to (1.9.1) and (1.9.2) and therefore, due to Lemma 1.3.5, is a solution to the evolution equation (1.3.13). For any $t \in [t_0, t_1]$, $\sigma > 0, \sigma < t_1 - t_0$ this yields the inclusion

$$(1.9.17) \qquad \mathcal{X}[t+\sigma] \subseteq \mathcal{X}[t] + \int_t^{t+\sigma} \mathcal{P}(\tau)d\tau + \int_t^{t+\sigma} f(\tau)d\tau$$
$$\subseteq \mathcal{X}[t] + \sigma\mathcal{P}(t) + \sigma f(t) + o(\sigma)\mathcal{S}$$

that follows from the definition of the Hausdorff semidistance h_+ as well as from relations (1.3.6) and (1.3.7). The inclusion (1.9.17) may be rewritten due to (1.9.16) as

$$(1.9.18) \qquad \mathcal{X}[t+\sigma] \subseteq \mathcal{X}[t] \cap \mathcal{Y}[t] + \sigma\mathcal{P}(t) + \sigma f(t) + o(\sigma)\mathcal{S} \ .$$

The latter relation indicates that (1.9.15) is true for $t \in [t_0, t_1]$. From the relations of the above it also follows that $\mathcal{X}[t]$ is absolutely h_+ continuous in the sense of Definition 1.3.2. We thus come to the proposition

Theorem 1.9.2 *The set-valued function $\mathcal{X}[t]$ is a solution to the evolution equation (1.9.15).*

It is not difficult to observe that an isolated trajectory $x[t] = x(t, t_0, x^0)$ that is viable relative to constraint (1.9.3) is also a solution to (1.9.15). Given $\mathcal{X}[t]$ and any other solution $\mathcal{Z}[t]$ to (1.9.15), the following assertion is true.

Lemma 1.9.1 *The set-valued function $\mathcal{X}[t]$ is a maximal solution to (1.9.15), namely,*

$$\mathcal{Z}[t] \subseteq \mathcal{X}[t]$$

for any other solution $\mathcal{Z}[t]$ to (1.9.15).

We leave to the reader to verify both Lemma 1.9.1 as well as the following assertion

Lemma 1.9.2 *The mapping $\mathcal{X}[t] = \mathcal{X}(t, t_0, \mathcal{X}^0)$ satisfies the semigroup property*

$$(1.9.19) \qquad \mathcal{X}(t, t_0, \mathcal{X}^0) = \mathcal{X}(t, \tau, \mathcal{X}(\tau, t_0, \mathcal{X}^0)) \ .$$

There is another, stronger form of an evolution equation which should be mentioned in this context. We will precede this with a definition

Definition 1.9.3 *A multivalued function $\mathcal{Y}(t) \in \mathrm{comp}\mathbb{R}^n$ is said to be absolutely continuous on an interval $T = [t_0, t_1]$ if its support function*

$$\rho(\ell|\mathcal{Y}(t)) = f(t,\ell)$$

is absolutely continuous on the interval T, for any $\ell \in \mathcal{S}$.

This definition ensures that $f(t,\ell)$ is absolutely continuous on T uniformly in $\ell \in \mathcal{S}$.

We may now formulate

Theorem 1.9.3 *Assume the multifunction $\mathcal{Y}(t)$ to be absolutely continuous on the interval T. Also assume that there exists a trajectory $x(t)$ of system (1.8.1),(1.8.2) such that $x(t) \in \mathrm{int}\mathcal{Y}(t)$. Then the multifunction $\mathcal{X}[t]$ is the unique solution to the evolution equation*

(1.9.20)
$$\lim_{\sigma\to 0} \sigma^{-1} h(\mathcal{X}[t+\sigma], (\mathcal{X}[t] + \sigma\mathcal{P}(t) + \sigma f(t)) \cap \mathcal{Y}(t+\sigma)) = 0$$

$$\mathcal{X}[t_0] = \mathcal{X}^0 \quad .$$

Equation (1.9.19) is somewhat different from (1.9.15), particularly in having involved the Hausdorff distance h rather than the semidistance h_+. The proof of Theorem 1.9.3 is given in papers [190], [193].

Let us now formally write the equation (1.9.19) with $A(t) \not\equiv 0$. This gives

(1.9.21)
$$\lim_{\sigma\to 0} \sigma^{-1} h(\mathcal{X}[t+\sigma], ((I + \sigma A(t))\mathcal{X}[t] + \sigma\mathcal{P}(t) + \sigma f(t)) \cap \mathcal{Y}(t+\sigma)) = 0 \quad .$$

Equation (1.9.19) can also be treated in backward time, namely, in the following form

(1.9.22)
$$\lim_{\sigma\to 0} \sigma^{-1} h(\mathcal{W}[t-\sigma], (\mathcal{W}[t] - \sigma\mathcal{P}(t) - \sigma f(t)) \cap \mathcal{Y}(t-\sigma)) = 0 \quad ,$$

$$\mathcal{W}[t_1] = \mathcal{M} \quad .$$

Theorem (1.9.3) obviously yields

Lemma 1.9.3 *With $\mathcal{Y}(t)$ absolutely continuous, equation (1.9.21) has a unique solution defined on the interval T.*

Set $\mathcal{W}[\tau]$ allows the following interpretation.

Definition 1.9.4 *A solvability set $\mathcal{W}_0[\tau]$ under state constraints (1.9.3), is the set of all states $\{x_\tau\} = \mathcal{W}_0[\tau]$ such that there exists a measurable function $u(t)$ that generates a trajectory $x(t,\tau,x_\tau) = x[t]$, $\tau \leq t \leq t_1$ of system (1.9.1) that satisfies the inclusion $x[t_1] \in \mathcal{M}$ together with restriction (1.9.3).*

Here $\mathcal{W}_0[\tau]$ is obviously the same as the attainability set, but is taken in backward time and *is clearly a solution to equation (1.9.21)*, so that $\mathcal{W}_0[\tau] \equiv \mathcal{W}[\tau]$. With $\mathcal{M} = \mathbb{R}^n$, the set $\mathcal{W}_0[\tau]$ is also known as *the viability kernel* (relative to constraint (1.9.3)), [17].

An alternative version of (1.9.21) is given by the equation

$$(1.9.23) \quad \lim_{\sigma\to 0} \sigma^{-1} h_+(\mathcal{Z}[t-\sigma], \mathcal{Z}[t] \cap \mathcal{Y}[t] - \sigma\mathcal{P}(t) - \sigma f(t)) = 0$$

$$\mathcal{Z}[t_1] \subseteq \mathcal{M} \ .$$

A solution to (1.9.23) exists under weaker assumptions than those for (1.9.22) ($\mathcal{Y}[t]$ may be assumed to be merely continuous and even upper semicontinuous). Its solution is nonunique, however. Thus any viable, single-valued trajectory of (1.8.1) and (1.8.2), if taken in backward time, satisfies (1.9.23), and the proof of the respective existence theorem is similar to that of Theorem 1.9.2.

We finally emphasize the following property that could be proved through standard procedures.

Lemma 1.9.4 *The multifunction $\mathcal{W}[t]$ is the maximal solution to equation (1.9.22).*

Function $\mathcal{W}_0[t], t_0 \leq t \leq t_1$ generates a *solvability tube* that is a crucial element for solving the problem of *control synthesis under state constraints.*

Remark 1.9.1 *With $\mathcal{M} = \mathbb{R}^n, \mathcal{X}^0 = \mathbb{R}^n$, the viability set $\mathcal{W}[t]$ is the collection of all positions $\{t, x\}$, from each of which there exists a control $u(\cdot) \in \mathcal{U}_{\mathcal{P}}^0$ that keeps the respective trajectory $x(t)$ within the state constraint (1.9.3.). A set $\mathcal{W}[t]$ with such a property is referred to as weakly invariant relative to constraint (1.8.3) (see [17]).*

1.10 Control Synthesis Under State Constraints

Given the solvability tube $\mathcal{W}[t]$ of the previous paragraph we may construct a multivalued synthesizing strategy $\mathcal{U}(t, x)$ that solves the problem of *control synthesis under state constraints.*

Definition 1.10.1 *Given a terminal set $\mathcal{M} \in \mathrm{comp}\mathbb{R}^n$, the problem of* **control synthesis under state constraints** *consists in specifying a* **solvability set** $\mathcal{W}(\tau, t_1, \mathcal{M}) = \mathcal{W}^0[\tau]$ *and a set-valued* **feedback control strategy** $u = \mathcal{U}(t, x), \mathcal{U}(\cdot, \cdot) \in U_{\mathcal{P}}^c$, *such that all the solutions to the differential inclusion*

$$\dot{x}(t) \in \mathcal{U}(t, x) + f(t) \tag{1.10.1}$$

that start from any given position $\{\tau, x_\tau\}$, $x_\tau = x[\tau]$, $x_\tau \in \mathcal{W}(\tau, t_1, \mathcal{M})$, $\tau \in [t_0, t_1]$ would satisfy the restrictions

$$x(t) \in \mathcal{Y}(t), \quad \tau \leq t \leq t_1 \ , \tag{1.10.2}$$

$$x(t_1) \in \mathcal{M} \ . \tag{1.10.3}$$

Definition 1.9.1 is nonredundant provided $\mathcal{W}^0[\tau] = \mathcal{W}(\tau, t_1, \mathcal{M}) \neq \emptyset$, where $\mathcal{W}^0[\tau]$ is *the largest set of states x_τ*, from which the solution to the problem of Definition 1.9.1 does exist at all.

Following the same reasoning as in the absence of state constraints (see Sections 1.3 and 1.4), it may be shown that the set $\mathcal{W}^0[\tau]$ will coincide with set $\mathcal{W}[\tau]$ of Section 1.1.9, so that $\mathcal{W}^0[\tau] \equiv \mathcal{W}_0[\tau] \equiv \mathcal{W}[\tau], \tau \in T$. We shall further use notation $\mathcal{W}[\tau]$ for this set and for the respective tube ($\tau \in T$).

Let us now consider the tube $\mathcal{W}[\tau]$, $\tau \leq t \leq t_1$, and define a feedback strategy

(1.10.4) $$\mathcal{U}(t,x) = \partial_l f(t, -l_W^0(t,x))$$

similar to that of (1.4.20) and (1.8.9). Here, as before, $f(t,l) = \rho(l|\mathcal{P}(t))$ and $l = l_W^0(t,x)$ is the maximizer for the expression

(1.10.5) $$d_W[t,x] = \max\{(l,x) - \rho(l|\mathcal{W}[t]) \,\Big|\, \|l\| \leq 1\}$$

or

$$d_W[t,x] = (l^0, x) - \rho(l^0|\mathcal{W}[t])$$

if $d_W[t,x] > 0$ (otherwise $l^0 = 0$).

Here

(1.10.6) $$d_W^2[t,x] = \min\{(x - z, x - z)|z \in \mathcal{W}[t]\} = V(t,x) \ .$$

To prove that $\mathcal{U}(t,x)$ is a solution to our problem we have to calculate the derivative

(1.10.7) $$\frac{d}{dt}V(t,x) = 2d_W[t,x]\frac{d}{dt}d_W[t,x] = 2d_W[t,x]\frac{d}{dt}d_W[t,x]$$

due to the inclusion

(1.10.8) $$\dot{x} \in \mathcal{U}(t,x) + f(t) \ .$$

We assume in this section that the support function $\rho(l|\mathcal{Y}(t))$ of the multifunction $\mathcal{Y}(t)$ is absolutely continuous.

In order to do that, let us first calculate the left partial derivative in t of the support function $\rho(l|W[t])$, namely,

$$\frac{\partial_- \rho(l|W[\tau])}{\partial \tau}$$

where

$$\frac{\partial_- \rho(l|W[\tau])}{\partial \tau} = \lim_{\sigma \to 0} \sigma^{-1}(\rho(l|\mathcal{W}[\tau - \sigma]) - \rho(l|\mathcal{W}[\tau]))$$

for a given direction $l \in \mathbb{R}^n$. We will further use the relation (1.9.21), particularly to calculate the increment

$$\rho(l|\mathcal{W}[\tau - \sigma]) - \rho(l|\mathcal{W}[\tau])$$

through the relation

$$\mathcal{W}[\tau-\sigma] = (\mathcal{W}[\tau] - \sigma\mathcal{P}(\tau) - \sigma f(\tau)) \cap \mathcal{Y}(\tau-\sigma) + r(\sigma)$$

where $\sigma^{-1}h(r(\sigma),0) \to 0$ with $\sigma \to 0$.

Since

$$h(\mathcal{W}',\mathcal{W}'') = \max\{\rho(l|\mathcal{W}') - \rho(l|\mathcal{W}'') \,\Big|\, \|l\| = 1\}$$

we observe that the increments

$$\Delta_1(\sigma) = \sigma^{-1}(\rho(l|\mathcal{W}[\tau-\sigma]) - \rho(l|\mathcal{W}[\tau]))$$

and

$$\Delta_2(\sigma) = \sigma^{-1}(\rho(l|(\mathcal{W}[\tau] - \sigma\mathcal{P}(\tau) - \sigma f(\tau)) \cap \mathcal{Y}(\tau-\sigma)) - \rho(l|\mathcal{W}[\tau]))$$

are such that

$$\lim_{\sigma\to 0} |\Delta_1(\sigma) - \Delta_2(\sigma)| = 0 \quad .$$

Therefore it suffices to calculate the derivative

$$\left.\frac{dg(\sigma)}{d\sigma}\right|_{\sigma=0}$$

for the function

$$\begin{aligned} g(\sigma) &= \rho(l|(\mathcal{W}[\tau] - \sigma\mathcal{P}(\tau) - \sigma f(\tau)) \cap \mathcal{Y}(\tau-\sigma)) \\ &= \min\{\rho(p|(\mathcal{W}[\tau] - \sigma\mathcal{P}(\tau) - \sigma f(\tau))) + \rho(l-p|\mathcal{Y}(\tau-\sigma))|p \in \mathbb{R}^n\} \end{aligned}$$

since

$$\left.\frac{dg(\sigma)}{d\sigma}\right|_{\sigma=0} = \frac{\partial_-\rho(l|\mathcal{W}[\tau])}{\partial\tau} \quad .$$

The calculation then follows the techniques of directional differentiation given, for example, in [89]. This finally yields

Lemma 1.10.1 *The following relation holds*

(1.10.9)
$$\left.\frac{\partial_-\rho(l|\mathcal{W}[\tau])}{\partial\tau}\right|_{\tau=t} =$$

$$\min\{\rho(-p|\mathcal{P}(t)) - (p, f(t)), -\frac{\partial}{\partial t}(\rho(l-p|\mathcal{Y}(t))\Big| p \in \mathcal{F}(t,l)\} \quad ,$$

where

$$(1.10.10)\quad \mathcal{F}(t,l) = \{p \in \mathbb{R}^n : k(t,l) - k(t,p) = \rho(l-p|\mathcal{Y}(t))\}$$

and

$$k(t,l) = \rho(l|\mathcal{W}[t]) \ .$$

Here the relations (1.10.9) and (1.10.10) reflect the fact that the infimal convolution that defines $g(0), t = \tau$, is *exact*, namely, the minimum in (1.0.9) is taken over all $p \in \mathbb{R}^n$ that satisfy the equality

$$\rho(p|\mathcal{W}[t]) + \rho(l-p|\mathcal{Y}(t)) = \rho(l|\mathcal{W}[t]) \ .$$

Let us elaborate on this result. Since the properties of $\mathcal{W}[t]$ imply $\mathcal{W}[t] \subseteq \mathcal{Y}(t)$, we have

$$\rho(l|\mathcal{W}[t] \cap \mathcal{Y}(t)) = \min\{\rho(l|\mathcal{W}[t]), \rho(l|\mathcal{Y}(t))\}$$

and therefore the minimum of $g(0)$ over p is attained at either $p = 0$ (which is when $\rho(l|\mathcal{W}[t]) \leq \rho(l|\mathcal{Y}(t))$ or $p = l$ (which is when $\rho(l|\mathcal{Y}(t) = \rho(l|\mathcal{W}[t]))$. Formula (1.10.9) actually yields

$$\left.\frac{\partial_- \rho(l|\mathcal{W}[\tau])}{\partial \tau}\right|_{\tau=t} =$$

$$(1.10.11)\qquad \min\{\rho(-l|\mathcal{P}(t)) - (l, f(t))\,,\ -\frac{\partial}{\partial t}\rho(l|\mathcal{Y}(t))\} \ .$$

Relation (1.10.9) allows to calculate the right directional derivative

$$d_+ d_{\mathcal{W}}[t,x]/dt$$

due to system (1.9.1) through formula (1.9.5).

In view of the equality

$$\left.\frac{\partial_- \rho(l|\mathcal{W}[\tau])}{\partial \tau}\right|_{\tau=t} = -\frac{\partial \rho(l|\mathcal{W}[\tau])}{\partial \tau}$$

we come to

$$\begin{aligned}
\frac{d_+ r_{\mathcal{W}}[t,x]}{dt} &= (l^0, u + f(t)) + \frac{\partial \rho(l^0|\mathcal{W}[t])}{\partial t} = \\
&= (l^0, u + f(t)) + \\
&\quad \min\{\rho(-l^0|\mathcal{P}(t)) - (l^0, f(t)), -\frac{\partial}{\partial t}\rho(l^0|\mathcal{Y}(t))\} \\
&\leq (l^0, u + f(t)) + \rho(-l^0|\mathcal{P}(t)) + (-l^0, f(t)) \ ,
\end{aligned}$$

which is true for almost all t. The last relation turns into an equality $d_+ d_{\mathcal{W}}[t,x]/dt = 0$, if $u \in \mathcal{U}(t,x)$, where $\mathcal{U}(t,x)$ is given by (1.10.4). This conclusion produces

Lemma 1.10.2 *Once $u \in \mathcal{U}(t,x)$, where $\mathcal{U}(t,x)$ is defined by (1.10.4), then almost everywhere the derivative*

$$d_+ d_{\mathcal{W}}[t,x]/dt \leq 0 \quad ,$$

and therefore

$$\frac{d_+}{dt} V(t,x) \Big|_{u \in \mathcal{U}(t,x)} \leq 0 \quad a.e. \tag{1.10.12}$$

Similar to Sections 1.1.4 and 1.1.8, it suffices to prove that strategy $\mathcal{U}(t,x)$ of (1.10.4) does solve the problem 1.10.1 of control synthesis under state constraints. We thus come to the proposition

Theorem 1.10.1 *The problem of control synthesis under state constraints of Definition 1.10.1 is solved by strategy $\mathcal{U}(t,x)$ of (1.10.4).*

The problem is obviously solvable if the starting position $\{t,x\}$ is such that $x \in \mathcal{W}[t]$, where $\mathcal{W}[t]$ is the solvability set given by the unique solution to equation (1.9.22) or by the unique maximal solution to equation (1.9.23). It is not difficult to prove though that $\mathcal{W}[t]$ is the largest set from which the solution does exist at all. The respective proof is similar to the one given in the last part of Section 1.8, so the last theorem may be complemented by

Lemma 1.10.3 *In order that the solution strategy $\mathcal{U}(t,x)$ of Definition 1.10.1 can be applied to position $\{\tau, x_\tau\}$, it is necessary and sufficient that $x_\tau \in \mathcal{W}[\tau]$.*

The results of this section may be explained through DP techniques.

Exercise 1.10.1. (a) Introducing the value function

$$V_0(t,x) = \min\left\{ \int_t^{t_1} h_+^2(x(\tau), \mathcal{Y}(\tau))d\tau \; + \; h_+^2(x(t_1), \mathcal{M}) \Big| u(\tau) \in \mathcal{P}(\tau) \right\} \; ,$$

check whether it satisfies the corresponding H-J-B equation for system (1.10.8) and what would be the relations between the solutions to the problem of Definition 1.10.1 achieved through $V_0(t,x)$ and through function $V(t,x)$ of (1.10.6).

(b) Taking Example 1.5.1, complement it by a state constraint

$$(x(t) - n(t), N(t)(x(t) - n(t))) \leq 1, \quad N(t) > 0 \; ,$$

and find the solvability set $\mathcal{W}^0[\tau]$ of Definition 1.10.1 by following the schemes of (1.5.15)–(1.5.17). Calculate the analogy of formula (1.5.17) for the given state constraint.

We finally come to the next topic which incorporates all the difficulties specific for the previous sections. This is the problem of control synthesis under both uncertainty and state constraints.

1.11 State Constrained Uncertain Systems: Viability Under Counteraction

Consider system (1.7.1) with terminal set $\mathcal{M}$, state constraint (1.10.2) and constraints (1.1.2), (1.6.1) on the control u and the uncertain input f.

Definition 1.11.1 *The problem of* **control synthesis under uncertainty and state constraint** *consists in specifying a solvability set* $\mathcal{W}[\tau] = \mathcal{W}_*(\tau, t_1, \mathcal{M})$ *and a set-valued control strategy* $\mathcal{U}(t,x)$ *such that all the solutions* $x[t] = x(t, \tau, x_\tau)$ *to the differential inclusion (1.8.1) that start at a given position* $\{\tau, x_\tau\}, x_\tau = x[\tau] \in \mathcal{W}_*(\tau, t_1, \mathcal{M}), \tau \in [t_0, t_1)$, *would reach the terminal set* $\mathcal{M}$ *at time* t_1, *so that* $x(t_1) \in \mathcal{M}$, *and would also satisfy the state constraint (1.10.2), namely,*

$$x[t] \in \mathcal{Y}(t), \quad \tau \leq t \leq t_1 \; .$$

Here the multivalued function $\mathcal{Y}(t)$ with values in comp$\mathbb{R}^n$ is again taken to be absolutely continuous.

It is clearly the strategy $\mathcal{U}(t,x)$ that is responsible for the solution $x[t]$ to satisfy the state constraint (1.10.2), no matter what is the disturbance $f(t)$.

In this section we will have to combine the schemes of Sections 1.1.7, 1.1.8 and Sections 1.1.9, 1.1.10. The technicalities of this combination require a more or less sophisticated mathematical treatment, the details of which are not directly relevant to the topics of this book. They are the subject of other publications (see [194], [195]). We will however give a concise presentation of the solution to this problem emphasizing the substantial interrelations important for the results.

The solution strategy $\mathcal{U}(t,x)$ will again be determined by a relation of type (1.10.4) and (1.10.5), where $\mathcal{W}[t]$ has to be substituted by $\mathcal{W}_*[t]$ – *the solvability set* of Definition 1.11.1. The basic evolution equation for $\mathcal{W}_*[t]$ now has the form

$$\lim_{\sigma\to 0}\sigma^{-1}h_+(\mathcal{Z}[t-\sigma]+\sigma\mathcal{Q}(t),\mathcal{Z}[t]\cap\mathcal{Y}(t)-\sigma\mathcal{P}(t))=0\ ,$$

$$\mathcal{Z}[t_1]\subseteq\mathcal{M}\ ,$$

so that the following assertion holds.

Lemma 1.11.1 *The solvability set $\mathcal{W}_*[t]$ for the problem of control synthesis under both uncertainty and state constraints as formulated in Definition 1.11.1 is the maximal solution to equation (1.11.1) with boundary condition $\mathcal{Z}[t_1]=\mathcal{M}$.*

The proof of this assertion follows the lines of Sections 1.1.7–1.1.10. As we have seen above, the property important for Control Synthesis is the behavior of the directional derivative

$$\frac{d}{dt}V(t,x)$$

$$V(t,x)=d^2_{\mathcal{W}_*}[t,x]=h_+^2(x,\mathcal{W}_*[t])$$

along the solutions to the differential inclusion (1.10.1) with $f(t)$ unknown, but bounded:

$$f(t) \in \mathcal{Q}(t), \quad t \in [t_0, t_1] \quad . \tag{1.11.1}$$

Combining the calculations of Sections 1.1.8 and 1.1.10 under Assumption 1.6.1 we come to

Lemma 1.11.2 *The derivative $dd(x, \mathcal{W}_*[t])/dt$ is given by*

$$\frac{d}{dt} d(x, \mathcal{W}_*[t]) = (l^0, u(t) + f(t)) +$$

$$+ \min\{\rho(l^0 | (-\mathcal{P}(t)) \dot{-} \mathcal{Q}(t)) \, , \, -\frac{\partial}{\partial t}\rho(l^0 | \mathcal{Y}(t))\} \leq$$

$$\leq (l^0, u(t) + f(t)) \; + \; \rho(l^0 | - \mathcal{P}(t)) \; - \; \rho(l^0 | \mathcal{Q}(t)) \quad . \tag{1.11.2}$$

The synthesizing strategy $\mathcal{U}^0_*(t, x)$ may now be defined in the same way as $\mathcal{U}^0(t, x)$ of Section 1.1.8, that is according to (1.8.9), but with $\mathcal{W}^*[t]$ substituted by $\mathcal{W}_*[t]$ of Definition 1.11.1, (the notation $\mathcal{U}^0(t, x)$ is also substituted by $\mathcal{U}^0_*(t, x)$).

Similarly to Section 1.1.8 (Theorem 1.8.1), the previous Lemma implies

Theorem 1.11.1 *The strategy $\mathcal{U}^0_*(t, x)$ defined by (1.8.9) (with $\mathcal{W}^*[t]$ substituted by $\mathcal{W}_*[t]$) resolves the problem of control synthesis under uncertainty and state constraints of Definition 1.11.1.*

Therefore every solution $x[t]$ to the system

$$\dot{x} \in \mathcal{U}^0_*(t, x) + f(t) \tag{1.11.3}$$

$$x(t_0) \in \mathcal{W}_*[t_0] \tag{1.11.4}$$

satisfies the constraint

$$x[t] \in \mathcal{W}_*[t], \quad t_0 \leq t \leq t_1 \tag{1.11.5}$$

and therefore the inclusion

$$x[t_1] \in \mathcal{M} \quad .$$

In this case we will say that system (1.11.3) is *viable relative to constraint (1.10.2) under counteraction (1.11.2)*, provided $x(t_0)$ satisfies (1.11.6).

In other terms we may say that $\mathcal{W}_*[t], t_0 \leq t \leq t_1$, is a tube of *strongly invariant sets* for system (1.11.3) *under counteraction (1.11.2)*. (The latter term indicates that *all the solutions to the differential inclusion (1.11.3) that start in $\mathcal{W}[t_0]$, do satisfy the state constraint (1.10.2).*)

The assertions of this section finalize the concise description of the solutions to the problems of evolution and control synthesis in the presence of uncertainty and state constraints. A topic for further discussion is the application of set-valued calculus to the problem of *state-estimation.*

1.12 Guaranteed State Estimation: The Bounding Approach

One of the basic problems of modelling and control is *to estimate the state* of an uncertain or incompletely defined dynamic system on the basis of *on- or off-line observations* corrupted by noise. Leaving aside the well-developed stochastic approach to these problems, and following the emphasis of the present book, we shall again assume the set-valued interpretations of the respective problems.

Namely, as in Section 1.6, an *uncertain system* is understood to be one of the following type

$$\dot{x}(t) \in A(t)x(t) + u(t) + f(t) \ , \tag{1.12.1}$$

$$t_0 \leq t \leq t_1, \qquad x(t_0) = x^0 \ ,$$

where $A(t) \in \mathcal{L}(\mathbb{R}^n, \mathbb{R}^n), u(t)$ is a given function (a preselected control) and $f(t) \in \mathbb{R}^n$ is the *unknown but bounded input* (disturbance).

It is presumed that the *initial state* $x^0 \in \mathbb{R}^n$ is also unknown but bounded, so that

$$f(t) \in \mathcal{Q}(t), \quad t_0 \leq t \leq t_1 \ , \tag{1.12.2}$$

$$x^0 \in \mathcal{X}^0 \ , \tag{1.12.3}$$

where the set $\mathcal{X}^0 \subset \text{conv}\mathcal{R}^n$ and the continuous set-valued function $\mathcal{Q}(t) \in \text{comp}\mathbb{R}^n$, are given in advance.

Equation (1.12.1) may be complemented, as we have seen earlier, in Section 1.9, by a *state constraint*

$$(1.12.4) \qquad G(t)x(t) \in \mathcal{Y}(t), \qquad t_0 \leq t \leq t_1$$

with $G(t) \in \mathcal{L}(\mathbb{R}^n, \mathbb{R}^m)$ and $\mathcal{Y}(t) \in$ conv $\mathbb{R}^m$, $m \leq n$. The constraint (1.12.3) may be particularly generated by a *measurement equation*

$$(1.12.5) \qquad y(t) = G(t)x(t) + v(t), \qquad t_0 \leq t \leq t_1 \ ,$$

with an *unknown but bounded error*

$$(1.12.6) \qquad v(t) \in \mathcal{K}(t), \qquad t_0 \leq t \leq t_1$$

where $\mathcal{K}(t) \in$ conv $\mathbb{R}^m$, $t_0 \leq t_1$ is an absolutely continuous set-valued map (recall Section 1.9).

With the realization $y(\cdot)$ being known, restriction (1.12.4), (1.12.5) turns into

$$(1.12.7) \qquad G(t)x(t) \in y(t) - \mathcal{K}(t), \quad t_0 \leq t \leq t_1 \ ,$$

so that

$$\mathcal{Y}(t) = \{x : G(t)x \in y(t) - \mathcal{K}(t)\} \ ,$$

(however, the whole function $y(\cdot)$ may not be known in advance, arriving *on-line*).[11]

Our objective will be to *estimate the system output*

$$(1.12.8) \qquad z(t) = Hx(t), \quad z \in \mathbb{R}^r, \quad r \leq n \ , \quad t_0 \leq t \leq t_1$$

at any prescribed instant of time t.

More precisely, the problem is to specify *the range of the output* $z(t)$ that is consistent with relations (1.12.1)–(1.12.4) (the *Attainability Problem under State Constraints*), or the set of all outputs $z(t)$ consistent with system (1.12.1)–(1.12.3), and measurement equation (1.12.5) and (1.12.6), with realization $y(t)$ of the measurement being given (the *Guaranteed State Estimation Problem*).

[11]The class of functions $y(t)$ admitted here clearly depends on the class of functions $v(t)$ – the error noise. We shall implicitly specify theses classes along the course of the presentation, through the properties of $\mathcal{Y}(t)$.

The solution to both problems is, therefore, given in the form of *a set* representing thus *the bounding approach to state estimation.*

Our aim here is not to repeat the well-known information [277], [181], [225]. on these issues, but to rewrite some theoretical results focusing them on the main objective, which is further to devise in Part IV some constructive algorithmic procedures based on ellipsoidal techniques that would allow a computer simulation with graphical representations.[12]

Let us specify the problems considered here, starting with the *Attainability Problem.* As indicated in Section 1.9, the *attainability domain* $\mathcal{X}(t, t_0, x^0)$ for (1.12.1) and (1.12.2) under state constraint (1.12.4) at time $t \in [t_0, t_1]$ from point $x^0 \in \mathbb{R}^n$ is the cross-section at $t \in [t_0, t_1]$ of the tube of all trajectories $x[\cdot] = x(\cdot, t_0, x^0)$ that satisfy (1.12.1), (1.2.2), and (1.12.4), namely,

$$\mathcal{X}(\cdot, t_0, x^0) = \bigcup \{x(\cdot, t_0, x^0) \mid x_0 \in \mathcal{X}^0\} \ . \tag{1.12.9}$$

Define the map $\mathcal{X}[t] = \mathcal{X}(t, t_0, \mathcal{X}^0)$ as

$$\mathcal{X}(t, t_0, \mathcal{X}^0) \ = \ \bigcup \{\mathcal{X}(t, t_0, x^0) | x^0 \in \mathcal{X}^0\} \ .$$

The multivalued map $\mathcal{X}[\cdot]$ generates a *generalized dynamic system.* Namely, the mapping

$$\mathcal{X} \ : \ [t_0, t_1] \times [t_0, t_1] \times \text{conv}\mathbb{R}^n \ \rightarrow \ \text{conv}\mathbb{R}^n$$

possesses a *semigroup property*, that is, whatever are the values $t_0 \leq t \leq \tau \leq \theta \leq t_1$, we have

$$\mathcal{X}(\theta, t, \mathcal{X}[t]) = \mathcal{X}(\theta, \tau, \mathcal{X}(\tau, t, \mathcal{X}[t])) \ .$$

Also, the set-valued map $\mathcal{X}$, or in other words, the tube $\mathcal{X}[t]$, $(t_0 \leq t \leq t_1)$ satisfies an evolution equation, the funnel equation of type (1.9.20) [190], [193], which is
(1.12.10)

$$\lim_{\sigma \to +0} \sigma^{-1} h(\mathcal{X}[t + \sigma], ((I + A(t)\sigma)\mathcal{X}[t] + \sigma\mathcal{P}(t)) \cap \mathcal{Y}(t + \sigma)) = 0 \ \ ,$$

[12] The first descriptions of state estimation (observation) problems under unknown but bounded errors may be traced to papers [166], [318], [277], [177]. The set-valued approach to such problems in continuous time appears to have independently started from publications [54], [178], [278], [181].

$$t_0 \le t \le t_1 \quad ,$$

$$\mathcal{X}[t_0] = \mathcal{X}_0 \quad ,$$

Equation (1.12.10) is correctly posed and has a unique solution that defines the tube $\mathcal{X}[\cdot] = \mathcal{X}(\cdot, t_0, \mathcal{X}_0)$ for system (1.12.1)–(1.12.4) if the map $\mathcal{Y}(\cdot)$ is such that the support function

$$\rho(\ell \mid \mathcal{K}(t)) = \max\{(\ell, p) \mid p \in \mathcal{K}(t)\}$$

and the function $y(t)$ are *absolutely continuous* in t [193].

Using only one of the Hausdorff semi-distances in (1.12.10) leads to the loss of *uniqueness* of the solutions, but allows us to relax the requirements on the multivalued function $\mathcal{Y}(t)$.

Consider the evolution equation of type (1.9.23) which, in our case, transforms into

$$(1.12.11) \qquad \lim_{\sigma \to +0} \sigma^{-1} h_+(\mathcal{Z}[t+\sigma], ((I + A(t)\sigma)\mathcal{Z}[t] \cap \mathcal{Y}(t)) + $$

$$+\sigma \mathcal{P}(t)) = 0 \quad ,$$

with

$$t_0 \le t \le t_1 \quad ,$$

and

$$\mathcal{Z}[t_0] = \mathcal{X}_0 \quad .$$

As we have observed earlier, the solution to this equation is nonunique. By complementing it with an extremality condition, we obtain alternative descriptions for the multivalued map $\mathcal{X}[\cdot]$.

A set-valued map $\mathcal{X}_+[\cdot]$ will be defined as a *maximal solution* to (1.12.11) if it satisfies (1.12.11) for almost all $t \in [t_0, t_1]$ and if there exists no other solution $\mathcal{Z}[\cdot]$, such that $\mathcal{X}_+[t] \subset \mathcal{Z}[t]$ for all $t \in [t_0, t_1]$ and $\mathcal{X}_+[\cdot] \neq \mathcal{W}[\cdot]$. Equation (1.12.11) has a *unique maximal solution* under relatively mild conditions (for example, if $\mathcal{K}(t) - y(t)$ is only *upper semicontinuous* in t) [194]. Particularly, it allows us to treat a reasonably large class of *discontinuous set-valued functions* $\mathcal{Y}(t)$.

Under the conditions required for the existence and uniqueness of the solutions to (1.12.10), one may also observe, that $\mathcal{X}[\cdot] = \mathcal{X}_+[\cdot]$. As mentioned in Section 1.9, equation (1.12.11) is an alternative version relative to (1.12.10).

The *Guaranteed State Estimation Problem* may now be formulated more precisely. Namely, suppose that the measurement $y(\cdot) = y^*(\cdot)$, due to system (1.12.1) and (1.12.4), is given and is generated by an unknown triplet

(1.12.12) $$\zeta^*(t) = \{x_0^*, f^*(t), v^*(t)\} \ ,$$
$$t_0 \leq t \leq t_1 \ ,$$

that complies with the constraints (1.12.2), (1.12.3), (1.12.5), and (1.12.6), that is:

(1.12.13) $$\dot{x}^*[t] = A(t)x^*[t] + u(t) + f^*(t) \qquad x_0^* \in \mathcal{X}_0^*,$$
(1.12.14) $$y^*(t) = G(t)x^*[t] + v^*(t), \qquad t_0 \leq t \leq t_1 \ .$$

Then the tube $\mathcal{X}[t] = \mathcal{X}^*[\cdot]$ of domains $\mathcal{X}^*[t] = \mathcal{X}[t] = \mathcal{X}(t, t_0, \mathcal{X}^0)$ generated by (1.12.1)–(1.12.3), (1.12.5), (1.12.6) and calculated due to the knowledge of the measurement $y[\cdot] = y^*[\cdot]$, does *always* contain the unknown actual trajectory $x^*[\cdot]$, generated by $\zeta^*(\cdot)$. Each set $\mathcal{X}^*[t]$ therefore gives a *guaranteed estimate* of the state $x^*[t]$ of system (1.12.1) on the basis of the available measurement $y^*(\tau)$, $t_0 \leq \tau \leq t$ under the constraints (1.12.2), (1.12.3), and (1.12.7).

Definition 1.12.1 *The set $\mathcal{X}[t] = \mathcal{X}(t, t_0, \mathcal{X}_0)$, of states $x = x(t)$ of system (1.12.1) that, with given $y(\tau), t_0 \leq \tau \leq t$, are consistent with the constraints (1.12.2), (1.12.3), and (1.12.7) is referred to as* **the information domain** *relative to measurement $y(\cdot)$).*

The information domain $\mathcal{X}[t]$, [181], is also referred to as the domain of consistency, or the feasibility domain [56], [278], [225]. As mentioned above, it is *the attainability domain* $\mathcal{X}[t]$ for system (1.12.1), (1.12.2), and (1.12.7).

The solution of the *guaranteed estimation problem* is to specify the tube $\mathcal{X}[t] = \mathcal{X}^*[t]$, $t_0 \leq t \leq t_1$, defined for a given measurement $y(t) = y^*(t)$. The results of Section 1.9 allow the following assertion.

Theorem 1.12.1

(i) With $\mathcal{K} - y(t)$ upper semicontinuous, the tube $\mathcal{X}^*[t]$ is the unique maximal solution to the evolution equation (1.2.11).

(ii) With $\rho(l|\mathcal{K}(t)-y(t))$ absolutely continuous in t, the tube (set-valued function)$\mathcal{X}^*[t]$ satisfies the evolution equation (1.12.10). It is also the unique maximal solution to (1.2.11).

(iii) Once $\mathcal{X}^*[t]$ is known, the estimate of the output $z(t)$ is the set $\mathcal{Z}(t) = H\mathcal{X}^*[t]$.

To conform with the assumptions on $\mathcal{Y}(t)$ of the above, one ought to presume, for example, that with $\mathcal{K}(t) = \text{const}$, the function $y(t)$ is piecewise-continuous (from the right), if we use equation (1.12.11), or absolutely continuous, if we use (1.12.10).

It is important to emphasize that in many applied problems the observed measurement output $y(t)$ is not obliged to be continuous. We shall therefore further allow it to be only Lebesgue-measurable. In order to imbed this situation in the given schemes, we shall apply the idea of *singular perturbation technique.* But one must of course realize that this time the object of application is a differential inclusion and that the propagation of well-known results [295], [296] in singular perturbation theory *to trajectory tubes* would require specific treatment.

Consider the system of differential inclusions ($\epsilon > 0$):

(1.12.15)
$$\dot{x} \in A(t)x + \mathcal{P}(t) \quad ,$$

(1.12.16)
$$\epsilon\dot{w} \in -G(t)x + \mathcal{K}(t) \quad ,$$

(1.12.17)
$$\{x(t_0), w(t_0)\} \in Z_0 \quad t_0 \le t \le \tau \quad .$$

Here $w \in \mathbb{R}^m, z_0 \in \text{conv}(\mathbb{R}^n \times \mathbb{R}^m)$. As in the earlier sections, by $\mathcal{X}[t] = \mathcal{X}(t, t_0, \mathcal{X}_0)$, we denote the *trajectory tube* of system (1.1.16) that consists of all those trajectories that start at $\mathcal{X}[t_0] = \mathcal{X}_0$ and satisfy the state constraint (1.12.4) for all $t \in [t_0, \tau]$ $\mathcal{X}[t]$ is obviously the tube of all viable trajectories relative to constraints (1.12.4) and (1.12.3).

Following this notation, symbol $Z[t] = \mathcal{Z}(\tau, t_0, Z_0, \epsilon)$ will denote the tube of solutions $z(t) = \{x(t), y(t)\}$ to the system (1.12.15)–(1.12.17) on the interval $[t_0, \tau]$. We will also use the notation $\Pi_x W$ for the projection of set $W \subset \mathbb{R}^n \times \mathbb{R}^m$ onto the space $\mathbb{R}^n$ of variables x.

Here the constraint (1.12.4) may particularly be generated by a measurement equation, as in (1.12.7) or (1.12.5) and (1.12.6), where function $y(t)$ – the realization of the observations – is allowed to be Lebesgue-measurable. The "bad" properties of $y(t)$ are then clearly due to the bad measurement noise $v(t)$ in (1.12.5).

Our aim is still to describe the tube $\mathcal{X}[t]$. However, in order to achieve this, we shall not study system (1.12.16),(1.12.4) directly, but shall rather deal with the perturbed system (1.12.16)–(1.12.18). The latter system may then be fully treated within the standard framework of Sections 1.9 and 1.10, and papers [192], [193]. The following assertion is true.

Theorem 1.12.1 *Assume that*

$$\mathcal{X}_0 \subseteq \Pi_x Z_0 \quad . \tag{1.12.18}$$

Then for every trajectory $x(\cdot) \in \mathcal{X}[\cdot]$ of (1.12.15),(1.12.3),(1.12.4) there exists a vector $w_0 \in \mathbb{R}^m$ such that $\{x(t_0), w_0\} \in Z_0$ and for every $\tau \in [t_0, t_1]$

$$z(\tau) = \{x(\tau), w_{(\tau)}\} \in \mathcal{Z}(\tau, t_0, \mathcal{Z}_0, \epsilon)$$

for all $\epsilon > 0$.

Corollary 1.12.1 *Assume (1.12.18) to be true. Then*

$$\mathcal{X}[\tau] \subseteq \Pi_x(\cap\{Z(\tau, t_0, Z_0, \epsilon) | \epsilon > 0\}) \quad .$$

Let us now introduce another system of differential inclusions of type (1.12.15) and (1.12.16), but with a time-dependent matrix $L(t)$ instead of the scalar $\epsilon > 0$:

$$\dot{x} \in A(t)x + \mathcal{P}(t) \quad , \tag{1.12.19}$$

$$L(t)\dot{y} \in -G(t)x + \mathcal{K}(t) \tag{1.12.20}$$

$$z_0 = \{x(t_0), w(t_0)\} \in Z_0, \quad t_0 \leq t \leq \tau \quad . \tag{1.12.21}$$

The class of all continuous invertible matrix functions $L(t) \in \mathcal{L}(\mathbb{R}^n, \mathbb{R}^n), t \in [t_0, t_1]$ will be denoted as $\mathbf{L}$ and the solution tube to system (1.12.19)–(1.12.21) will be denoted as $\mathcal{Z}[t, L] = \mathcal{Z}(t, t_0, X_0, L)$.

The following analogy of Theorem 1.12.2 is true.

Theorem 1.12.2 *Assume relation (1.12.18) to be true. Then for every $x(\cdot) \in \mathcal{X}[\cdot]$ there exists a vector $w_0 \in \mathbb{R}^m$ such that*

$$\{x(t_0), w_0\} \in Z_0 \quad ,$$

and for every $\tau \in [t_0, t_1]$

$$z(\tau) = \{x(\tau), y(\tau)\} \in Z[\tau, L] \quad ,$$

whatever is the function $L(\cdot) \in \mathbf{L}$.

Corollary 1.12.2 *Assume relation (1.12.18) to be true. Then*

$$\mathcal{X}[\tau] \subseteq \Pi_x(\cap\{Z[\tau, L] | L(\cdot) \in \mathbf{L}\}) \quad . \tag{1.12.22}$$

The principal result of the singular perturbations method applied to the guaranteed estimation problem discussed here is formulated as follows

Theorem 1.12.3 *Let us assume*

$$\Pi_x Z_0 \subseteq X_0 \quad .$$

Then for every $\tau \in [t_0, t_1]$ the following inclusion is true

$$\Pi_x(\cap\{Z[\tau, L] | L(\cdot) \in \mathbf{L}\}) \subseteq \mathcal{X}[\tau] \quad . \tag{1.12.23}$$

This result may be proved within the techniques of Sections 1.9 and 1.10. Its details may be found in [193].

Relations (1.12.22) and (1.12.23) yield an exact description of the set $X[\tau]$ through the solution of the perturbed differential inclusions (1.12.19)–(1.12.21) that are without any state constraints:

Theorem 1.12.4 *Under the assumption*

$$\Pi_x Z_0 = \mathcal{X}_0$$

the following formula is true

$$\mathcal{X}[\tau] = \Pi_x(\cap\{Z[\tau, L] | L(\cdot) \in \mathbf{L}\}) \tag{1.12.24}$$

for any $\tau \in [t_0, t_1]$.

The application of this theorem to the calculation of information sets will be illustrated in Section 4.6, where it will be further modified to suit the related ellipsoidal techniques.

The conventional theory of guaranteed state estimation as introduced in [181], [225], may require us to find the *worst-case estimate* of $x(t)$ *as a vector* $x^0(t)$, which is usually taken to be the *"Chebyshev center"* of set $\mathcal{X}(t)$, namely, as the solution to the problem

$$\max_z\{\| x^0(t) - z \| \,|z \in \mathcal{X}(t)\} = \tag{1.12.25}$$

$$= \min_x \max_z\{\| x - z \| \,|x \in \mathcal{X}(t), z \in \mathcal{X}(t)\} \ .$$

The Chebyshev center of a set $\mathcal{X}$ is the center of the smallest Euclidean ball that includes $\mathcal{X}$. Its calculation leads to mathematical programming problems of special type [139], [86], [88], [69], [209]. The approximate calculation of Chebyshev centers is generating an increasing literature [225]. A less investigated problem is to find *the Steiner center* [275] of set $\mathcal{X}[t]$.

The interested reader who has managed to reach these lines may be curious to know whether the results of the last few sections could be interpreted in some conventional way, in terms of DP, as in Section 5, for example. These questions are discussed further in the first sections of Part IV.

1.13 Synopsis

We shall now summarize the results of the previous sections. Recall that we have considered the system

$$\dot{x}(t) = u(t) + f(t) \ , t_0 \leq t \leq t_1 \ , \tag{1.13.1}$$

with constraints on *the controls*

$$u(t) \in \mathcal{P}(t) \ , \tag{1.13.2}$$

the unknown inputs

$$f(t) \in \mathcal{Q}(t) \ , \tag{1.13.3}$$

the initial state

(1.13.4) $$x^0 \in X^0,\ x(t_0) = x_0 \ ,$$

and *the state space variables*

(1.13.5) $$G(t)x \in \mathcal{Y}(t)\ ,\ G(t) \in \mathcal{L}(\mathbb{R}^n, \mathbb{R}^m)\ ,$$

and with continuous in time set-valued functions

$$\mathcal{P}(t) \in conv\mathbb{R}^n,\ \mathcal{Q}(t) \in conv\mathbb{R}^n,\ \mathcal{Y}(t) \in conv\mathbb{R}^m$$

and matrix-valued function $G(t)$ taken to be continuous in t. (Recall that the presumed property of $Q(t)$ being continuous is translated into the presumption that $f(t)$ is continuous whenever $Q(t) = f(t)$ is reduced to a singleton $f(t)$, otherwise $f(t)$ is allowed to be measurable in t.)

Among the problems of control and estimation for this system we have singled out five for detailed treatment to demonstrate the suggested approach. These are the following

I *System with no input uncertainty and no state constraints* (Sections 1.2 and 1.3):

$f(t)$ – given ; $Q(t) = f(t)$ - single-valued, $\mathcal{Y}(t) \equiv \mathbb{R}^m$.

II *System with input uncertainty and with no state constraints* (Sections 1.6 and 1.7):

$f(t)$ – unknown, but bounded, due to (1.14.3), $\mathcal{Y}(t) \equiv \mathbb{R}^m$.

III *System with state constraint but no uncertainty* (Section 1.9):

$f(t)$ – given; $Q(t) \equiv f(t); \mathcal{Y}(t) \in conv\mathbb{R}^m$ – absolutely continuous in t.

IV *System with uncertainty and with state constraints* (Section 1.11):

$f(t)$ – unknown but bounded, due to (1.13.3); $\mathcal{Y}(t) \in conv\mathbb{R}^m$ – same as in **III**.

IV' *System with measurement output* (Section 1.12), with uncertainty in the inputs, initial states and measurement noise: control $u(t)$ – given, input $f(t)$ – unknown, but bounded, due to (1.13.3), state constraint given in the form

$$y(t) \in G(t)x + \mathcal{K}(t)$$

or, particularly,

$$y(t) \in \mathcal{Y}(t)\ ,\ \mathcal{Y}(t) \ = \ y(t) - \mathcal{K}(t) \ .$$

Here $y(t)$ is the available measurement, $\mathcal{K}(t)$ is the bound on the measurement error.

The first issue discussed was the calculation of the *attainability domains* and the *attainability tubes*. These were given through the solutions of the following evolution funnel equations with set-valued solutions, namely,

for case **I**

$$\lim_{\sigma\to 0} \sigma^{-1} h(\mathcal{X}[t+\sigma], \mathcal{X}[t] + \sigma\mathcal{P}(t) + \sigma f(t)) = 0 \ ,$$

(1.13.6) $$\mathcal{X}[t_0] = \mathcal{X}^0 \ ;$$

for case **II**

$$\lim_{\sigma\to 0} \sigma^{-1} h_+(\mathcal{X}[t+\sigma] - \sigma\mathcal{Q}(t), \mathcal{X}[t] + \sigma\mathcal{P}(t)) = 0 \ ,$$

or equivalently,

$$\lim_{\sigma\to 0} \sigma^{-1} h_+(\mathcal{X}[t+\sigma], (\mathcal{X}[t] + \sigma\mathcal{P}(t)) \dot{-} (-\mathcal{Q}(t))) = 0 \ ,$$

under condition (1.13.6);

for case **III**

$$\lim_{\sigma\to 0} \sigma^{-1} h(\mathcal{X}[t+\sigma], (\mathcal{X}[t] + \sigma\mathcal{P}(t) + f(t)) \cap \mathcal{Y}(t+\sigma)) = 0 \ ,$$

or

$$\lim_{\sigma\to 0} \sigma^{-1} h_+(\mathcal{X}[t+\sigma], \mathcal{X}[t] \cap \mathcal{Y}(t) + \sigma\mathcal{P}(t) + f(t)) = 0 \ ,$$

under (1.13.6);

for case **IV**

$$\lim_{\sigma\to 0} \sigma^{-1} h_+(\mathcal{X}[t+\sigma] - \sigma\mathcal{Q}(t), \mathcal{X}[t] \cap \mathcal{Y}(t) + \mathcal{P}(t) + f(t)) \ = \ 0 \ ,$$

under condition (1.13.6);

for case **IV'**

$$\lim_{\sigma\to 0} \sigma^{-1} h(\mathcal{X}[t+\sigma], (\mathcal{X}[t] + \sigma(u(t) + y(t)) + \sigma\mathcal{Q}(t)) \cap \mathcal{K}(t)) = 0 \ ,$$

(if the function $\mathcal{Y}(t) = y(t) - \mathcal{K}(t)$ is absolutely continuous in t) or

$$\lim_{\sigma \to 0} \sigma^{-1} h_+(\mathcal{X}[t+\sigma], \mathcal{X}[t] \cap (y(t) - \mathcal{K}(t)) + \sigma(u(t) + \mathcal{Q}(t))) = 0 \quad ,$$

(if the function $\mathcal{Y}(t)$ is upper semicontinuous in t, particularly, if $\mathcal{K}(t)$ is continuous and $y(t)$ is piecewise continuous from the right).

Both equations are considered under conditions (1.13.6).

The respective attainability domains are given through the respective *unique solutions* to the evolution equations when these are written in terms of the Hausdorff *distance* $h(\cdot,\cdot)$ and through the *maximal solutions* (with respect to inclusion) for the equations written in terms of the Hausdorff *semidistance* $h_+(\cdot,\cdot)$. The attainability domain for case **IV'** is the *information domain* for the guaranteed state estimation problem of Section 1.12.

The second group of issues consists of problems of *goal-oriented nonlinear control synthesis*. Here the objective is to reach a preassigned terminal target set $\mathcal{M}$ at given time $t = t_1$ by selecting a feedback control strategy $\mathcal{U}(t,x) \in U_{\mathcal{P}}^{c}$ which in general turns out to be nonlinear, as the controls are bounded here by *magnitude bounds*. The overall synthesized system is then described by a nonlinear differential inclusion.

For each of the system types **I–IV** this strategy $\mathcal{U} = \mathcal{U}_0(t,x)$ is selected in a standard way by minimizing the derivative

$$(1.13.7) \quad \frac{d}{dt}V(t,x)\Big|_{u=\mathcal{U}_0} = \min\left\{ \frac{d}{dt}V(t,x)\Big|_{u=\mathcal{U}} : \mathcal{U} \in U_{\mathcal{P}}^{c} \right\} \quad ,$$

where

$$V(t,x) = d^2(x, \mathcal{W}[t])$$

and $W[t]$ is the cross-section of the respective *solvability tube*.[13]

The strategy $\mathcal{U}(t,x)$ may also be calculated directly, without introducing the tube $\mathcal{W}[t]$, but, as indicated in Sections 1.5, 1.8, 1.10, and 1.11, by solving for the respective problems the respective H-J-B equations with value functions

$$V^*(t,x) = \min\{h_+^2(x(t_1,t_0,x), \mathcal{M}) | \mathcal{U}(\cdot,\cdot) \in U_{\mathcal{P}}^{c}\}$$

[13]This type of solution was introduced by N.N. Krasovski under the name of *extremal aiming strategy* with solution tubes $W[t]$ being referred to as bridges, see [168], [169] and also [171].

for case **I**,

$$V^*(t,x) = \min_{\mathcal{U}} \max_{f} \{h_+^2(x(t_1,t_0,x),\mathcal{M}) | \mathcal{U}(\cdot,\cdot) \in U_{\mathcal{P}}^c, f(\cdot) \in \mathcal{Q}(\cdot)\}$$

for case **II**,

$$V_0(t,x) = \min \left\{ \int_t^{t_1} h_+^2(x(\tau),\mathcal{Y}(\tau))d\tau + h_+^2(x(t_1,t_0,x),\mathcal{M}) \middle| u(\cdot) \in \mathcal{P}(\cdot) \right\}$$

for case **III**,

$$V_0(t,x) = \min_{\mathcal{U}} \max_{f}$$

$$\left\{ \int_t^{t_1} h_+^2(x(\tau),\mathcal{Y}(\tau))d\tau + h_+^2(x(t_1,t_0,x),\mathcal{M}) \middle| \mathcal{U}(\cdot,\cdot) \in U_{\mathcal{P}}^c, f(\cdot) \in \mathcal{Q}(\cdot) \right\}$$

for case **IV**,

and with further application of (1.14.6) to $V = V^*(t,x)$ or $V = V_0(t,x)$.

For systems of type (1.14.1) the respective solvability sets $\mathcal{W}[t]$ will turn out to be *level sets* for the corresponding value functions, namely,

$$\mathcal{W}[t] = \{x : V^*(t,x) \leq 0\}$$

for cases **I, II** and

$$\mathcal{W}[t] = \{x : V_0(t,x) \leq 0\}$$

for cases **III, IV**.

The control strategies are then determined from relation (1.13.6) taken for the corresponding value functions.

The ability to calculate the solvability tubes $\mathcal{W}(\cdot)$ eliminates the necessity to solve the H-J-B equation. The specific emphasis is that these tubes may be calculated through evolution equations which are precisely the ones introduced for the attainability domains but should be now taken *in backward time*. Namely, we have introduced the following equations:

for case **I**

(1.13.8) $$\lim_{\sigma \to 0} \sigma^{-1} h(W[t-\sigma], W[t] - \sigma\mathcal{P}(t) - \sigma f(t)) = 0 \ ,$$

for case **II**

(1.13.9) $$\lim_{\sigma\to 0} \sigma^{-1} h_+(W[t-\sigma] + \sigma\mathcal{Q}(t), W[t] - \sigma\mathcal{P}(t)) = 0 \ ,$$

or

(1.13.10) $$\lim_{\sigma\to 0} \sigma^{-1} h_+(W[t-\sigma], (W[t] - \sigma\mathcal{P}(t)) \dot{-} (-\mathcal{Q}(t))) = 0 \ ,$$

for case **III**

(1.13.11) $$\lim_{\sigma\to 0} \sigma^{-1} h(W[t-\sigma], (W[t] - \sigma\mathcal{P}(t) - \sigma f(t)) \cap \mathcal{Y}(t-\sigma)) = 0 \ ,$$

or

(1.13.12) $$\lim_{\sigma\to 0} \sigma^{-1} h_+(W[t-\sigma], W[t] \cap \mathcal{Y}(t) - \sigma\mathcal{P}(t) - \sigma f(t)) = 0 \ ,$$

for case **IV**

(1.13.13) $$\lim_{\sigma\to 0} \sigma^{-1} h_+(W[t-\sigma] + \sigma\mathcal{Q}(t), W[t] \cap \mathcal{Y}(t) - \sigma\mathcal{P}(t)) = 0 \ ,$$

All of these equations have to be solved with boundary condition

(1.13.14) $$W[t_1] = \mathcal{M} \ .$$

The unique solutions to equations (1.13.7) and (1.13.10) with boundary condition (1.13.13) and the maximal solutions to equations (1.13.8), (1.13.9), (1.13.11), and (1.13.12) with the same boundary condition give us the respective solvability tubes $\mathcal{W}[\cdot]$ that produce the the crucial elements $\mathcal{W}[t]$ for calculating the required control strategies $\mathcal{U}(t,x)$.

Needless to say, equations (1.13.9), (1.13.10), and (1.3.12) are particular cases of equations (1.13.13) and (1.13.8) is a particular case of (1.13.11). The solution to (1.13.8) is also the maximal solution to a modification of this equation which is (1.3.9), where distance h is substituted by semidistance h_+.

Now it should be probably clear that equations (1.13.7)–(1.13.12) may serve to be the motivation and the basis for introducing discretized schemes with set-valued elements. In other words, we may loosely assume:

for case **I**

(1.13.15) $$W[t-\sigma] \simeq W[t-\sigma] - \sigma\mathcal{P}(t) - \sigma f(t) \ ,$$

for case **II**

(1.13.16) $$W[t-\sigma] \simeq (W[t] - \sigma\mathcal{P}(t)) \dot{-} (-\mathcal{Q}(t)) \ ,$$

for case **III**

(1.13.17) $$W[t-\sigma] \simeq (W[t] - \sigma\mathcal{P}(t) - \sigma f(t)) \cap \mathcal{Y}(t-\sigma)$$

or

(1.13.18) $$W[t-\sigma] \simeq W[t] \cap \mathcal{Y}(t) - \sigma\mathcal{P}(t) - \sigma f(t) \ ,$$

for case **IV**

(1.13.19) $$W[t-\sigma] \simeq (W[t] \cap \mathcal{Y}(t) - \sigma\mathcal{P}(t)) \dot{-} (-\mathcal{Q}(t)) - \sigma f(t) \ .$$

The equalities (1.13.14)–(1.13.18) are true relative to an error of order $\gamma(\sigma)$, where $\sigma^{-1}\gamma(\sigma) \to 0, \sigma \to 0$.[14]

The last relations indicate that the basic set-valued operations for the topics of this book are the *geometrical* (Minkowski) sums $(+,-)$ and differences$(\dot{-})$ of convex compact sets as well as their intersections $(\cap)$. Since an arbitrary convex compact set is an infinite-dimensional element (that may be identified with its support function, for example), the respective numerical calculations require finite-dimensional approximations. This book indicates *ellipsoidal approximations* as an appropriate technique.

1.14 Why Ellipsoids?

The aim of this book is to indicate some constructive techniques for solving problems of estimation and feedback control under set-membership uncertainty and state constraints with the hope that these techniques will allow effective algorithmization and computer animation. As we have seen in the above, the basic mathematical tool for describing the class of problems raised here is *set-valued calculus*. It is probably not unnatural, therefore, that the specific methods selected in the sequel are based on an *ellipsoidal technique* that would allow us to approximate the

[14]The indicated relations describe *first order approximations* to the exact set-valued solutions of the above. The theory of second-order approximations to solution tubes for differential inclusions was discussed in paper [309].

set-valued solutions of the above by ellipsoidal-valued solutions. Particularly, the set-valued attainability and solution tubes of the previous sections will be further approximated by ellipsoidal-valued functions.

Technically one of the basic justifications for such an approach is that the crossections $\mathcal{X}[t], \mathcal{W}[t]$ of functions $\mathcal{X}[\cdot], \mathcal{W}[\cdot]$ with convex compact values may be presented (for all the cases **I–IV** under consideration) in the form of intersections

$$\mathcal{X}[t] = \cap \mathcal{E}_+^{(\alpha)}(t) \ , \ \mathcal{W}[t] = \cap \mathcal{E}_+^{(\beta)}(t) \tag{1.14.1}$$

over a parametrized infinite variety of ellipsoidal-valued functions $\mathcal{E}_+^{(\alpha)}(t), \mathcal{E}_+^{(\beta)}(t)$ (which may even be assumed denumerable). Each of these, in its turn, may be calculated by solving a system of ordinary differential equations (ODE's). The calculation of $\mathcal{X}[t], \mathcal{W}[t]$ would thus be *parallelized into an array of identical problems each of which would consist in solving an ODE* that describes an ellipsoidal-valued function $\mathcal{E}_+^{(\alpha^*)}(t)$ or $\mathcal{E}_+^{(\beta^*)}(t)$.

The ellipsoidal representations (1.14.1) for $\mathcal{X}[t], \mathcal{W}[t]$ are *exact* and are true for the solutions of each of the evolution equations indicated in Section 1.14, (that is, *for all the cases* **I–IV** indicated in this section). Moreover, in the absence of state constraints (cases **I, II**), the ellipsoidal calculus used here also allows effective *internal ellipsoidal approximations* in the form of

$$\mathcal{X}[t] = \overline{\cup \mathcal{E}_-^{(\gamma)}(t)}, \mathcal{W}[t] = \overline{\cup \mathcal{E}_-^{(\delta)}(t)} \ , \tag{1.14.2}$$

where the dash stands for the closure of the respective set and where $\mathcal{E}_-^{(\gamma)}(t), \mathcal{E}_-^{(\delta)}(t)$, again stands for the elements of an infinite denumerable variety of ellipsoidal-valued functions described by *ordinary differential equations.*

The ellipsoidal calculus suggested further yields, among others, the ability to address the following issues:

(i) The exact representation and approximation of attainability domains for linear systems with or without state constraints through both external and internal ellipsoids.

(ii) The treatment of attainability and solvability tubes $\mathcal{X}[t], \mathcal{W}[t]$ under set-membership uncertainty (counteraction) in the inputs. These, for the linear systems considered here, may be particularly described by alternated integrals of L.S. Pontryagin – an object far more complicated than the standard set-valued (Aumann) integral that represents similar tubes in the absence of input uncertainty. Nevertheless, the respective tubes, given by alternated integrals or by corresponding evolution equations of the funnel type, still allow exact internal and external ellipsoidal representations.

(iii) The exact ellipsoidal representation or external approximation of the *information domains* for guaranteed (set-membership) state estimation under unknown but bounded errors.

(iv) The possibility to single out individual external or internal approximating ellipsoids that are *optimal* relative to some given optimality criterion (trace, volume, diameter, etc.) or a combination of such criteria. They also allow to apply vector-valued criteria to the approximation problem.[15]

Loosely speaking, the representations of type (1.14.1) and (1.14.2) mean that the more ellipsoids are allowed to approximate $\mathcal{X}[t], \mathcal{W}[t]$ (in practice this depends on the number of available processors), the more accurate will be the approximation, so that, in theory, an infinite (denumerable) variety of ellipsoids would produce the exact relations (1.14.1) and (1.14.2). Thus, each ellipsoidal-valued function could be treated through a single processor which solves an ODE of fixed dimension. The number of available processors would then determine the accuracy of the solution.

The application of ellipsoidal techniques will further allow us to devise relatively simple control strategies for a control synthesis that will ensure guaranteed results for the related problems. The strategies will then be given in the form of *analytic designs rather than algorithms* as it is in the exact case. The important feature that allows such ellipsoidal-based analytical designs is that the multivalued mappings that generate the internal ellipsoidal tubes $\mathcal{E}_-^{\delta}(t)$ for the solvability sets $\mathcal{W}[t]$ satisfy *a generalized semigroup property* on one hand, and, on the other, the tubes

[15]The possibility of exact representations and of vector-valued criteria for approximating attainability domains under state constraints by external ellipsoids was indicated in monograph [181]. The minimal volume criteria for these problems was thoroughly studied in [278], [73].

are nondominated (inclusion-maximal) among all such ellipsoidal-valued functions. These two properties allow us to demonstrate that the tubes $\mathcal{E}_{-}^{\delta}(t)$ possess the property of being an ellipsoidal bridge similar to the "Krasovski bridge" of the exact solution, (see Sections 3.5 and 3.8).

It is obvious, of course, that one of the options is to approximate the set-valued functions $\mathcal{X}[t], \mathcal{W}[t]$ by using boxes or, more generally, by *polyhedral-valued functions.* This approach, which, of course, has its advantages and disadvantages, lies beyond the scope of the present book. (We address the reader to [187], [175], [225], [75].) However, it appears that the main difficulty lies in the fact that computational complexity is such that the number of elementary operations here increases exponentially with the number of steps in the sampled problem. A natural desire will then be to parallelize the polyhedral approximation into problems of smaller dimensions.

(v) Another motivation for using ellipsoids comes from Section 4.2. There the nondifferentiable solution $V(t, x)$ to the H-J-B equation (its level set is the attainability domain) is approximated by quadratic functions whose level sets are nondegenerate ellipsoids. At the same time, these functions are precisely the test functions used in defining the generalized viscosity solutions of the H-J-B equation.

(vi) The ellipsoidal approaches described in this book allow direct links to related stochastic problems of estimation and control with Gaussian models of uncertainty.

It also appears useful to remark that any convex compact set $\mathcal{Q}$ in $\mathbb{R}^n$ may be presented as an intersection of ellipsoids

$$\mathcal{Q} = \cap \mathcal{E}^{(\sigma)}$$

(This fact is a consequence of an *ellipsoidal separation theorem* – the property that every point $x \notin \mathcal{Q}$ may be separated from $\mathcal{Q}$ by an ellipsoidal surface.)

The last fact justifies that we further take all the sets $\mathcal{X}^0, \mathcal{P}(t), \mathcal{Q}(t), \mathcal{Y}(t)$, X^0 that define the preassigned constraints (1.2.1), (1.1.2), (1.6.1), and (1.9.3) on the system to be ellipsoidal-valued.

Part II. THE ELLIPSOIDAL CALCULUS

Introduction

This part is a separate text on *ellipsoidal calculus* – a technique of representing basic operations on ellipsoids. The operations treated here are motivated by the requirements of Part I of the present book. However, the results given here may be applied, of course, to a substantially broader class of problems that arise in mathematical modelling, particularly, in optimization and approximation, identification and experiment planning, probability and statistics, stabilization, adaptive control, mathematical morphology, and other areas.

The operations on ellipsoids are discussed in the following order. First, these are the geometrical (Minkowski) sums and differences of two nondegenerate ellipsoids with the difference having a nonvoid interior. Each sum and difference is approximated – both *externally and internally* – by a corresponding parametrized variety of ellipsoids. With the number of approximating ellipsoids increasing to infinity, the approximations converge, in the limit, to exact representations. The external representations are given by *intersections* and the internal by *unions* (or their closures) over the respective varieties each of which is infinite, denumerable, at least. Taking intersections or unions over some finite subsets of these varieties, we come to external and internal *approximations* of the sums and differences (Sections 2.2 and 2.3). Particularly, we take *only one element* of the respective variety, which is *optimal* in some sense (an array of possible optimality criteria is discussed at the end of Section 2.1). Then the sums and differences will be approximated (internally or externally) by an optimal ellipsoid (Section 2.5). These criteria may include its diameter, the sum of its axes or their squares (the trace of the matrix defining the ellipsoid) [111], [181], [263], [225], etc. A widely studied criterion is the volume of the ellipsoid (see [278], [73]).

A certain reciprocity consists in the fact that *the external ellipsoids for the sums and the internal ones for the differences* are given by the same type of parametrization which differs in both cases only in some signs in the representation formula. A similar fact is true for *the internal ellipsoidal approximations of the sums and the external ones of the differences* (Sections 2.2 and 2.3).

The obtained representations are then propagated to finite sums of nondegenerate ellipsoids and to set-valued integrals of ellipsoidal-valued functions $\mathcal{E}[t]$ (which are not obliged to be ellipsoids). These sums and integrals are again approximated externally and internally. Moreover, if the upper limit of the set-valued integral

$$\mathcal{X}[t] = \int_{t_0}^{t} \mathcal{E}[\tau]d\tau$$

varies, then the parameters of the ellipsoidal functions that approximate $\mathcal{X}[t]$ may be described by *ordinary differential equations* (Sections 2.7 and 2.8). The important element here is that these ellipsoidal-valued functions that approximate $\mathcal{X}[t]$ are nondominated with respect to inclusion. Namely, they are inclusion minimal for the external ellipsoids and inclusion-maximal for the internals.

Intersections of ellipsoids are the topic of Section 2.6. Several types of external ellipsoidal approximations are described here with exact representations in the limit. An indication is finally given on how to construct varieties of internal ellipsoidal approximations of intersections of ellipsoids. The construction of internal ellipsoidal approximations to polyhedral and other types of convex sets are important particularly in algorithmic problems of mathematical programming [152], [282]. The solution to these is usually given in the form of an algorithm. (Also note the theory of analytical centers [287].) However, to be consistent with the approach presented here, one must be able to indicate a variety of internal ellipsoids whose union (or its closure) would approximate a nondegenerate intersection of ellipsoids (from inside) with any desired degree of accuracy.

In Section 2.4 we also mention a direction toward the calculation of approximation errors (depending on the number of approximating ellipsoids). Effective algorithms for estimating the errors as well as the computational complexity of these problems are among the issues that present a further challenge. We also believe that one should not drop the problem of finding perhaps rough but simple error estimates.

2.1 Basic Notions: The Ellipsoids

As we have seen in Part I, the basic set-valued operations involved in the calculation of solutions to the control problems of the above are the following:

- the geometrical (Minkowski) sum of convex sets,
- the geometrical (Minkowski) difference of convex sets,
- the intersection of convex sets,
- the affine transformations of convex sets.

Let us elaborate on the first two of these operations presuming that the sets involved are convex and compact.

Definition 2.1.1 *Given sets* $\mathcal{H}_1, \mathcal{H}_2 \in \text{comp } \mathbb{R}^n$, *the geometrical (Minkowski) sum* $\mathcal{H}_1 + \mathcal{H}_2$ *is defined as*

$$\mathcal{H}_1 + \mathcal{H}_2 = \bigcup_{h_1 \in \mathcal{H}_1} \bigcup_{h_2 \in \mathcal{H}_2} \{h_1 + h_2\} \ .$$

Obviously, the support function

$$\rho(\ell | \mathcal{H}_1 + \mathcal{H}_2) = \rho(l | \mathcal{H}_1) + \rho(l | \mathcal{H}_2) \ .$$

Definition 2.1.2 *Given sets* $\mathcal{H}_1, \mathcal{H}_2 \in \text{comp } \mathbb{R}^n$, *the geometrical (or Minkowski or also internal) difference* $\mathcal{H}_1 \dot{-} \mathcal{H}_2$ *is defined as*

$$\mathcal{H}_1 \dot{-} \mathcal{H}_2 = \{h \in \mathbb{R}^n : h + \mathcal{H}_2 \subseteq \mathcal{H}_1\} \ .$$

This means $\mathcal{H}_1 \dot{-} \mathcal{H}_2 \neq \emptyset$ if there is an element $h \in \mathbb{R}^n$, such that

$$h + \mathcal{H}_2 \subseteq \mathcal{H}_1 \ .$$

Clearly

$$\mathcal{H}_1 \dot{-} \mathcal{H}_2 = \{h \in \mathbb{R}^n : h \in \mathcal{H}_1 - h_2, \text{ for all } h_2 \in \mathcal{H}_2\} \ .$$

What follows from here are the assertions

Lemma 2.1.1 *The set* $\mathcal{H}_1 \dot{-} \mathcal{H}_2$ *may be presented as*

$$\mathcal{H}_1 \dot{-} \mathcal{H}_2 = \bigcap_{h_2 \in \mathcal{H}_2} \bigcup_{h_1 \in \mathcal{H}_1} \{h_1 - h_2\} \ . \tag{2.1.1}$$

Lemma 2.1.2 *The geometrical difference* $\mathcal{H} = \mathcal{H}_1 \dot{-} \mathcal{H}_2$ *is the maximal convex set (with respect to inclusion) among those that satisfy the relation*

$$\mathcal{H} + \mathcal{H}_2 \subseteq \mathcal{H}_1 \ , \tag{2.1.2}$$

namely,

$$\mathcal{H} = \mathcal{H}_1 \dot{-} \mathcal{H}_2$$

if and only if

$$\mathcal{H} + \mathcal{H}_2 \subseteq \mathcal{H}_1 \quad \textit{and} \quad \mathcal{H}' + \mathcal{H}_2 \subseteq \mathcal{H}_1 \quad \textit{imply} \quad \mathcal{H}' \subseteq \mathcal{H} \ .$$

In terms of support functions the inclusion (2.1.2) yields

$$\rho(l|\mathcal{H}) \leq \rho(l|\mathcal{H}_1) - \rho(l|\mathcal{H}_2) = f(l), \quad \forall l \in \mathbb{R}^n, \tag{2.1.3}$$

so that, if $f(l)$ were convex, we would have

$$\rho(l|\mathcal{H}) = \rho(l|\mathcal{H}_1) - \rho(l|\mathcal{H}_2)$$

(since $\mathcal{H}$ is the *inclusion-maximal* set that satisfies (2.1.3)).

Otherwise,

$$\rho(l|\mathcal{H}) = (\text{ co } f)(l) = f^{**}(l) \ ,$$

where co f is the "lower envelope" of $f(l)$ [100].

The following properties may happen to be useful

Lemma 2.1.3 *With* $\mathcal{H}_1, \mathcal{H}_2, \mathcal{H}_3 \in \text{comp}\mathbb{R}^n$ *we have*

$$\mathcal{H}_1 \dot{-} (\mathcal{H}_2 + \mathcal{H}_3) = (\mathcal{H}_1 \dot{-} \mathcal{H}_2) \dot{-} \mathcal{H}_3 \ , \tag{2.1.4}$$

$$(\mathcal{H}_1 + \mathcal{H}_2) \dot{-} \mathcal{H}_3 \supseteq \mathcal{H}_1 + (\mathcal{H}_2 \dot{-} \mathcal{H}_3) \ . \tag{2.1.5}$$

Let us indicate examples that would illustrate that in the last relation both an equality and a strict inclusion are possible.

Example 2.1.1

Assume

$$\begin{aligned} \mathcal{S}_r(0) &= \{(x,y) \in \mathbb{R}^2 : x^2 + y^2 \leq r^2\}, \\ \mathcal{H}_1 &= \mathcal{S}_1(0), \\ \mathcal{H}_2 &= \mathcal{S}_1(0), \\ \mathcal{H}_3 &= \{(x,y) \in \mathbb{R}^2 : -1 \leq x \leq 1, y = 0\} \ . \end{aligned}$$

Then

$$\begin{aligned} \mathcal{H}_1 + \mathcal{H}_2 &= \mathcal{S}_2(0), \\ (\mathcal{H}_1 + \mathcal{H}_2) \dot{-} \mathcal{H}_3 &= \mathcal{S}_2(0) \dot{-} \mathcal{H}_3 = \mathcal{H} \ , \end{aligned}$$

where, obviously

$$\mathcal{H} = \{(x,y) \in \mathbb{R}^2 : (x+z)^2 + y^2 \leq 4, |z| \leq 1\} \ .$$

In other words, the set $\mathcal{H}$ is the intersection of the sets

$$\{(x,y) \in \mathbb{R}^2 : (x+1)^2 + y^2 \leq 4\},$$

$$\{(x,y) \in \mathbb{R}^2 : (x-1)^2 + y^2 \leq 4\} \ .$$

On the other hand, clearly, $\mathcal{H}_2 \dot{-} \mathcal{H}_3 = \{0\}$, according to the definition of the geometric difference. Therefore, $\mathcal{H}_1 + (\mathcal{H}_2 \dot{-} \mathcal{H}_3) = \mathcal{H}_1 + \{0\} = \mathcal{S}_1(0)$ and

$$\mathcal{S}_1(0) \subset \mathcal{H} \ .$$

The relation (2.1.5) is therefore a strict inclusion.

Example 2.1.2

Take

$$\begin{aligned} \mathcal{H}_1 &= \{(x,y) \in \mathbb{R}^2 : x = 0, |y| \leq 1\}, \\ \mathcal{H}_2 &= \mathcal{S}_1(0), \\ \mathcal{H}_3 &= \{(x,y) \in \mathbb{R}^2 : |x| \leq 1, y = 0\}. \end{aligned}$$

Then

$$\mathcal{H}_1 + \mathcal{H}_2 = \bigcup\{(x,y) \in \mathbb{R}^2 : x^2 + (y+z)^2 \le 1 \, , \; |z| \le 1\}$$

and one may observe

$$(\mathcal{H}_1 + \mathcal{H}_2) \dot{-} \mathcal{H}_3 = \mathcal{H}_1 \quad .$$

On the other hand, clearly,

$$\mathcal{H}_2 \dot{-} \mathcal{H}_3 = \{0\} \quad ,$$

and

$$\mathcal{H}_1 + (\mathcal{H}_2 \dot{-} \mathcal{H}_3) = \mathcal{H}_1 \quad .$$

In this case the inclusion (2.1.5) is an equality.

The convex compact set $\mathcal{H} = \mathcal{H}_1 \dot{-} \mathcal{H}_2$ is defined to be *inclusion-maximal* relative to those convex compact sets that satisfy the relation

$$\mathcal{H}_2 + \mathcal{H} \subseteq \mathcal{H}_1 \quad ,$$

being therefore an *internal difference.* It is not difficult to prove that *the internal difference is unique.*

Definition 2.1.3 *A set $\mathcal{H}_e$ will be defined as* **an external difference** *$\mathcal{H}_e = \mathcal{H}_1 \div \mathcal{H}_2$ if it is inclusion-minimal relative to those convex compact sets that satisfy the relation*

$$\mathcal{H}_2 + \mathcal{H} \supseteq \mathcal{H}_1 \quad . \tag{2.1.6}$$

The external difference may also be defined as the class $\mathbf{H} = \{\mathcal{H}\}$ of all sets $\mathcal{H} \in$ conv $(\mathbb{R}^n)$ that satisfy the inclusion (2.1.6).

Then

$$\rho(l|\mathcal{H}_2) + \rho(l|\mathcal{H}) \ge \rho(l|\mathcal{H}_1)$$
$$\forall \mathcal{H} \in \mathbf{H}, \quad \forall l \in \mathbb{R}^n$$

and it is possible to demonstrate that

$$\inf\{\rho(l|\mathcal{H})|\mathcal{H} \in \mathbf{H}\} = \rho(l|\mathcal{H}_1) - \rho(l|\mathcal{H}_2) \le \rho(\ell|\mathcal{H}_e)$$

(since $\mathcal{H}_1, \mathcal{H}_2 \in$ comp $(\mathbb{R}^n)$ and $l \in \mathbb{R}^n$ is finite-dimensional).

Our further aim is to introduce an *ellipsoidal calculus* that allows us to approximate the above relations for *convex compact sets* through *ellipsoidal-valued* relations.

Definition 2.1.4 *An ellipsoid $\mathcal{E}(a,Q)$ with center $a \in \mathbb{R}^n$ and configuration matrix Q (symmetric and nonnegative) is defined as the set*

$$\mathcal{E}(a,Q) = \{x \in \mathbb{R}^n : (l,x) \leq (l,a) + (l,Ql)^{\frac{1}{2}}, \forall l \in \mathbb{R}^n\} \quad ,$$

where its support function $\rho(l|\mathcal{E}(a,Q))$ is defined by the equation

$$\rho(l|\mathcal{E}(a,Q)) = (l,a) + (l,Ql)^{\frac{1}{2}} \quad .$$

With Q *nondegenerate*, the ellipsoid $\mathcal{E}(a,Q)$ could also be presented otherwise, in terms of the inequality

$$\mathcal{E}(a,Q) = \{x \in \mathbb{R}^n : (x-a)'Q^{-1}(x-a) \leq 1\} \quad .$$

which gives a direct, conventional description.[16]

We will now proceed with the following basic set-valued operations, applying them to ellipsoids.

The geometrical (Minkowski) sum. Given two ellipsoids $\mathcal{E}_1 = \mathcal{E}(a_1,Q_1)$,

$\mathcal{E}_2 = \mathcal{E}(a_2,Q_2)$, their sum $\mathcal{E}_1+\mathcal{E}_2$ may obviously not be an ellipsoid (find an example). We will therefore be interested in *ellipsoidal approximations* $\mathcal{E}_+^{(+)}, \mathcal{E}_+^{(-)}$ of the sum $\mathcal{E}_1+\mathcal{E}_2$ where

$$\mathcal{E}_+^{(+)} \supseteq \mathcal{E}_1 + \mathcal{E}_2$$

is an *external approximation* and

$$\mathcal{E}_+^{(-)} \subseteq \mathcal{E}_1 + \mathcal{E}_2$$

is an *internal approximation.*

As we shall see, it is not difficult to observe that $\mathcal{E}_+^{(+)}, \mathcal{E}_+^{(-)}$ are not unique. Therefore we shall further describe a rather complete parametrised variety of such ellipsoids.

[16]This representation is also true for degenerate matrices Q but then Q^{-1} does not exist and has to be substituted by the *Moore-Penrose pseudoinverse* for Q [120].

Definition 2.1.5 *Let* $\mathbf{E}_0$ *denote a certain variety of ellipsoids* $\mathcal{E}$. *An ellipsoid* $\mathcal{E}_0$ *will be* **inclusion-maximal** *relative to the variety* $\mathbf{E}_0$, *if the inclusions* $\mathcal{E}_0 \in \mathbf{E}_0$, $\mathcal{E}_0 \subseteq \mathcal{E}$ *imply the equality* $\mathcal{E}_0 = \mathcal{E}$. *Inclusion-minimality is defined similarly.*

We will further indicate the *inclusion-minimal external* approximations and the *inclusion-maximal internal* approximations of $\mathcal{E}_1 + \mathcal{E}_2$.

The geometrical (Minkowski) difference. For two given ellipsoids $\mathcal{E}_1, \mathcal{E}_2 \in$ comp $\mathbb{R}^n$ the geometrical (internal) difference

$$\mathcal{E}_1 \dot{-} \mathcal{E}_2 = \{x : x + \mathcal{E}_2 \subseteq \mathcal{E}_1\}$$

is unique. However it may not be an ellipsoid (give an example).

Definition 2.1.6 *An external ellipsoidal estimate of the difference* $\mathcal{E}_1 \dot{-} \mathcal{E}_2$ *will be defined as an ellipsoid* $\mathcal{E}_-^{(+)}$ *that satisfies the inclusion*

$$\mathcal{E}_-^{(+)} \supseteq \mathcal{E}_1 \dot{-} \mathcal{E}_2 \ ,$$

while an internal ellipsoidal approximation of the difference $\mathcal{E}_1 \dot{-} \mathcal{E}_2$ *is an ellipsoid* $\mathcal{E}_-^{(-)}$ *that satisfies the inclusion*

$$\mathcal{E}_-^{(-)} + \mathcal{E}_2 \subseteq \mathcal{E}_1 \ .$$

Obviously

$$\mathcal{E}_-^{(-)} \subseteq \mathcal{E}_1 \dot{-} \mathcal{E}_2 \ .$$

We will further be interested in the inclusion-maximal internal and the inclusion-minimal external ellipsoidal approximations $\mathcal{E}_-^{(-)}$, $\mathcal{E}_-^{(+)}$ of the difference $\mathcal{E}_1 \dot{-} \mathcal{E}_2$.

As we shall observe in the sequel, the inclusion-maximal approximations $\mathcal{E}_-^{(-)}$ and the inclusion-minimal approximations $\mathcal{E}_-^{(+)}$ are not unique. They could also be interpreted as *nondominated elements* of a (partially) ordered family. (It is now the family of sets in $\mathbb{R}^n$ and the ordering is inclusion.) This interpretation naturally follows from Definition 2.1.5.

The Intersections. Given $\mathcal{E}_1, \mathcal{E}_2 \in$ comp $\mathbb{R}^n$, its intersection $\mathcal{E}_1 \cap \mathcal{E}_2$ in general is not an ellipsoid. We are interested first of all in its external approximations

$$\mathcal{E}^+ \supseteq \mathcal{E}_1 \cap \mathcal{E}_2$$

seeking, for example, the inclusion-minimal (nondominated) ellipsoids. We will again discover that these are not unique and will try to describe a rather complete variety of such ellipsoids.

A more difficult problem is to find an internal ellipsoidal approximation

$$\mathcal{E}^- \subseteq \mathcal{E}_1 \cap \mathcal{E}_2$$

to the intersection. Indeed the intersection may easily turn out to be a convex set of rather general nature, namely, a nonsymmetrical set relative to any point or plane or even a degenerate convex set in the sense that if taken in $\mathbb{R}^n$ it will have no interior point.

A relatively simpler situation occurs when the centers of $\mathcal{E}_1$ and $\mathcal{E}_2$ coincide.

Affine Transformations. As an exercise one may easily check that the inclusion

$$x \in \mathcal{E}(a, Q)$$

is equivalent to the following

$$Ax + b \in \mathcal{E}(Aa + b, AQA') \ .$$

Let us now indicate two useful properties of symmetrical sets.

Lemma 2.1.4 *Suppose that a set $\mathcal{H} \in$ comp $\mathbb{R}^n$ is symmetrical, namely, $\mathcal{H} = -\mathcal{H}$. Then*

(a) the inclusion $\mathcal{H} \subseteq \mathcal{E}(a, Q)$ implies $\mathcal{H} \subseteq \mathcal{E}(0, Q)$ and

(b) $\mathcal{H} \supseteq \mathcal{E}(a, Q)$ implies $\mathcal{H} \supseteq \mathcal{E}(0, Q)$.

Proof. Suppose $\mathcal{H} \subseteq \mathcal{E}(a,Q)$, but the inclusion $\mathcal{H} \subseteq \mathcal{E}(0,Q)$ does not hold. Then there exists a set $\mathcal{L} \neq \emptyset$ such that

$$\mathcal{L} = \{l \in \mathbb{R}^n : \rho(l|\mathcal{H}) > (l,Ql)^{\frac{1}{2}}, \|l\| = 1\} \ .$$

Since $\mathcal{H}$, $\mathcal{E}(0,Q)$ are symmetrical, the property

$$\rho(l|\mathcal{H}) > (l,Ql)^{\frac{1}{2}}$$

will also hold for $l \in -\mathcal{L}$.

Since $\mathcal{H} \subseteq \mathcal{E}(a,Q)$, we have

$$\rho(l|\mathcal{H}) \leq (a,l) + (l,Ql)^{\frac{1}{2}}, \quad l \in \mathbb{R}^n \ .$$

Due to the previous supposition this implies

$$(a,l) > 0, \qquad l \in \mathcal{L} \ , \tag{2.1.7}$$

and also the inequality

$$(a,l) > 0, \qquad l \in -\mathcal{L} \ ,$$

which contradicts (2.1.7). The assertion (a) is thus proved.

To prove assertion (b) we observe

$$\rho(l|\mathcal{H}) \geq (l,a) + (l,Ql)^{\frac{1}{2}}, \quad l \in \mathbb{R}^n \ . \tag{2.1.8}$$

Since $\mathcal{H} = -\mathcal{H}$ we have $\rho(l|\mathcal{H}) = \rho(-l|\mathcal{H})$ for all $l \in \mathbb{R}^n$. Therefore

$$\rho(l|\mathcal{H}) \geq (-l,a) + (l,Ql)^{\frac{1}{2}}, \quad l \in \mathbb{R}^n \ .$$

Together with (2.1.8) the latter inequality yields

$$\rho(l|\mathcal{H}) \geq (l,Ql)^{\frac{1}{2}}, \quad l \in \mathbb{R}^n \ .$$

The assertion is thus proved.

As we have already mentioned, our objective is to add and subtract ellipsoids (in the geometrical sense) and also to intersect them and to apply affine transformations. The results of these operations are convex sets which may either again be ellipsoids or, what is more common, may not be ellipsoidal at all. In the latter case we will introduce *internal and external ellipsoidal approximations* of these sets. Out of all the possible approximating ellipsoids we will prefer to select the inclusion minimal or maximal ellipsoids observing that these extremal ellipsoids are the *nondominated elements* (relative to inclusion) of the respective varieties.

Among the nondominated varieties of inclusion-minimal or inclusion-maximal ellipsoids we may then want to single out some individual elements that would be *optimal* relative to some prescribed *optimality criterion.* We will therefore indicate a class $\Psi = \{\psi(\mathcal{E}(a,Q))\}$ of criteria functions $\psi(\mathcal{E}(a,Q))$ that would be

(a) defined on the set of all nondegenerate ellipsoids $\{\mathcal{E}(a,Q)\}$ and nonnegative-valued,

(b) monotonous by increasing with respect to inclusion:

$$\psi(\mathcal{E}_1) \leq \psi(\mathcal{E}_2) \text{ if } \mathcal{E}_1 \subseteq \mathcal{E}_2 \ .$$

(We shall generally also require the monotonicity property (b) to be invariant relative to affine transformations of ellipsoids.)

Let Q stand for a symmetric positive matrix and $\varphi_k(Q)$, $k = 1,\ldots,n$, for the coefficient of the $(n-k)$-th degree term of its characteristic polynomial,

$$\chi(\lambda) = \sum_{k=0}^{n} \chi_k \lambda^k \ ,$$

so that $\chi_k = \varphi_{n-k}(Q)$.

Let $\sigma(Q) = \{\lambda_1,\ldots,\lambda_n\}$ denote the set of eigenvalues of Q.

Lemma 2.1.5 *Suppose* $\mathcal{E}(a_1,Q_1) \supseteq \mathcal{E}(a_2,Q_2)$*.Then for all* $m \in N$ *we have*

(i) $\varphi_k(Q_1^m) \geq \varphi_k(Q_2^m)$, $\quad (k = 1,\ldots,n)$,

(ii) $\max\{\sigma(Q_1^m)\} \geq \max\{\sigma(Q_2^m)\}$,

(iii) $\min\{\sigma(Q_1^m)\} \geq \min\{\sigma(Q_2^m)\}$.

Proof. From Lemma 2.1.4 it follows that if $\mathcal{E}(a_1, Q_1) \supseteq \mathcal{E}(a_2, Q_2)$, then

$$\mathcal{E}(0, Q_1) \supseteq \mathcal{E}(0, Q_2) \ .$$

The latter inclusion yields $Q_1 \geq Q_2$.

Having two positive $n \times n$ matrices Q_1, Q_2 with respective eigenvalues

$$\lambda_1^{(1)} \leq \lambda_2^{(1)} \leq \ldots \leq \lambda_n^{(1)}$$

and

$$\lambda_1^{(2)} \leq \lambda_2^{(2)} \leq \ldots \leq \lambda_n^{(2)} \ ,$$

we observe that if $Q_1 \geq Q_2$, then [120],

$$\lambda_i^{(1)} \geq \lambda_i^{(2)}, \quad i = 1, \ldots, n \ .$$

The latter property yields the assertions of the Lemma.

Let us now indicate some common type measures

$$\psi[Q] = \psi(\mathcal{E}(0, Q))$$

for the size of an ellipsoid $\mathcal{E}(0, Q)$.

(a) The trace:

$$\psi[Q] = \mathrm{tr}(Q) = \varphi_{n-1}(Q) = \lambda_1 + \ldots + \lambda_n \ ,$$

is actually the sum of the squares of the semiaxes of $\mathcal{E}(0, Q)$. Given $\mathcal{E}(0, Q) = \{x \in \mathbb{R}^n : (Q^{-1}x, x) \leq 1\}$, with support function $\rho(l|\mathcal{E}(0, Q)) = (l, Ql)^{\frac{1}{2}}$, a canonic orthogonal transformation $Tx = z$ ($|T| \neq 0$) transforms $\mathcal{E}(0, Q)$ into $\mathcal{E}(0, TQT')$, where TQT' is diagonal, with diagonal elements λ_i. (Here the transformation $Q \to TQT'$ keeps the eigenvalues of TQT' the same as of Q and the lengths of the semiaxes of $\mathcal{E}(0, TQT')$ the same as of $\mathcal{E}(0, Q)$). Thus

$$\rho(l|\mathcal{E}(0, TQT')) = (\sum_{i=1}^{n} \lambda_i l_i^2)^{\frac{1}{2}} \ ,$$

so that the length of the i-th semiaxis of $\mathcal{E}(0,Q)$ is

$$\rho(e^{(i)}|\mathcal{E}(0,TQT')) = \sqrt{\lambda_i} \ ,$$

where $e^{(i)} = (e_1^{(i)},\ldots,e_n^{(i)})$, $e_j^{(i)} = \delta_{ij}$, is the i-th orth in the orthogonal coordinate space of $\mathbb{R}^n$.

Therefore $\mathrm{tr}(Q)$ is equal to the sum of the squares of the semiaxes of $\mathcal{E}(0,Q)$.

(b) *The trace of the square* yields a criterion

$$\psi[Q] = \mathrm{tr}(Q^2) = \varphi_{n-1}(Q^2) \ .$$

(c) *The product* $\psi[Q] = \lambda_1 \cdot \lambda_2 \cdot \ldots \cdot \lambda_n = \varphi_0(Q)$ is proportional *to the volume* vol $(\mathcal{E}(0,Q))$ of $\mathcal{E}(0,Q)$.

Indeed a direct calculation yields [213],

$$\mathrm{vol}\ (\mathcal{E}(0,Q)) = \pi^{\frac{n}{2}}(\ \det\ Q)^{1/2}(\Gamma(\frac{n}{2}+1))^{-1} \ ,$$

where Γ stands for *the gamma-function* (see [213], [276]). One just has to recall that the determinant $\det Q$ of Q is equal to the product $\varphi_0(Q) = \lambda_1 \cdot \ldots \cdot \lambda_n$.

(d) *The diameter :* $\psi[Q] = d(\mathcal{E}(0,Q))$. Here the value

$$\max\{\lambda_i \in \mathbb{R} : i = 1,\ldots,n\} = \left(\frac{d}{2}\right)^2 \ ,$$

where $d = d(\mathcal{E}(0,Q))$ is *the diameter* of $\mathcal{E}(0,Q)$, so that $d/2$ is the radius of the smallest n-dimensional ball that includes $\mathcal{E}(0,Q)$.

This follows from the fact that $d/2$ is equal to the length of the largest semiaxis of $\mathcal{E}(0,Q)$.

It is obvious that monotonous functions of those appearing in Lemma 2.1.5 as well as their positive combinations are also monotonous with respect to inclusion. This indicates the range of cases that we are able to handle. However we shall formulate our results primarily for vol $\mathcal{E}(0,Q), \mathrm{tr}(Q)$, and $\mathrm{tr}(Q^2)$.

We shall now specify some parametrized varieties of ellipsoids that allow us to approximate the geometrical sums and differences of ellipsoids and even to give an exact representation of these.

2.2 External Approximations: The Sums Internal Approximations: The Differences

In this section we will deal only with *nondegenerate* ellipsoids. Given two such ellipsoids $\mathcal{E}_1 = \mathcal{E}(a, Q_1)$ and $\mathcal{E}_2 = \mathcal{E}(a, Q_2)$, denote the roots of the equation

$$\det(Q_1 - \lambda Q_2) = 0$$

as

$$\lambda_{\min} = \lambda_1 \leq \lambda_2 \leq \ldots \leq \lambda_n = \lambda_{\max}, \quad (\lambda_1 > 0, \lambda_n < \infty) \ .$$

These roots are also said to be the relative eigenvalues of the matrices $Q_1, Q_2 \in \mathcal{L}(\mathbb{R}^n, \mathbb{R}^n)$.

Consider a parametric family of matrices

$$Q(p) = (1 + p^{-1})Q_1 + (1 + p)Q_2 \ .$$

We will also be interested in the family $Q(-p)$.

Denote

$$\prod^+ = [\lambda_{\min}^{1/2}, \lambda_{\max}^{1/2}] \ ,$$

$$\prod^- = \prod^+ \bigcap (1, \lambda_{\min}) \ .$$

Lemma 2.2.1 *(a) The ellipsoid $\mathcal{E} = \mathcal{E}(a_1 + a_2, Q(p)), p > 0$, is properly defined and is an external approximation of the sum $\mathcal{E}_1 + \mathcal{E}_2$, i.e.,*

$$\mathcal{E}_1 + \mathcal{E}_2 \subseteq \mathcal{E}(a_1 + a_2, Q(p)) \tag{2.2.1}$$

for any $p > 0$.

(b) With vector $l \in \mathbb{R}^n, \|l\| = 1$, given, the equality

$$p = (Q_1 l, l)^{1/2}(Q_2 l, l)^{-1/2} \tag{2.2.2}$$

defines a scalar parameter $p \in \prod^+$, such that

$$\rho(l|\mathcal{E}(a_1 + a_2, Q(p))) = \rho(l|\mathcal{E}(a_1, Q_1) + \mathcal{E}(a_2, Q_2)) \ . \tag{2.2.3}$$

Conversely, with parameter $p \in \prod^+$ given, there exists a vector $l \in \mathbb{R}^n$ with $\|l\| = 1$, such that equalities (2.2.2) and (2.2.3) are true.

Proof. The inequality

$$p^{-1}(Q_1l,l) + p(Q_2l,l) \geq 2(Q_1l,l)^{1/2}(Q_2l,l)^{1/2}$$

is obviously true for any $p > 0$.

Adding $(Q_1l,l) + (Q_2l,l)$ to both sides, we obtain

$$(Q(p)l,l)^{1/2} \geq (Q_1l,l)^{1/2} + (Q_2l,l)^{1/2} \quad , \tag{2.2.4}$$

where $Q(p) > 0$ for any $p > 0$.

With a further addition of $(l, a_1 + a_2)$ to both sides this implies

$$\rho(l|\mathcal{E}(a_1 + a_2, Q(p))) \geq \rho(l|\mathcal{E}(a_1, Q_1)) + \rho(l|\mathcal{E}(a_2, Q_2))$$

for any $l \in \mathbb{R}^n$ and therefore, implies the inclusion (2.2.1).

To prove the assertion (b), with $l \in \mathbb{R}^n$ given, we select the parameter p due to (2.2.2), observing that $p \in \prod^+$ (check the latter inclusion as an exercise, using the extremal properties of matrix eigenvalues, see, e.g., [120]). After a substitution of (2.2.2) into (2.2.4), the latter turns into an equality for the given l (this can be verified through direct calculation). The equality (2.2.3) is therefore true for any given l with p and l related through (2.2.2).

Conversely, with $p \in \prod^+$ given, there exists a vector $l \in \mathbb{R}^n, \|l\| = 1$, such that (2.2.2) and therefore (2.2.3) do hold. This follows from Theorem 7.10, Chapter X of reference [120], due to the continuity in l of the right-hand side of (2.2.2).

Q.E.D.

A similar reasoning passes through for geometrical differences.

Lemma 2.2.2 *Suppose* int $\mathcal{E}(0, Q_1) \supset \mathcal{E}(0, Q_2)$. *Then*

(a) $\mathcal{E} = \mathcal{E}(a_1 - a_2, Q(-p))$ *is a nondegenerate ellipsoid if and only if*

$$p \in (1, \lambda_{\min}). \tag{2.2.5}$$

For these values of p *the ellipsoid* $\mathcal{E}$ *is an internal approximation of the difference* $\mathcal{E}_1 \dot{-} \mathcal{E}_2$, *i.e.,*

$$\mathcal{E}(a_1 - a_2, Q(-p)) \subseteq \mathcal{E}_1 \dot{-} \mathcal{E}_2 \quad .$$

(b) *With vector $l \in \mathbb{R}^n, \|l\| = 1$, given, the equality (2.2.2) defines a scalar parameter p. If $p \in \Pi^-$, then*

$$\rho(l|\mathcal{E}(a_1 - a_2, Q(-p))) = \rho(l|\mathcal{E} \dot{-} \mathcal{E}_2) = \rho(l|\mathcal{E}_1) - \rho(l|\mathcal{E}_2) \ .$$

Conversely, with parameter $p \in \Pi^-$ given, there exists a vector $l \in \mathbb{R}^n, \|l\| = 1$, such that the last equality is true together with (2.2.2).

Proof. Consider the inclusion

$$int\mathcal{E}(0, Q_1) \supset \mathcal{E}(0, Q_2) \ ,$$

which implies

$$(l, Q_1 l) > (l, Q_2 l)$$

for any $l \in \mathbb{R}^n$ and therefore implies the condition

$$p = \frac{(l, Q_1 l)^{1/2}}{(l, Q_2 l)^{1/2}} > 1 \ .$$

Taking

$$Q(-p) = (1 - p^{-1})Q_1 + (1 - p)Q_2 = (p - 1)(Q_1 p^{-1} - Q_2) \ ,$$

we observe, in view of the condition $p > 1$ that $Q(-p)$ is positive definite if and only if $Q_1 p^{-1} - Q_2 > 0$, which means $p < \lambda_{\min}$. This yields $p \in (1, \lambda_{\min})$.

Following the proof along a scheme similar to that of Lemma 2.2.1 with p substituted by $(-p)$, we come to the inequality

(2.2.6) $$(Q(-p)l, l)^{1/2} \leq (Q_1 l, l)^{1/2} - (Q_2 l, l)^{1/2},$$

which is true for any $l \in \mathbb{R}^n$ and equivalent to

$$\rho(l|\mathcal{E}(a_1 - a_2, Q(-p))) \leq \rho(l|\mathcal{E}(a_1, Q_1)) - \rho(l|\mathcal{E}(a_2, Q_2)) \ .$$

The latter inequality is further equivalent to the inclusion

$$\mathcal{E}(a_1 - a_2, Q(-p)) + \mathcal{E}(a_2, Q_2) \subseteq \mathcal{E}(a_1, Q_1)$$

which, due to the definition of the geometrical difference and the conditions of the Lemma, implies

$$\mathcal{E}(a_1 - a_2, Q(-p)) \subseteq \mathcal{E}(a_1, Q_1) \dot{-} \mathcal{E}(a_2, Q_2) \tag{2.2.7}$$

for any $p \in (1, \lambda_{\min})$. As shown above, the latter condition ensures that $Q(-p) > 0$.

To prove assertion (b) with $l \in \mathbb{R}^n$ given, we suppose the parameter p to be defined due to (2.2.2) and such that $p \in \prod^-$. Under these conditions a direct substitution of p into (2.2.6) turns the latter into an equality. The inclusion (2.2.8) together with the relation

$$\rho(l|\mathcal{E}(a_1, Q_1) \dot{-} \mathcal{E}(a_2, Q_2)) \leq \rho(l|\mathcal{E}(a_1, Q_1)) - \rho(l|\mathcal{E}(a_2, Q_2))$$

then yields equality (2.2.6) for the given values of l and p.

On the other hand, once $p \in \prod^-$ is given, there exists a vector $l \in \mathbb{R}^n$, such that equality (2.2.2) is fulfilled (due to Theorem 7.10, Chapter X of [120], and the continuity of the right-hand side of (2.2.2) in l).

This also yields (2.2.6) for the given p and l.

Q.E.D.

Consider a positive definite, symmetric matrix C with elements $\{c_{ij}\}$ where i stands for the row and j for the column of C. Also assume the symbol $\overline{1,0} = \emptyset$.

Lemma 2.2.3 *Fix a vector $l \in \mathbb{R}^n$, $\|l\| = 1$ and suppose that for some $m \in [0, n]$, we have*

$$\begin{aligned} l_j &= 0 \ \ \text{if} \ j \in \overline{1, m}, \\ l_j &\neq 0 \ \ \text{if} \ j = \overline{m+1, n} \ . \end{aligned} \tag{2.2.8}$$

Suppose in addition that $\mathcal{E}_1 = \mathcal{E}(0, Q_1), \mathcal{E}_2 = \mathcal{E}(0, Q_2)$ and that the matrices Q_1, Q_2, Q are diagonal. Then the following implications hold:

(a) If

$$\mathcal{E}(0, Q) \supseteq \mathcal{E}(0, C) \supseteq \mathcal{E}_1 + \mathcal{E}_2$$

and

$$\rho(l|\mathcal{E}(0,Q)) = \rho(l|\mathcal{E}_1 + \mathcal{E}_2)$$

then

$$c_{ij} = 0 \quad \textit{for all } i \neq j, \quad i \in \overline{m+1,n}$$

(b) If

$$\mathcal{E}(0,Q) \subseteq \mathcal{E}(0,C) \subseteq \mathcal{E}_1 \dot{-} \mathcal{E}_2$$

and

$$\rho(l|\mathcal{E}(0,Q)) = \rho(l|\mathcal{E}_1 \dot{-} \mathcal{E}_2)$$

then

$$c_{ij} = 0 \quad \textit{for all } i \neq j, \quad i \in \overline{m+1,n} \ .$$

Proof. We shall now prove assertion (a). Assertion (b) will be left as an exercise. Its proof is similar to that of (a).

For the given vector l define an array of vectors $h^{(k)}, k \in \overline{m+1,n}$, where

$$h_j^{(k)} = \begin{cases} l_j & \text{if } j \neq k \\ -l_j & \text{if } j = k \ . \end{cases}$$

By diagonality of the respective matrices and the equality of supports we have

$$(2.2.9) \quad (Qh^{(k)}, h^{(k)})^{1/2} = (Q_1 h^{(k)}, h^{(k)})^{1/2} + (Q_2 h^{(k)}, h^{(k)})^{1/2} \ .$$

Combined with the inclusion relations this implies

$$(2.2.10) \qquad (Qh^{(k)}, h^{(k)})^{1/2} = (C\, h^{(k)}, h^{(k)})^{1/2} \ .$$

Assuming

$$Q = \text{diag}\,\{q_{11}, \ldots, q_{nn}\} \ ,$$

$$Q_j = \text{diag}\,\{q_{11}^{(j)}, \ldots, q_{nn}^{(j)}\} \ , \quad j = 1, 2.$$

Define the function $\chi : \mathbb{R}^n \to \mathbb{R}$ by the equality

$$\chi(z) = \sum_{i=1}^{n} (q_{ii} - c_{ii}) z_i^2 \ .$$

Here we have for all $k \in \overline{m+1,n}$

$$\chi(l) = \chi(h^{(k)}) \tag{2.2.11}$$

and due to (2.2.10) we come to

$$\chi(l) = \sum_{i,j=m+1;i\neq j}^{n} c_{ij} l_i l_j \ . \tag{2.2.12}$$

Substituting l for $h^{(k)}$ with a fixed value of k into (2.2.11), (2.2.12) and taking into account the symmetry of C, we have

$$2 \sum_{j=1;j\neq k}^{n} c_{kj} l_k l_j = 0 \ .$$

The respective terms may now be cancelled out from the right-hand side of (2.2.12) for the respective value of k. Taking equation (2.2.12) in its reduced form we can cancel out similar terms for a new value $k^* \neq k$. Repeating this procedure for all the values $k^* \in \overline{m+1,n}$ except for a previously fixed pair r, s, $r \neq s, r, s \in \overline{m+1,n}$, we finally come to

$$\chi(l) = 2c_{rs} l_r l_s \ ,$$

so that the last cancellation yields $\chi(l) = 0$. This directly implies $c_{rs} = 0$. Since r, s were chosen arbitrarily, what follows is that $c_{rs} = 0$, for any $r, s \in \overline{m+1,n}; r \neq s$. The proof is therefore complete for $m = 0$.

If $m > 0$, then take

$$\varphi(l) = \frac{1}{2}(\rho^2(z|\mathcal{E}(0,Q)) - \rho^2(z|\mathcal{E}(0,C))) \ . \tag{2.2.13}$$

The function $\varphi(z)$ has a *local* minimum at $z = l$, $z = h^{(k)}$ for any $k \in \overline{m+1,n}$.

By differentiability we necessarily have

$$\left.\frac{\partial \varphi(z)}{\partial z}\right|_{z=h^{(k)}} = 0, \quad \left.\frac{\partial \varphi(z)}{\partial z}\right|_{z=l} = 0 \ .$$

For all the values of $i \in \overline{1,m}, k \in \overline{m+1,n}$ this yields

$$\sum_{j=m+1}^{n} c_{ij} h_j^{(k)} = 0, \quad \sum_{j=m+1}^{n} c_{ij} l_j = 0 \ .$$

Substracting the first relation from the second one for ech $k = \overline{m+1,n}$ and recalling that $h_j^{(k)} \neq 0$ for any $j,k \in \overline{m+1,n}$, we conclude that $c_{ij} = 0,\ i \neq j$, for any $i \in \overline{1,m}, j \in \overline{m+1,n}$.

Q.E.D.

Lemma 2.2.4 *Consider an ellipsoid $\mathcal{E}(0,C)$ together with ellipsoids $\mathcal{E}_1 = \mathcal{E}(0,Q_1), \mathcal{E}_2 = \mathcal{E}(0,Q_2)$, assuming that Q_1, Q_2 are diagonal. Further assume the vector $l \in \mathbb{R}^n, \|l\| = 1$ to be given and the parameter p to be defined due to relation (2.2.2) with given l.*

Then the following implications hold:
If

$$\mathcal{E}(0,Q(p)) \supseteq \mathcal{E}(0,C) \supseteq \mathcal{E}_1 + \mathcal{E}_2$$

and

$$\rho(l|\mathcal{E}(0,Q(p))) = \rho(l|\mathcal{E}_1 + \mathcal{E}_2) \ ,$$

then

$$\mathcal{E}(0,Q(p)) = \mathcal{E}(0,C) \quad \text{and} \quad p \in \Pi^+ \ .$$

Proof. Denote $Q_j = \text{diag } \{q_{11}^{(j)}, \ldots, q_{nn}^{(j)}\},\ j = 1,2,\ Q(p) = \text{diag } \{q_{11}, \ldots, q_{nn}\},\ C = \{c_{ij}\}$. Also keep the notation of (2.2.9) and the definition of $h^{(k)}$ of the previous Lemma.

Due to the previous Lemma 2.2.3 (a) and the inclusion $\mathcal{E}(0,Q(p)) \supseteq \mathcal{E}(0,C)$, we have

(2.2.14) $$c_{ii} \leq q_{ii}, \qquad i \in \overline{m+1,n} \ .$$

The equality (2.2.3) with value of p from (2.2.2) yields

$$\sum_{i=m+1}^{n} q_{ii} l_i^2 = \sum_{i=m+1}^{n} c_{ii} l_i^2 \ ,$$

where $l_i \neq 0$. Together with (2.2.14) this implies $c_{ii} = q_{ii}$ for $i \in \overline{m+1,n}$.

By the conditions of the Lemma both nonnegative functions

$$\xi(l) = \rho(l|\mathcal{E}(0,Q(p))) - \rho(l|\mathcal{E}(0,C))$$

and

$$\zeta(l) = \rho(l|\mathcal{E}(0,C)) - \rho(l|\mathcal{E}_1 + \mathcal{E}_2)$$

have local minima at $l = h^{(k)},\ k \in \overline{m+1,n}$ with $\xi(h^{(k)}) = 0, \zeta(h^{(k)}) = 0$. Due to the differentiability of these functions, the second order necessary conditions of optimality imply that the matrices of the second order partial derivatives are nonnegative, namely

$$\left(\frac{\partial}{\partial l}\right)^2 \xi(l)\Big|_{l=h^{(k)}} \geq 0$$

and

$$\left(\frac{\partial}{\partial l}\right)^2 \zeta(l)\Big|_{l=h^{(k)}} \geq 0 \ .$$

In particular, this implies that the diagonal elements of the respective matrices are nonnegative, or, after a direct calculation,

$$\frac{q_{ii}}{(Q(p)h^{(k)}, h^{(k)})^{1/2}} \geq \frac{c_{ii}}{(Ch^{(k)}, h^{(k)})^{1/2}} - \frac{\left(\sum_{j=m+1}^{n} c_{ij} h_j^{(k)}\right)^2}{(Ch^{(k)}, h^{(k)})^{3/2}}$$

for all $i = \overline{1,m}, k \in \overline{m+1,n}$. Here the second term on the right-hand side disappears due to Lemma 2.2.3. If we now observe that the denominators in the resulting inequality are equal (due to the condition of the Lemma and being the values of the support functions at $l = h^{(k)}$) we may conclude that $q_{ii} \geq c_{ii}$ for all $i \in \overline{1,m}$.

Due to Lemma 2.2.3 and the definition of $h^{(k)}$ a similar condition for the matrix of second derivatives of ζ yields the following inequality for the diagonal elements $i = \overline{1,m}$:

$$\text{(2.2.15)} \quad \frac{c_{ii}}{(Ch^{(k)}, h^{(k)})^{1/2}} \geq \frac{q_{ii}^{(1)}}{(Q_1 h^{(k)}, h^{(k)})^{1/2}} + \frac{q_{ii}^{(2)}}{(Q_2 h^{(k)}, h^{(k)})^{1/2}}$$

Take the right-hand side of (2.2.16), multiply and divide it by $(Q_1 h^{(k)}, h^{(k)})^{1/2} + (Q_2 h^{(k)}, h^{(k)})^{1/2}$, recalling that

$$Q(p) = (1 + p^{-1})Q^{(1)} + (1 + p)Q^{(2)} \ ,$$

where

$$p = (l, Q^{(1)} l)^{1/2} (l, Q^{(2)} l)^{-1/2}, \quad l = h^{(k)} \ .$$

Substituting the obtained relations into (2.2.15) and using the equality relation for the support functions at $l = h^{(k)}$, we come to the condition $q_{ii} \leq c_{ii}$ and therefore to the equality $q_{ii} = c_{ii}, i \in \overline{1,m}$.

The inclusion $\mathcal{E}(0, Q(p)) \supseteq \mathcal{E}(0, C)$ implies that the matrix $Q(p) - C$ is nonnegative. Since it was just established that its diagonal elements are all equal to zero, what follows is that all the rest of the elements must also be zero. The Lemma is thus proved.

However, before proving an assertion similar to Lemma 2.2.4, but for the differences $\mathcal{E}_1 \dot{-} \mathcal{E}_2$, we will first prove the following essential

Theorem 2.2.1 *Suppose that for the ellipsoids $\mathcal{E}_1 = \mathcal{E}(a_1, Q_1)$, $\mathcal{E}_2 = \mathcal{E}(a_2, Q_2)$ the matrices Q_1, Q_2 are positive definite and that $Q(p)$ is defined due to formula (2.2.1).*

Then the set of inclusion-minimal external estimates of the sum $\mathcal{E}_1 + \mathcal{E}_2$ will consist of the ellipsoids of the form $\mathcal{E}(a_1 + a_2, Q(p))$, with $p \in \prod^+$.

Proof. Without loss of generality, referring also to Lemma 2.1.4, we may assume all the centers of the ellipsoids considered here to be zero, particularly $a_1 = a_2 = 0$.

Given an ellipsoid $\mathcal{E}(0, Q) \supseteq \mathcal{E}_1 + \mathcal{E}_2$ let us indicate that there exists a value p such that the ellipsoid $\mathcal{E}(0, Q(p))$ could be squeezed in between $\mathcal{E}(0, Q)$ and $\mathcal{E}_1 + \mathcal{E}_2$, so that we would have

$$\mathcal{E}_1 + \mathcal{E}_2 \subseteq \mathcal{E}(0, Q(p)) \subseteq \mathcal{E}(0, Q) \ .$$

We may obviously consider $\mathcal{E}(0, Q)$ to be tangential to $\mathcal{E}_1 + \mathcal{E}_2$, assuming the existence of a vector $l = \bar{l} \in \mathbb{R}^n$, $\|\bar{l}\| = 1$, such that

$$\rho(\bar{l} | \mathcal{E}(0, Q)) = \rho(\bar{l} | \mathcal{E}_1 + \mathcal{E}_2) \ . \tag{2.2.16}$$

Let us now select an invertible matrix T such that the matrices $Q_1^* = T'Q_1T$, $Q_2^* = T'Q_2T$ would both be diagonal. The existence of such a transformation T follows from results in Linear Algebra and the theory of matrices (see, for example, [120]).

The transformation T obviously does not violate the inclusion $\mathcal{E}(0,Q) \supseteq \mathcal{E}_1 + \mathcal{E}_2$, so that with $Q^* = T'QT$ we still have

$$\mathcal{E}(0,Q^*) \supseteq \mathcal{E}(0,Q_1^*) + \mathcal{E}(0,Q_2^*) \ .$$

Taking the mapping $l = Tz$ one may transform the equality (2.2.16), which is

$$(\bar{l}, Q\bar{l})^{1/2} = (\bar{l}, Q_1\bar{l})^{1/2} + (\bar{l}, Q_2\bar{l})^{1/2}$$

into

$$(\bar{z}, Q^*\bar{z})^{1/2} = (\bar{z}, Q_1^*\bar{z})^{1/2} + (\bar{z}, Q_2^*\bar{z})^{1/2}$$

where $\bar{z} = T^{-1}\bar{l}$. Following (2.2.2) we may now select

$$\bar{p} = (\bar{z}, Q_1^*\bar{z})^{1/2}(\bar{z}, Q_2^*\bar{z})^{-1/2}$$

and further take

$$Q^*(\bar{p}) = (1 + \bar{p}^{-1})Q_1^* + (1 + \bar{p})Q_2^* \ .$$

We then come to the relations

$$\mathcal{E}(0,Q_1^*) + \mathcal{E}(0,Q_2^*) \subseteq \mathcal{E}(0,Q^*(\bar{p})) \ ,$$

(2.2.17)
$$\begin{aligned} \rho(\bar{z}|\mathcal{E}(0,Q_1^*)) \ &+ \ \rho(\bar{z}|\mathcal{E}(0,Q_2^*)) \\ &= \ \rho(\bar{z}|\mathcal{E}(0,Q^*(\bar{p}))) = \rho(\bar{z}|\mathcal{E}(0,Q^*)). \end{aligned}$$

From Lemma 2.1.4, part (a) it now follows that $\mathcal{E}(0,Q^*(\bar{p})) \subseteq \mathcal{E}(0,Q^*)$.

Indeed with this inclusion being false, there would have existed a vector z^* such that

(2.2.18)
$$\rho(z^*|\mathcal{E}(0,Q^*(\bar{p}))) > \rho(z^*|\mathcal{E}(0,Q^*)) \ .$$

The vector z^* is obviously noncollinear with $\bar{z}$. Define Z to be the 2-dimensional space generated by $\bar{z}, z^*$ and $\mathcal{E}_z(0,Q)$ to be the projection of the ellipsoid $\mathcal{E}(0,Q)$ on space Z. In view of (2.2.17) and the inequality (2.2.18) we have

$$\mathcal{E}_z(0,Q^*(\bar{p})) \supseteq \mathcal{E}_z(0,Q^*)$$

and

$$\begin{aligned}\rho(\bar{z}|\mathcal{E}_z(0,Q^*)) &= \rho(\bar{z}|\mathcal{E}_z(0,Q^*(\bar{p})))\\ &= \rho(\bar{z}|\mathcal{E}_z(0,Q_1^*)+\mathcal{E}_z(0,Q_2^*)) \ .\end{aligned}$$

From Lemma 2.2.4 it then follows that $\mathcal{E}_z(0,Q^*) = \mathcal{E}_z(0,Q^*(p))$ is in contradiction with (2.2.18).

Q.E.D.

The following proposition is similar to Lemma 2.2.4, but is applied to geometrical differences.

Lemma 2.2.5 *Consider a nondegenerate ellipsoid $\mathcal{E}(0,C)$ together with ellipsoids $\mathcal{E}_1 = \mathcal{E}(0,Q_1)$, $\mathcal{E}_2 = \mathcal{E}(0,Q_2)$, assuming that Q_1, Q_2 are diagonal. Further assume the vector $l = \bar{l}$, $\|l\| = 1$ to be given and the parameter $p = \bar{p}$ to be defined due to relation (2.2.2), $l = \bar{l}$.*

Then, if the ellipsoid $\mathcal{E}(0,Q(-\bar{p}))$ is properly defined ($Q(-\bar{p}) > 0$), the relations

$$\mathcal{E}(0,Q(-\bar{p})) \subseteq \mathcal{E}(0,C) \subseteq \mathcal{E}_1 \dot{-} \mathcal{E}_2 \tag{2.2.19}$$

$$\rho(\bar{l}|\mathcal{E}(0,Q(-\bar{p}))) = \rho(\bar{l}|\mathcal{E}_1 \dot{-} \mathcal{E}_2) \tag{2.2.20}$$

imply

$$\mathcal{E}(0,Q(-\bar{p})) = \mathcal{E}(0,C) \qquad \text{and} \qquad \bar{p} \in \prod{}^{-} \ .$$

Proof. Let us start with the indication that the inclusion $\mathcal{E}(0,C) \subseteq \mathcal{E}_1 \dot{-} \mathcal{E}_2$ implies the existence of an $\varepsilon > 0$ such that

$$\varepsilon\|l\| \leq (l,Cl)^{1/2} \leq (l,Q_1 l)^{1/2} - (l,Q_2 l)^{1/2}$$

and therefore

$$\bar{p} = (\bar{l},Q_1\bar{l})^{1/2}(\bar{l},Q_2\bar{l})^{-1/2} > 1 \ .$$

Let us further proceed with all the formal procedures, presuming also

$$\bar{p} < \lambda_{\min} = \min\{(l,Q_1 l)(l,Q_2 l)^{-1} \in \mathbb{R} : \|l\| = 1\} \ ,$$

so that altogether $\bar{p} \in (1,\lambda_{\min})$ and therefore $Q(-\bar{p}) > 0$.

Suppose that $\bar{p}, \bar{l}$ is the pair given in the formulation of the theorem. Then for this pair the relations (2.2.20), (2.2.21) are fulfilled, and the first of them is equivalent to the inequalities

$$\rho(l|\mathcal{E}(0, Q(-\bar{p}))) \leq \rho(l|\mathcal{E}(0, C)) \leq \rho(l|\mathcal{E}_1 \dot{-} \mathcal{E}_2)$$

for any $l \in \mathbb{R}^n$.

Further on, the inclusion $\mathcal{E}(0, C) \subseteq \mathcal{E}_1 \dot{-} \mathcal{E}_2$ implies

$$\mathcal{E}(0, C) + \mathcal{E}_2 \subseteq \mathcal{E}_1 \ .$$

By Lemma 2.2.1 there exists an ellipsoid $\mathcal{E}(0, C(p))$ with

(2.2.21) $$C(p) = (1 + p^{-1})C + (1 + p)Q_2 \ ,$$

which satisfies the inclusion

(2.2.22) $$\mathcal{E}(0, C) + \mathcal{E}_2 \subseteq \mathcal{E}(0, C(p))$$

for any $p > 0$. With $l = \bar{l}$ given and $p = p^*$ taken as

(2.2.23) $$p^* = (C\bar{l}, \bar{l})^{1/2}(Q_2\bar{l}, \bar{l})^{-1/2}$$

it also satisfies the equality

(2.2.24) $$\rho(\bar{l}|\mathcal{E}(0, C(p^*))) = \rho(\bar{l}|\mathcal{E}(0, C)) + \rho(\bar{l}|\mathcal{E}_2) \ .$$

According to Theorem 2.2.1 we then have

(2.2.25) $$\mathcal{E}(0, C) + \mathcal{E}_2 \subseteq \mathcal{E}(0, C(p^*)) \subseteq \mathcal{E}_1 \ .$$

(Note that here

$$\mathcal{E}(0, C) \subseteq \mathcal{E}(0, C(\bar{p})) \dot{-} \mathcal{E}_2 \subseteq \mathcal{E}_1 \dot{-} \mathcal{E}_2 \ ) \ .$$

Rearranging (2.2.21) and taking $p^0 = 1 + p^*$, we obtain

(2.2.26) $$C = (1 - (p^0)^{-1})C(p^*) + (1 - p^0)Q_2 \ .$$

Being defined through (2.2.23), the value p^* is positive (as $C > 0$) and $p^0 > 1$. From (2.2.25) we also observe $C(p^*) \leq Q_1$ and due to (2.2.23), (2.2.24) we have

$$\begin{aligned} p^0 = p^* + 1 &= ((C\bar{l},\bar{l})^{1/2} + (Q_2\bar{l},\bar{l})^{1/2})(Q_2\bar{l},\bar{l})^{-1/2} \\ &= ((C(p^*),\bar{l},\bar{l})^{1/2})(Q_2\bar{l},\bar{l})^{1/2} \leq (Q_1\bar{l},\bar{l})^{1/2}(Q_2\bar{l},\bar{l})^{-1/2} = \bar{p} \ , \end{aligned}$$

so that $p^0 \leq \bar{p}$.

Combining (2.2.25), (2.2.26) we come to the inequality

$$C \leq Q(-\bar{p}) = (1 - \bar{p}^{-1})Q_1 + (1 - \bar{p})Q_2$$

which may be checked by direct calculation, observing that $(l, Cl) \leq (l, Q(-p)l)$, $\forall l \in \mathbb{R}^n$.

Together with (2.2.19) this produces

$$C = Q(-\bar{p}) \ .$$

Since C is nondegenerate, we have $Q(-\bar{p}) > 0$ and therefore $\bar{p}$ indeed lies within the domain $p \in (1, \lambda_{\min})$.

Q.E.D.

Let us now prove the analogy of Theorem 2.2.1 for geometrical differences.

Theorem 2.2.2 *Suppose that* int $\mathcal{E}(0, Q_1) \supseteq \mathcal{E}(0, Q_2)$ *holds. Then the set of maximal internal estimates of the difference* $\mathcal{E}_1 \dot{-} \mathcal{E}_2$ *consists of ellipsoids of the form*

$$\mathcal{E}(a_1 - a_2, Q(-p)), \ p \in \Pi^- \ .$$

Proof. Without loss of generality suppose again that $a_1 = 0, a_2 = 0$. Given a nondegenerate ellipsoid $\mathcal{E}(0, Q) \subseteq \mathcal{E}_1 \dot{-} \mathcal{E}_2$ let us indicate that there exists a value p such that the ellipsoid $\mathcal{E}(0, Q(-p))$ could be squeezed in between $\mathcal{E}(0, Q)$ and $\mathcal{E}_1 \dot{-} \mathcal{E}_2$, so that we would have

$$\mathcal{E}(0, Q) \subseteq \mathcal{E}(0, Q(-p)) \subseteq \mathcal{E}_1 \dot{-} \mathcal{E}_2 \ . \qquad (2.2.27)$$

We may consider $\mathcal{E}(0,Q)$ to be tangential to $\mathcal{E}_1 \dot{-} \mathcal{E}_2$, assuming the existence of a vector $\bar{l} \in \mathbb{R}^n$, $\|\bar{l}\| = 1$, such that

$$\rho(\bar{l}|\mathcal{E}(0,Q)) = \rho(\bar{l}|\mathcal{E}_1 \dot{-} \mathcal{E}_2) \quad . \tag{2.2.28}$$

Let us first define $\mathcal{E}(0,Q(-\bar{p}))$ with

$$\bar{p} = (Q_1\bar{l},\bar{l})^{1/2}(Q_2\bar{l},\bar{l})^{-1/2}$$

and prove that $Q(-\bar{p})$ is positive definite. To do this, define the matrix

$$D(p^*) = (1+(p^*)^{-1})Q + (1+p^*)Q_2$$

or

$$Q = (1-p^{-1})D(p^*) + (1-p)Q_2 \quad , \tag{2.2.29}$$

where

$$p^* = (\bar{l},Q\bar{l})^{1/2}(\bar{l},Q_2\bar{l})^{-1/2}$$

and

$$p = p^* + 1 \quad .$$

Due to Theorem 2.2.1, we further observe

$$\mathcal{E}(0,Q) + \mathcal{E}_2 \subseteq \mathcal{E}(0,D(p^*)) \subseteq \mathcal{E}_1$$

or, in other words, that $D(p^*) \leq Q_1$. Moreover,

$$\rho(\bar{l}|\mathcal{E}(0,D(p^*))) = \rho(\bar{l}|\mathcal{E}(0,Q)) + \rho(\bar{l}|\mathcal{E}(0,Q_2)) \quad . \tag{2.2.30}$$

By (2.2.28) we have

$$\rho(\bar{l}|\mathcal{E}(0,Q)) + \rho(\bar{l}|\mathcal{E}_2) \leq \rho(\bar{l}|\mathcal{E}_1)$$

or

$$(\bar{l},Q\bar{l})^{1/2} + (\bar{l},Q_2\bar{l})^{1/2} \leq (\bar{l},Q_1\bar{l})^{1/2} \quad ,$$

so that

$$\begin{aligned} p = p^* + 1 &= ((\bar{l},Q\bar{l})^{1/2} + (\bar{l},Q_2\bar{l})^{1/2})(\bar{l},Q_2\bar{l})^{-1/2} \\ &= ((\bar{l},D(p^*)\bar{l})^{1/2}(\bar{l},Q_2\bar{l})^{1/2} \leq (\bar{l},Q_1\bar{l})^{1/2}(\bar{l},Q_2\bar{l})^{-1/2} = \bar{p} \quad . \end{aligned}$$

Together with (2.2.29) and the inequality $p \leq \bar{p}$ this yields $Q \leq Q(-\bar{p})$, also proving that $Q(-\bar{p}) > 0$, which means $\bar{p} \in (1,\lambda_{\min})$. On the other hand, due to Lemma 2.2.2(a), we have $\mathcal{E}(0, Q(-\bar{p})) \in \mathcal{E}(0, Q_1) \dot{-} \mathcal{E}(0, Q_2)$. We thus come to the desired inclusion (2.2.27).

Q.E.D.

To conclude this section we shall summarize its results in the following

Theorem 2.2.3 *Given nondegenerate ellipsoids $\mathcal{E}_1, \mathcal{E}_2$, the following relations are true*

$$\mathcal{E}_1 + \mathcal{E}_2 = \bigcap \{\mathcal{E}(a_1 + a_2, Q(p)) : p \in \Pi^+\} \ , \tag{2.2.31}$$

and with $\text{int}\ \mathcal{E}_1 \supseteq \mathcal{E}_2$,

$$\mathcal{E}_1 \dot{-} \mathcal{E}_2 = \overline{\bigcup \{\mathcal{E}(a_1 - a_2, Q(-p)) : p \in \Pi^-\}} \ , \tag{2.2.32}$$

where $\bar{Q}$ stands for the closure of set Q.

Proof. It is clearly sufficient to prove the theorem for $a_1 = a_2 = 0$.

From Lemma 2.2.1 it follows

$$\mathcal{E}_1 + \mathcal{E}_2 \subseteq \bigcap \{\mathcal{E}(0, Q(p)) : p \in \Pi^+\} \ .$$

To prove the exact equality, assume the existence of a point x^* such that

$$x^* \in \bigcap \{\mathcal{E}(0, Q(p)) : p \in \Pi^+\} \ , \tag{2.2.33}$$

$$x \notin \mathcal{E}_1 + \mathcal{E}_2 \ . \tag{2.2.34}$$

The last condition ensures the existence of a vector $l = l^*$ that yields

$$(l^*, x^*) > \rho(l^* | \mathcal{E}_1 + \mathcal{E}_2) \ . \tag{2.2.35}$$

Selecting

$$p = p^* = (l^*, Q_1 l^*)^{1/2} (l^*, Q_2 l^*)^{-1/2}$$

and following Lemma 2.2.1 (a), we come to

$$\rho(l^* | \mathcal{E}_1 + \mathcal{E}_2) = \rho(l^* | \mathcal{E}(0, Q(p^*))) \ .$$

Together with (2.2.35) this implies $x^* \notin \mathcal{E}(0, Q(p^*))$ in contradiction with (2.2.33). The equality (2.2.31) is thus proved.

To prove (2.2.32), we recall that $\text{int}\,(\mathcal{E}_1 \dot{-} \mathcal{E}_2) \neq \emptyset$ and follow Lemma 2.2.2 which immediately yields

$$\bigcup\{\mathcal{E}(0, Q(-p)) : p \in \Pi^-\} \subseteq \mathcal{E}_1 \dot{-} \mathcal{E}_2 \ .$$

To indicate that there is actually an exact equality, assume the existence of such a vector x^* that

$$x^* \in \text{int}\,(\mathcal{E}_1 \dot{-} \mathcal{E}_2) \ , \tag{2.2.36}$$

$$x^* \notin \bigcup\{\mathcal{E}(0, Q(-p)) : p \in \Pi^-\} \ . \tag{2.2.37}$$

Since $x^* \in \text{int}\,(\mathcal{E}_1 \dot{-} \mathcal{E}_2)$ there exists an $\varepsilon > 0$ for which

$$\mathcal{S}_\varepsilon(x^*) = \{(x - x^*, x - x^*) \leq \varepsilon^2\} \subseteq \text{int}\,(\mathcal{E}_1 \dot{-} \mathcal{E}_2) \ .$$

As $\mathcal{E}_1 = -\mathcal{E}_1$, $\mathcal{E}_2 = -\mathcal{E}_2$ (these sets are symmetrical around the origin), we obviously have $\mathcal{E}_1 \dot{-} \mathcal{E}_2 = -(\mathcal{E}_1 \dot{-} \mathcal{E}_2)$ and therefore the whole set

$$\mathcal{S} = \{x : x \in \mathcal{S}_\varepsilon(z),\ z = -x^* + 2\alpha x^*, \alpha \in [0, 1]\}$$

satisfies $\mathcal{S} \subseteq \text{int}\,(\mathcal{E}_1 \dot{-} \mathcal{E}_2)$. What follows is that there exists a nondegenerate ellipsoid $\mathcal{E}(0, C^*) \subseteq \mathcal{S} \subseteq \text{int}\,(\mathcal{E}_1 \dot{-} \mathcal{E}_2)$. (Give an example by explicit calculation of C^*, assuming set

$$\mathcal{X}^* = \{z \ : \ z \ = \ -x^* + 2\alpha x^*, \ \alpha \in [0, 1]\}$$

to be its largest axis.) From Theorem 2.2.2 it now follows that for some $p^* \in \Pi^-$ there exists an ellipsoid $\mathcal{E}(0, Q(-p^*))$ that satisfies $x^* \subseteq \mathcal{E}(0, C^*) \subseteq \mathcal{E}(0, Q(-p^*)) \in \mathcal{E}_1 \dot{-} \mathcal{E}_2$ in contradiction with (2.2.36), (2.2.37).

Q.E.D.

Theorem 2.2.3 may be illustrated on a 2-dimensional example. In the center of Figure 2.2.1(a) we see two ellipsoids whose sum is the nonellipsoidal set that is the intersection of the nondominated (inclusion-minimal) ellipsoids that approximate it *externally* and are constructed

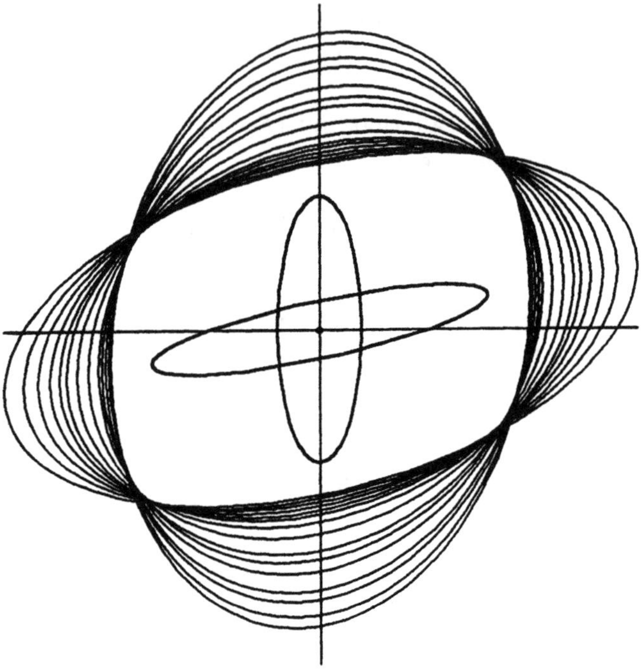

Figure 2.2.1(a).

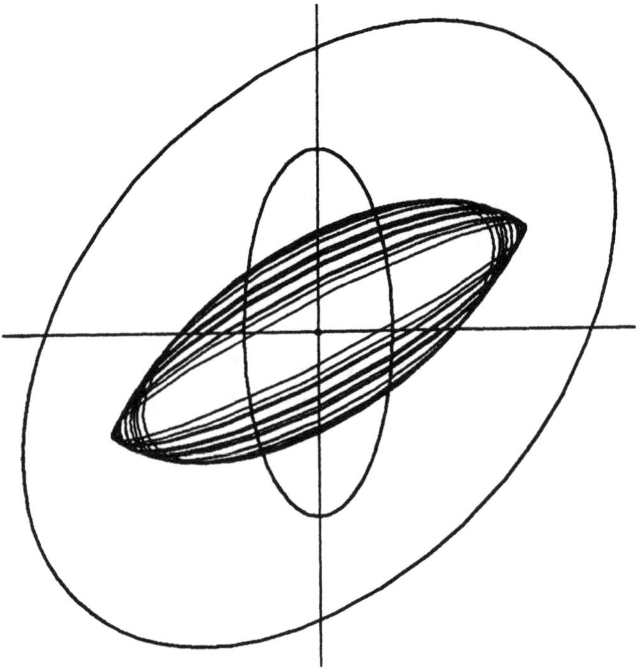

Figure 2.2.1(b).

from formula (2.2.31). Figure 2.2.1(b) shows a nondegenerate geometric difference of two ellipsoids (the set with two kinks) that also arrives as the (closure of the) union of the nondominated (inclusion-maximal) ellipsoids that approximate it *internally* and are constructed from formula (2.2.32). In both examples the parameters $p \in \Pi^+$, $S \in \Sigma$ are chosen randomly but give a good illustration of the nature of the approximations.

2.3 Internal Approximations: The Sums External Approximations: The Differences

We shall now introduce a representation that will allow an *internal approximation of the sum* of two nondegenerate ellipsoids by a parametrized variety of ellipsoids, and an *external approximation of the geometrical difference* of these. It will be demonstrated that this approximation may be exact.

Given $\mathcal{E}_1 = \mathcal{E}(a_1, Q_1), \mathcal{E}_2 = \mathcal{E}(a_2, Q_2)$, where $Q_1 > 0$, $Q_2 > 0$, we introduce a parametric family of matrices $Q_+[S]$, where

$$Q_+[S] = S^{-1}[(SQ_1S')^{1/2} + (SQ_2S')^{1/2}]^2 S'^{-1} \tag{2.3.1}$$

and $S \in \Sigma$ with

$$\Sigma = \{S \in \mathcal{L}(\mathbb{R}^n, \mathbb{R}^n) : S' = S,\ |S| \neq 0\}\ .$$

The matrix S is therefore selected from the set Σ of symmetrical nondegenerate matrices.

In a similar way we define the variety

$$Q_-[S] = S^{-1}[(SQ_1S')^{1/2} - (SQ_2S)^{1/2}]^2 S'^{-1} \tag{2.3.2}$$

with $S \in \Sigma$.

The variety $Q_+[S]$ will be used for approximating the sums $\mathcal{E}_1 + \mathcal{E}_2$ (internally), while the variety $Q_-[S]$ for approximating the differences $\mathcal{E}_1 \dot{-} \mathcal{E}_2$ (externally).

Let us start from the first case.

Lemma 2.3.1

(a) *The ellipsoid* $\mathcal{E} = \mathcal{E}(a_1 + a_2,\ Q_+[S])$ *is an internal approximation of the (Minkowski) sum* $\mathcal{E}_1 + \mathcal{E}_2$, *namely, for any* $S \in \Sigma$, *one has*

$$\mathcal{E}[S] = \mathcal{E}(a_1 + a_2, Q_+[S]) \subseteq \mathcal{E}_1 + \mathcal{E}_2 \quad . \tag{2.3.3}$$

For each $S \in \Sigma$ *there exists a vector* $l = l^*$, $\|l\| = 1$, *such that the equality*

$$\rho\left(l|\mathcal{E}(a_1 + a_2, Q_+[S])\right) = \rho(l|\mathcal{E}_1) + \rho(l|\mathcal{E}_2) \tag{2.3.4}$$

is true with $l = l^*$. *Conversely, for any* $l \in \mathbb{R}^n$, $\|l\| = 1$, *there exists a matrix* $S^* \in \Sigma$ *such that (2.3.4) is true with* $S = S^*$.

(b) *The ellipsoid* $\mathcal{E} = \mathcal{E}(a_1 - a_2,\ Q_-[S])$ *is an external approximation of the geometrical difference* $\mathcal{E}_1 \dot{-} \mathcal{E}_2$, *namely, for any* $S \in \Sigma$, *one has*

$$\mathcal{E}[S] = \mathcal{E}(a_1 - a_2, Q_-[S]) \supseteq \mathcal{E}_1 \dot{-} \mathcal{E}_2 \quad .$$

For each $S \in \Sigma$ *there exists a vector* $l = l^*$, $\|l\| = 1$, *such that the equality*

$$\rho\left(l|\mathcal{E}(a_1 - a_2, Q_-[S])\right) = \rho(l|\mathcal{E}_1) - \rho(l|\mathcal{E}_2)$$

is true with $l = l^*$. *Conversely, for any* $l \in \mathbb{R}^n$, $\|l\| = 1$, *there exists a matrix* $S^* \in \Sigma$ *such that the last relation is true with* $S = S^*$.

Proof. As in the previous sections it is clearly sufficient to consider the case, when $a_1 = 0,\ a_2 = 0$.

For any matrix $S \in \mathcal{L}(\mathbb{R}^n, \mathbb{R}^n)$ we have

$$\begin{aligned}(\rho(l|\mathcal{E}[S]))^2 = (l, Q_+[S]l) = & \\ = & \quad (l, Q_1 l) + (l, Q_2 l) + \\ + & \quad 2\left((SQ_1S')^{1/2}S'^{-1}l, (SQ_2S')^{1/2}S'^{-1}l\right) \quad .\end{aligned}$$

By Hölder's inequality

$$\begin{aligned}(\rho(l|\mathcal{E}[S]))^2 \leq & \quad (l, Q_1 l) + (l, Q_2 l) + \\ + & \; 2\left((SQ_1S')^{1/2}S'^{-1}l, (SQ_1S')^{1/2}S'^{-1}\right)^{1/2} \cdot \\ \cdot & \; \left((SQ_2S')^{1/2}S'^{-1}l, (SQ_2S')^{1/2}S'^{-1}l\right)^{1/2}\end{aligned}$$

or

$$(\rho(l|\mathcal{E}[S]))^2 \leq (l, Q_1 l) + (l, Q_2 l) + 2(l, Q_1 l)^{1/2}(l, Q_2 l)^{1/2} \quad ,$$

which proves the inclusion (2.3.3).

To prove that for a given S there exists an $l = l^*$ that ensures the equality (2.3.4), we observe, by direct substitution that this would be possible if there existed a number $\lambda > 0$ and a vector $l = l^*, \|l^*\| = 1$, that would ensure the relation

$$\left[(SQ_1S')^{1/2} - \lambda(SQ_2S')^{1/2}\right] S'^{-1}l^* = 0 \quad . \tag{2.3.5}$$

Denote

$$D = (Q_2^{1/2}Q_1Q_2^{-1/2})^{1/2}, \; z = Q_2^{-1/2}l^*, \; T = SQ_2^{1/2} \quad ,$$

then (2.3.5) reduces to

$$(TDDT')^{1/2}T'^{-1} \; z = \lambda(TT')^{1/2}T'^{-1}z \quad .$$

Suppose, in addition, that the matrix T is symmetrical. Then the last relation takes the form

$$(TD \cdot DT)^{1/2} \cdot T^{-1}z = \lambda z \quad . \tag{2.3.6}$$

Taking the polar decomposition [120], of the matrix TD, we obtain an orthogonal matrix U and a symmetrical (here also nonsingular) matrix H such that

$$TD = UH \quad . \tag{2.3.7}$$

The condition of symmetricity for H means that

$$TDU^{-1} = UDT \quad . \tag{2.3.8}$$

Substituting (2.3.7) into (2.3.6), we finally transform the original equation (2.3.4) to

$$U \cdot Dz = \lambda z \quad . \tag{2.3.9}$$

We now have to solve the system (2.3.8), (2.3.9) for $\lambda \in \mathbb{R}$, $U \in \mathcal{L}(\mathbb{R}^n, \mathbb{R}^n)$ orthogonal and $T \in \mathcal{L}(\mathbb{R}^n, \mathbb{R}^n)$ symmetrical and nonsingular, where the symmetrical, positive definite matrix $D \in \mathcal{L}(\mathbb{R}^n, \mathbb{R}^n)$ and the nonzero vector $z \in \mathbb{R}^n$ are given in advance.

With vector $Dz \neq 0$ given, there obviously exists an orthogonal matrix U such that vector UDz is directed along with z, hence there is a $\lambda > 0$

that ensures (2.3.9) with the U selected as above. What remains is to find a solution to the equation

$$TB' = BT$$

for symmetrical and nonsingular T where

$$B = UD \ .$$

This can be done by using a well known result of matrix theory. The proof given in Chapter VIII of [120] needs a slight modification which we leave to the reader.

For matrix S defined by the solution T obtained this way, equation (2.3.5) will hold. Using the polar decomposition theorem, we find an orthogonal matrix O such that $S^* = OS$ is symmetrical and therefore $S^* \in \sum$. From formula (2.3.1) it follows that

$$Q_+[S] = Q_+[OS]$$

and in such a way also that (2.3.5) is valid for $S = S^*$.

The proof of case (b) is analogous. Q.E.D.

The proof of Lemma 2.3.1 implies

Corollary 2.3.1 *The following equality is true*

$$\mathcal{E}_1 + \mathcal{E}_2 = \bigcup \{\mathcal{E}(a_1 + a_2, Q_+[S]) | S \in \sum\} \ ,$$

as well as

$$\mathcal{E}_1 \dot{-} \mathcal{E}_2 = \bigcup \{\mathcal{E}(a_1 - a_2, Q_-[S]) | S \in \sum\} \ .$$

The next step is to prove that the ellipsoids $\mathcal{E}(0, Q_+[S]), \mathcal{E}(0, Q_-[S])$ are the inclusion-maximal internal and the inclusion-minimal external estimates for $\mathcal{E}_1 + \mathcal{E}_2$ and $\mathcal{E}_1 \dot{-} \mathcal{E}_2$, respectively.

Theorem 2.3.1 *Consider the parametrized varieties of ellipsoids $\mathcal{E}(a_1 + a_2, Q_+[S])$, $\mathcal{E}(a_1 - a_2, Q_-[S])$, $S \in \Sigma$, generated, respectively, by the varieties of matrices $Q_+[S], Q_-[S]$, due to (2.3.1), (2.3.2). Then the following assertions are true*

(a) *the set of inclusion-maximal internal estimates of the sum* $\mathcal{E}_1 + \mathcal{E}_2$ *consists of ellipsoids of the form* $\mathcal{E}(a_1 + a_2, Q_+[S])$ *where* $S \in \Sigma$.

(b) *Assuming* $\text{int } \mathcal{E}_1 \supseteq \mathcal{E}_2$, *the set of inclusion-minimal external estimates of the difference* $\mathcal{E}_1 \dot{-} \mathcal{E}_2$ *consists of the ellipsoids of the form* $\mathcal{E}(a_1 - a_2, Q_-[S])$, $S \in \Sigma$.

Proof. As previously, we assume $a_1 = a_2$.

In order to prove the maximality of $\mathcal{E}(0, Q_+[S])$ we shall demonstrate that for any ellipsoid $\mathcal{E}(0, Q)$ the inclusions

$$\mathcal{E}(0, Q_+[S]) \subseteq \mathcal{E}(0, Q) \subseteq \mathcal{E}_1 + \mathcal{E}_2$$

imply

$$Q_+[S] = Q \quad .$$

According to Lemma 2.3.1 there is a condition in which there exists a vector $\ell \in \mathbb{R}^n$, $\|\ell\| = 1$, such that (2.3.4) is true. This is actually (2.3.6) (using the notations of the Lemma), where the matrix $(TDDT)^{1/2}$ is positive definite and symmetrical and the matrix T^{-1} is symmetrical. It is left to the reader to prove as an exercise that, under the above conditions, their product has simple structure, namely, a complete set $\{z_i \in \mathbb{R}^n \ : \ i = \overline{1, n}\}$ of linearly independent eigenvectors (that are not necessarily orthogonal). From this it follows that there is an invertible matrix $B \in \mathcal{L}(\mathbb{R}^n, \mathbb{R}^n)$ that maps the i-th unit vector $e_i \in \mathbb{R}^n$ into $Q_2^{-1/2} z_i \in \mathbb{R}^n$, for all $i \in \overline{1, n}$.

This leads to the relation

$$\rho(\ell|\mathcal{E}(0, B'Q_+[S]B)) \leq \rho(\ell|\mathcal{E}(0, B'QB)) \leq \tag{2.3.10}$$

$$\leq \rho(\ell|B'\mathcal{E}_1) + \rho(\ell|B'\mathcal{E}_2)$$

for all $\ell \in \mathbb{R}^n$ with equality holding for $\ell = e_i$, $i \in \overline{1, n}$. This implies that the diagonal elements of $B'Q_+[S]B$ and $B'QB$ coincide.

Substituting $\ell = e_i + e_j$, $i \neq j$ into (2.3.10) we obtain

$$q_{ii}^{(+)} + 2q_{ij}^{(+)} + q_{jj}^{(+)} \ \leq \ q_{ii} + 2q_{ij} + q_{jj}$$

for arbitrary fixed i and j, where $q_{kr}^{(+)}$ and q_{kr} denote the element in the k-th row and r-th column of the matrix $B'Q_+[S]B$ and $B'QB$, respectively. By the equality of the diagonal elements, this implies

$$q_{ij}^{(+)} \leq q_{ij} \quad .$$

Carrying out the substitution of $\ell = e_i - e_j$ into (2.3.10), we arrive at the reverse inequality. Taken together this means

$$B'Q_+[S]B = B'QB$$

and by the invertibility of B, the equality

$$Q_+[S] = Q \quad .$$

Part (a) is thus proved. The proof of part (b) is similar, and is left to the reader. However, one should bear in mind that part (b) is true only if the difference $\mathcal{E}_1 \dot{-} \mathcal{E}_2$ has a nonvoid interior which implies that matrix $Q_-[S] > 0$.

Q.E.D.

The second part of the Theorem implies the following assertion

Corollary 2.3.2 *The following representation is true*

$$\mathcal{E}_1 \dot{-} \mathcal{E}_2 = \bigcap \{\mathcal{E}(a_1 - a_2, Q_-[S]) | S \in \Sigma\} \quad . \tag{2.3.11}$$

A 2-dimensional illustration of Theorem 2.3.1 is given in Figures 2.3.1(a) and 2.3.1(b). The first one shows the nonellipsoidal sum of two ellipsoids and the variety of nondominated (inclusion-maximal) ellipsoids that approximate it internally, due to formula (2.3.3). The sum then arrives as the union of the internal ellipsoids (over all the variety of these, see Lemma 2.3.1). The second one shows a nondegenerated geometrical difference of two ellipsoids (the set with two kinks) and the variety of nondominated (inclusion-minimal) ellipsoids that approximate it externally, due to formula (2.3.11). This difference then appears as the intersection of the external ellipsoids (over all the variety). The parameters $S \in \Sigma$ are chosen randomly but give a good illustration of the nature of the representations.

Exercise 2.3.1. Check, whether in this section the class Σ of symmetrical nondegenerate matrices S may be reduced to the class of only positive matrices $S > 0, S = S'$.

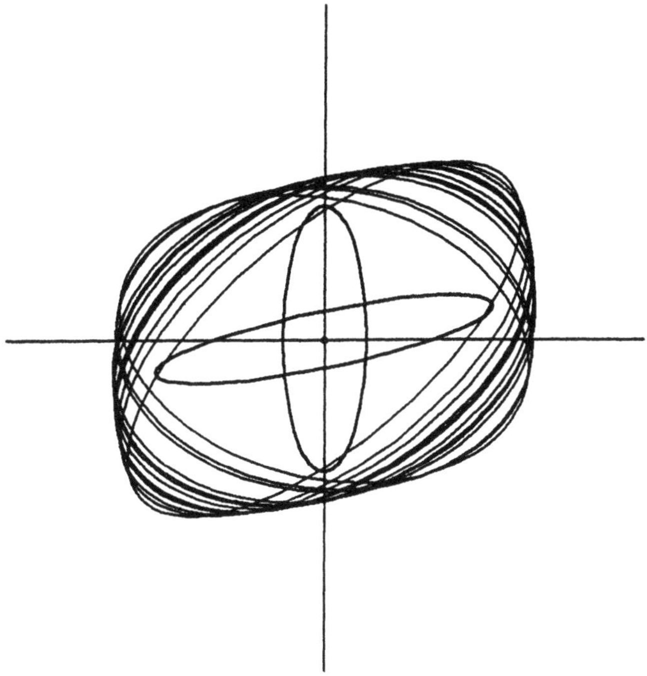

Figure 2.3.1(a).

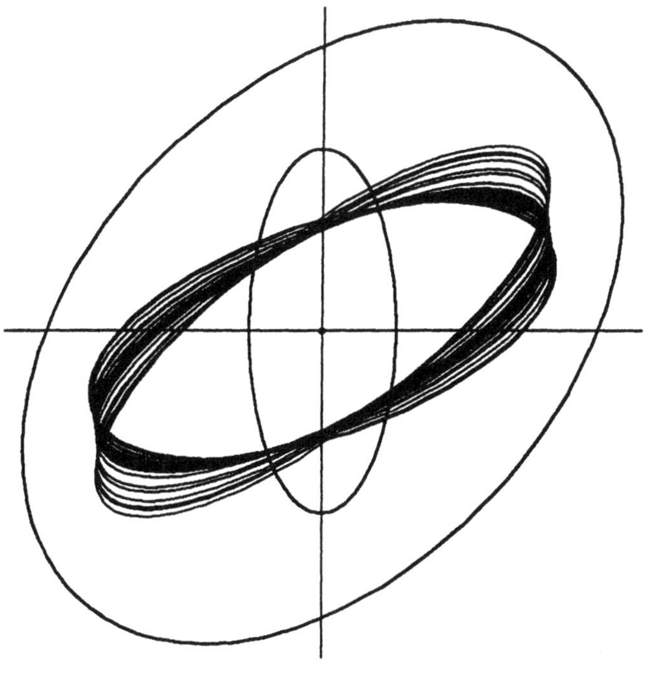

Figure 2.3.1(b).

2.4 Sums and Differences: The Exact Representation

The results of the previous sections indicate that the sums $\mathcal{E}_1 + \mathcal{E}_2$ and differences $\mathcal{E}_1 \dot{-} \mathcal{E}_2$ of ellipsoids could be exactly represented through the unions and intersections of the elements of certain parameterized families of ellipsoids. Let us once more indicate this result collecting all the facts in one proposition. When calculating $\mathcal{E}_1 \dot{-} \mathcal{E}_2$ we also assume $\mathrm{int}(\mathcal{E}_1 \dot{-} \mathcal{E}_2) \neq \emptyset$.

Theorem 2.4.1 *(The Representation Theorem).*

Let $\mathcal{E}_1 = \mathcal{E}(a_1, Q_1), \mathcal{E}_2 = \mathcal{E}(a_2, Q_2)$ be a pair of nondegenerate ellipsoids. Let $Q(p)$ be a parameterized family of ellipsoids

$$Q(p) = (1 + p^{-1})Q_1 + (1 + p)Q_2,$$

$$p \in \prod^+ = [\lambda_{\min}^{1/2}, \lambda_{\max}^{1/2}] \quad ,$$

where $\lambda_{\min} > 0$, $\lambda_{\max} < \infty$ are the roots of the equation

$$\det(Q_1 - \lambda Q_2) = 0$$

(the relative eigenvalues of Q_1, Q_2).

Let

$$\prod^- = \prod^+ \cap (1, \lambda_{\min}) \quad .$$

Also let $Q_+[S]$, $Q_-[S]$ denote the following parametrized families of ellipsoids

$$Q_+[S] = S^{-1}[(SQ_1S')^{1/2} + (SQ_2S')^{1/2}]^2 S'^{-1}$$

$$Q_-[S] = S^{-1}[(SQ_1S')^{1/2} - (SQ_2S')^{1/2}]^2 S'^{-1}$$

where

$$S \in \sum = \{S \in \mathcal{L}(\mathbb{R}^n, \mathbb{R}^n) : S' = S, |S| \neq 0\} \quad .$$

Then the following inclusions are true

$$\mathcal{E}_1 + \mathcal{E}_2 \subseteq \mathcal{E}(a_1 + a_2, Q(p)), \quad \forall p \in \prod^+ \quad , \tag{2.4.1}$$

$$\mathcal{E}_1 + \mathcal{E}_2 \supseteq \mathcal{E}(a_1 + a_2, Q_+[S]), \quad \forall S \in \sum, \tag{2.4.2}$$

$$(2.4.3) \qquad \mathcal{E}_1 \dot{-} \mathcal{E}_2 \subseteq \mathcal{E}(a_1 - a_2, Q_-[S]), \quad \forall S \in \sum \ ,$$

$$(2.4.4) \qquad \mathcal{E}_1 \dot{-} \mathcal{E}_2 \supseteq \mathcal{E}(a_1 - a_2, Q(-p)), \quad \forall p \in \prod\nolimits^- \ .$$

Moreover, the following exact representations are valid:

$$(2.4.5) \qquad \mathcal{E}_1 + \mathcal{E}_2 = \bigcap \{\mathcal{E}(a_1 + a_2, Q(p)) | p \in \prod\nolimits^+\} \ ,$$

$$(2.4.6) \qquad \mathcal{E}_1 + \mathcal{E}_2 = \overline{\bigcup \{\mathcal{E}(a_1 + a_2, Q_+[S]) | S \in \sum\}} \ ,$$

$$(2.4.7) \qquad \mathcal{E}_1 \dot{-} \mathcal{E}_2 = \bigcap \{\mathcal{E}(a_1 - a_2, Q_-[S]) | S \in \sum\} \ ,$$

$$(2.4.8) \qquad \mathcal{E}_1 \dot{-} \mathcal{E}_2 = \overline{\bigcup \{\mathcal{E}(a_1 - a_2, Q(-p)) | p \in \prod\nolimits^-\}} \ .$$

The facts given in this theorem may be treated as being related to *integral geometry*, particularly to the representations of ellipsoidal sets (bodies) in $\mathbb{R}^n$. The specific properties formulated in (2.4.1)–(2.4.4) and (2.4.5)–(2.4.8) reflect a certain type of *geometrical duality* in treating the geometrical sums and differences of ellipsoids. Namely, *the external* representations (2.4.1) for approximating *the sum* yields, with a change in the sign of the parameter p, *the internal* representation (2.4.4) *for the difference* and *the internal* representation (2.4.2) *for the sum* yields, with a change of sign (from $Q_+[S]$ to $Q_-[S]$), *the external* representation (2.4.3) *for the difference.* As it was also demonstrated in the previous sections, Theorem 2.4.1 also indicates that the parametrized varieties involved are also the varieties of inclusion-minimal external and inclusion-maximal internal estimates for $\mathcal{E}_1 + \mathcal{E}_2$, $\mathcal{E}_1 \dot{-} \mathcal{E}_2$. This can be summarized in

Theorem 2.4.2 *(i) Given $\mathcal{E}_1 + \mathcal{E}_2$ and an ellipsoid $\mathcal{E} \supseteq \mathcal{E}_1 + \mathcal{E}_2$, there exists a value $p \in \prod^+$ such that*

$$(2.4.9) \qquad \mathcal{E}_1 + \mathcal{E}_2 \subseteq \mathcal{E}(a_1 + a_2, Q(p)) \subseteq \mathcal{E} \ .$$

(ii) Given $\mathcal{E}_1 + \mathcal{E}_2$ and an ellipsoid $\mathcal{E} \subseteq \mathcal{E}_1 + \mathcal{E}_2$ there exists an $S \in \sum$, such that

$$(2.4.10) \qquad \mathcal{E} \subseteq \mathcal{E}(a_1 + a_2, Q_+[S]) \subseteq \mathcal{E}_1 + \mathcal{E}_2 \ .$$

*(iii) Given $\mathcal{E}_1 \dot{-} \mathcal{E}_2$, (*int $(\mathcal{E}_1 \dot{-} \mathcal{E}_2) \neq \emptyset$*) and an ellipsoid $\mathcal{E} \supseteq \mathcal{E}_1 \dot{-} \mathcal{E}_2$, there exists an $S \in \sum$, such that*

$$(2.4.11) \qquad \mathcal{E}_1 \dot{-} \mathcal{E}_2 \subseteq \mathcal{E}(a_1 - a_2, Q_-[S]) \subseteq \mathcal{E} \ .$$

(iv) *Given* $\mathcal{E}_1 \dot{-} \mathcal{E}_2$, *(*$\text{int}\,(\mathcal{E}_1 \dot{-} \mathcal{E}_2) \neq \emptyset$*) and an ellipsoid* $\mathcal{E} \supseteq \mathcal{E}_1 \dot{-} \mathcal{E}_2$, *there exists an* $p \in \prod^-$, *such that*

$$\mathcal{E} \subseteq \mathcal{E}(a_1 - a_2, Q(-p)) \subseteq \mathcal{E}_1 \dot{-} \mathcal{E}_2 \quad . \tag{2.4.12}$$

The variety $\{\mathcal{E}(a_1 + a_2, Q(p)),\ p \in \prod^+\}$ is therefore the set of non-dominated (inclusion-minimal) upper ellipsoidal estimates for $\mathcal{E}_1 + \mathcal{E}_2$. The variety of nondominated (inclusion-maximal) internal estimates for $\mathcal{E}_1 + \mathcal{E}_2$ is therefore $\{\mathcal{E}(a_1 + a_2, Q_+[S]),\ S \in \sum\}$.

Similarly, the varieties of non-dominated (inclusion-minimal) external and nondominated (inclusion-maximal) external estimates for $\mathcal{E}_1 \dot{-} \mathcal{E}_2$ are $\mathcal{E}(a_1 - a_2, Q_-[S])$ and $\mathcal{E}(a_1 - a_2, Q(-p))$.

The mentioned relations allow us to say that the nondominated ellipsoids, as described above, posses a certain type of *"Pareto" property.* The important fact is that *the mentioned Pareto property is invariant under linear transformations.* This means that after a linear transformation the nondominated ellipsoids remain nondominated. It is precisely this fact that allows us to propagate the static schemes of this section to systems with linear dynamics.

One of the further problems in ellipsoidal approximations to be discussed is to *to estimate the number of ellipsoids that would give a desired accuracy of approximation.*

Without going into the details of this problem, we shall briefly discuss it for the case of external approximation of the sum of two ellipsoids $\mathcal{E}_1 = \mathcal{E}(a_1, Q_1)$ and $\mathcal{E}_2 = \mathcal{E}(a_2, Q_2)$. Taking k arbitrary external ellipsoids $\mathcal{E}(a_1 + a_2, Q(p_k))$ of the type given in (2.4.1), we have

$$\mathcal{E}_1 + \mathcal{E}_2 \subset \cap\{\mathcal{E}(a_1 + a_2, Q(p_i)) | i = 1, ...k\} \quad ,$$

where $p_i \in \Pi^+$. Without loss of generality we further set $a_1 + a_2 = 0$.

Calculating the Hausdorff semidistance $(i = 1, ..., k)$

$$h_+(\cap\mathcal{E}(0, Q(p_i)), \mathcal{E}_1 + \mathcal{E}_2) = \zeta(p[k]) \quad ,$$

where $p[k] = \{p_1, ..., p_k\}$ is a k-dimensional vector, we come to

$$\zeta(p[k]) = \max\{\rho(l | \cap_i \mathcal{E}(0, Q(p_i))) - \rho(l | \mathcal{E}_1 + \mathcal{E}_2) | (l, l) \leq 1\} \quad ,$$

where, in its turn,

$$\rho(l|\cap_i \mathcal{E}(0,Q(p_i))) = \min\{\sum_{i=1}^{k}(l^{(i)},Q(p_i)l^{(i)})^{1/2}|\Sigma_{i=1}^{k}l^{(i)} = l\} \ .$$

Presuming that the best k ellipsoids are those that give the smallest Hausdorff semidistance $\zeta(p[k])$, we may specify them by solving the problem

$$\zeta^0(k) = \min\{\zeta(p[k])|p[k] : p_i \in \prod^+, i = 1, ...k\} \ .$$

The best ellipsoids are those that are generated by the optimalizing vector $p[k] = p^0[k]$ of the previous problem. It is not difficult to observe that $\zeta^0(k) \geq \zeta^0(k+1)$ and $\zeta^0(k) \to 0$ if $k \to \infty$. We thus assert the following

Lemma 2.4.1 *The minimal number of ellipsoids that approximate the sum $\mathcal{E}_1 + \mathcal{E}_2$ with given accuracy $\epsilon > 0$ may be determined as the smallest integer $k = k(\epsilon)$ that satisfies the inequality $\zeta^0(k) \leq \epsilon$.*

The respective optimal ellipsoids are those that are generated by the vector $p[k(\epsilon)] = p^0[k(\epsilon)]$.

It is also important, of course, to obtain the estimate in a more explicit form or to obtain estimates that are perhaps less precise, but simpler than the exact one, specified by the previous Lemma. An appropriate issue is also to describe effective algorithms for calculating $\zeta^0(k)$ or its estimates.

A similar reasoning may be applied to the other approximation problems. However, speaking in general, we should note that the nonsimple problem of *the accuracy and computational complexity* of the ellipsoidal approximation requires special treatment and in its full detail spreads beyond the scope of this book.

It may be also interesting to single out, among the variety of approximating ellipsoids, a single ellipsoid that is optimal in some sense.

2.5 The Selection of Optimal Ellipsoids

We shall now proceed with describing the *optimal ellipsoidal estimates* (external or internal) for $\mathcal{E}_1 + \mathcal{E}_2$ and $\mathcal{E}_1 \dot{-} \mathcal{E}_2$, selected through some cost function (optimality criterion). If the cost function $\psi(\mathcal{E})$ is *monotonous* by increasing *with respect to inclusion* ($\psi(\mathcal{E}') \geq \psi(\mathcal{E}'')$ or $\psi(\mathcal{E}') \leq \psi(\mathcal{E}'')$ if $\mathcal{E}' \supseteq \mathcal{E}''$), then, due to Theorem 2.4.2, the solution may be sought only from the parametrized varieties of this theorem. Some rather simple necessary conditions of optimality could then be applied for this purpose.

Lemma 2.5.1 *Suppose the function $\psi(\mathcal{E})$ is continuous and monotonously increasing with respect to inclusion. Then*

(a) The ψ-minimal external ellipsoidal estimate of the sum $\mathcal{E}_1 + \mathcal{E}_2$ is $\mathcal{E}(a_1 + a_2, Q(p^*))$, where p^* is the value for which the minimum of the function $f : (0, \infty) \longrightarrow \mathbb{R}$,

$$f(p) = \psi[Q(p)], \quad p \in \Pi^+ \quad ,$$

is attained.

(b) Suppose that $\operatorname{int}(\mathcal{E}_1) \supset \mathcal{E}_2$ holds. Then the ψ-maximal internal estimate of the difference $\mathcal{E}_1 \dot{-} \mathcal{E}_2$ is $\mathcal{E}(a_1 - a_2, Q(-p^{**}))$, where p^{**} is the value for which the maximum of the function $g : (1, \lambda_{\min}) \longrightarrow \mathbb{R}$

$$g(p) = \psi[Q(-p)], \quad p \in \Pi^- \quad ,$$

is attained.

Proof. By the monotonicity of ψ, follows from Theorem 2.4.2. Q.E.D.

We shall now apply Lemma 2.5.1 to find the minimal external estimates of the sum of two ellipsoids with respect to three important parameters of ellipsoids: the volume, the sum of squares of semiaxes (or trace), and $\operatorname{tr}(Q^2)$, and do the same for the maximal external estimates of the difference. Using our technique, analogous results can be obtained for other functions, like $\operatorname{tr}(Q^3)$, etc.

Although the case of the difference is similar to the above, in our test cases we are not in such a favorable position because the existence of a unique stationary point cannot always be guaranteed. However, as we shall see later, this still can be done up to $o(\varepsilon)$ in the important special case when $Q_2 = \varepsilon^2 Q_0$.

Lemma 2.5.2

(a) *There exists a unique ellipsoid with minimal sum of squares of semiaxes, (which is* $\mathrm{tr}(Q)$*) that contains the sum* $\mathcal{E}_1 + \mathcal{E}_2$*. It is of the form* $\mathcal{E}(a_1 + a_2, Q(p^*))$*, where*

$$p^* = \frac{(\mathrm{tr}Q_1)^{1/2}}{(\mathrm{tr}Q_2)^{1/2}} \quad . \tag{2.5.1}$$

(b) *Suppose that* $\mathrm{int}(\mathcal{E}_1) \supset \mathcal{E}_2$ *holds, and also that there exists an internally estimating ellipsoid for the difference* $\mathcal{E}_1 \dot{-} \mathcal{E}_2$ *such that its sum of squares of semiaxes (which is* $\mathrm{tr}(Q)$*) is maximal. Then it is of the form* $\mathcal{E}(a_1 - a_2, Q(-p^*))$*. Here* p^* *is defined by the equality (2.5.1) and* $p^* \in \Pi^-$*.*

Proof. The function $f : (0, \infty) \longrightarrow \mathbb{R}$

$$\begin{aligned}\frac{d}{dp} f(p) = \frac{d}{dp}\mathrm{tr}[Q(p)] &= \frac{d}{dp}\mathrm{tr}[(1+p^{-1})Q_1 + (1+p)Q_2] \\ &= \mathrm{tr}\frac{d}{dp}[(1+p^{-1})Q_1 + (1+p)Q_2]\end{aligned}$$

has one single root at $p = p^*$, therefore it has to be an element of Π^+.

In case (b), considering the function defined by $g(p) = f(-p)$, we have $g(p) = f(p)$, and therefore the root is the same as in the above case and unique. The matrix $Q(-p^*)$ being positive definite implies $p^* < \lambda_{\min}$.
Q.E.D.

Lemma 2.5.3

(a) *There exists a unique ellipsoid of minimal volume that contains the sum* $\mathcal{E}_1 + \mathcal{E}_2$*. It is of the form* $\mathcal{E}(a_1 + a_2, Q(p^*))$ *where* $p^* \in (0, \infty)$ *is the unique solution of the equation*

$$\sum_{i=1}^{n} \frac{1}{\lambda_i + p} = \frac{n}{p(p+1)} \quad . \tag{2.5.2}$$

For it we also have $p^* \in \Pi^+$*.*

(b) Suppose that $\mathcal{E}(a_2, Q_2) \subset \mathrm{int}(\mathcal{E}(a_1, Q_1))$, then there exists a unique ellipsoid $\mathcal{E}$ of maximal volume contained in the difference $\mathcal{E}_1 \dot{-} \mathcal{E}_2$. It is of the form $\mathcal{E} = \mathcal{E}(a_1 - a_2, Q(-p^{**}))$ where p^{**} is the single root of equation (2.5.2) falling into the set Π^-.

Proof. By Lemma 2.5.1 we only have to find $p^* \in (0, \infty)$ that minimizes the volume of $\mathcal{E}(a, Q(\cdot))$. This is obviously equivalent to the minimization of $\log\det(Q(\cdot))$, and $\det(Q(\cdot))$ depends on the product of the eigenvalues of Q. Since the eigenvalues of Q do not change with nondegenerate linear transformations, the minimal property of the estimating ellipsoid is not changed if a nondegenerate-affine transformation is applied. Because of the way $Q(p)$ is derived from Q_1 and Q_2, so is the value of $p^* \in (0, \infty)$. Therefore, we are allowed to suppose that

$$Q_j = \mathrm{diag}\{q^{(j)}{}_1, \ldots q^{(j)}{}_n\},\ j \in \overline{1,2}$$

and hence

$$Q(p) = \mathrm{diag}\{q_1(p), \ldots q_n(p)\} \ .$$

All this means that we have to find the roots of the function $f : (0, \infty) \longrightarrow \mathbb{R}$

$$f(p) = \frac{d}{dp} \log\det(Q(p)) \ .$$

Using the relationship

$$\frac{d}{dp} \log\det(Q(p)) = \mathrm{tr}\left[Q(p)^{-1} \frac{d}{dp} Q(p)\right]$$

we obtain

$$\mathrm{tr}[(p^{-1} Q_2{}^{-1} Q_1 + I)^{-1}] = \frac{n}{p+1} \ .$$

By diagonality this means that

$$\sum_{i=1}^{n} \frac{p}{\lambda_i + p} = \frac{n}{p+1} \ ,$$

where λ_i, $i \in \overline{1,n}$ are the eigenvalues of the pencil $Q_1 - \lambda Q_2$, and they are again invariant with respect to the affine transformation used for diagonalization.

The left hand side of this equality strictly increases from 0 to n and the right hand side strictly decreases from n to 0 while p increases from 0 to ∞. Therefore it has one single root corresponding to a minimum, as we have

$$\lim_{p \longrightarrow 0} \det(Q(p)) = \lim_{p \longrightarrow \infty} \det(Q(p)) = \infty \ .$$

The proof of (a) now is complete.

In the case of (b) we need the roots in $(-\infty, 0)$. If here the parameter p corresponding to the maximal volume fell onto any of the endpoints of the interval $(1, \lambda_{\min})$ then the matrix would be semidefinite, i.e., the ellipsoid would have zero volume. But this is excluded by the condition we set. After having established this, a similar argument shows that there is one single local maximum. Q.E.D.

Lemma 2.5.4

(a) *There exists a unique ellipsoid with minimal* $\mathrm{tr}(Q^2)$ *that contains the sum* $\mathcal{E}_1 + \mathcal{E}_2$. *It is of the form* $\mathcal{E}(a_1 + a_2, Q(p^*))$, *where* p^* *is the unique positive root of the polynomial*

$$f(p) = \gamma_{22}p^3 + \gamma_{12}p^2 - \gamma_{12}p - \gamma_{11} \qquad (2.5.3)$$

$$\gamma_{ij} = \mathrm{tr}(Q_i Q_j) \ ,$$

for $i, j \in \overline{1,2}$. *The value* $p^* \in \prod^+$.

(b) *Suppose that* $\mathrm{int}(\mathcal{E}_1) \supset \mathcal{E}_2$ *holds, and also that there exists an internally estimating ellipsoid for the difference* $\mathcal{E}_1 \dot{-} \mathcal{E}_2$ *with maximal* $\mathrm{tr}(Q^2)$. *Then it is of the form* $\mathcal{E}(a_1 - a_2, Q(-p^{**}))$. *Here* p^{**} *is a root of the polynomial defined by the equality* $g(p) = f(-p)$ *that falls into* Π^-.

Proof. Direct calculations indicate that

$$2(1+p)p^{-3} \cdot f(p) = \frac{d}{dp}\mathrm{tr}[Q(p)^2] \ .$$

All the coefficients γ_{ij}-s are positive, as we have the equality $\mathrm{tr}(Q_i Q_j) = \mathrm{tr}(R_j Q_i R_j)$ where R_j is the square root of Q_j. The right hand side is positive here, because we take the trace of a positive definite matrix. Hence $\gamma_{22} > 0$ and $\gamma_{11} > 0$, implying the existence of a positive root.

The equality $f(p) = 0$ is equivalent to

$$p^2 = \frac{\gamma_{12}p + \gamma_{11}}{\gamma_{22}p + \gamma_{12}} .$$

Here the left hand side is a convex strictly increasing function, while the right hand side is a continuous function with value $\gamma_{11}\gamma_{12}^{-1}$ at $p = 0$, which tends to $\gamma_{12}\gamma_{22}^{-1}$ with $p \to \infty$. Therefore their graphs may have no more than one intersection. The proof of (a) is complete.

The proof of the statement of part (b) is obvious. Q.E.D.

For the sake of approximating the solutions to differential inclusions we may need to treat a particular case, when the parameters of $\mathcal{E}_2 = \mathcal{E}(a_2, Q_2)$ may be presented as

$$a_2 = \varepsilon a_0, \quad Q_2 = \varepsilon^2 Q_0,$$

so that

$$\begin{aligned}\rho(l|\mathcal{E}(a_2, Q_2)) &= (l, a_2) + (l, Q_2 l)^{1/2} \\ &= \varepsilon(l, a_0) + \varepsilon(l, Q_0 l)^{1/2} \\ &= \varepsilon\rho(l|\mathcal{E}(a_0, Q_0)) ,\end{aligned}$$

or in other terms

$$\mathcal{E}(a_2, Q_2) = \varepsilon\mathcal{E}(a_0, Q_0) ,$$

i.e.,

(2.5.4) $$\mathcal{E}_2 = \varepsilon\mathcal{E}_0, \quad \mathcal{E}_0 = \mathcal{E}(a_0, Q_0) .$$

Lemma 2.5.5 *Let us have for $\varepsilon > 0$ the relation (2.5.4).*

(a) Then the ellipsoid with minimal sum of squares of semiaxes, (which is $\operatorname{tr}(Q)$) that contains the $\mathcal{E}_1 + \mathcal{E}_2$ is of the form $\mathcal{E}(a_1+a_2, Q_\varepsilon(q^*))$, where

(2.5.5) $$Q_\varepsilon(q) = Q_1 + \varepsilon(q^{-1}Q_1 + qQ_0) + \varepsilon^2 Q_0$$

and

(2.5.6) $$q^* = \frac{\operatorname{tr}^{1/2}(Q_1)}{\operatorname{tr}^{1/2}(Q_0)} .$$

(b) *Suppose that* $\mathrm{int}(\mathcal{E}_2) \neq \emptyset$. *Then the ellipsoid with the maximal sum of squares of semiaxes, (which is* $\mathrm{tr}(Q)$*) that is contained in the difference* $\mathcal{E}_1 \dot{-} \mathcal{E}_2$ *is of the form* $\mathcal{E}(a_1 - \varepsilon a_0, Q_\varepsilon(-q^*))$.

Proof. Writing formally

$$Q(p) = (1 + p^{-1})Q_1 + \varepsilon^2(1 + p)Q_0 \quad , \tag{2.5.7}$$

and calculating the optimal $p = p^*$ that gives $\mathrm{tr}(Q(p)) = \min$, we obtain

$$\varepsilon^2(p^*) = \frac{\mathrm{tr}Q_1}{\mathrm{tr}Q_0} \quad . \tag{2.5.8}$$

Introducing the notation

$$Q_\varepsilon(q) = Q(q\varepsilon^{-1}), \quad q = p\varepsilon \quad , \tag{2.5.9}$$

(2.5.5) follows. By (2.5.8) now the proof of part (a) is complete.

The proof of case (b) is similar to the above; however, here we also need the inequality

$$\varepsilon^{-1}q^* < \lambda_{\min} = \varepsilon^{-2}\mu_{\min} \quad ,$$

that will automatically hold for small ε. Here $\mu_{\min}$ is the minimal relative eigenvalue of Q_1 and Q_0. The Lemma is now proved for both cases. Q.E.D.

Lemma 2.5.6 *Let us have for* $\varepsilon > 0$ *the relation (2.5.4).*

(a) *Then the ellipsoid with minimal volume that contains the sum* $\mathcal{E}_0 + \mathcal{E}_2$ *is of the form* $\mathcal{E}(a_1 + \varepsilon a_0, Q_\varepsilon(q^*))$, *with* $Q_\varepsilon(q)$ *given by (2.5.5) and*

$$q^* = \frac{n^{1/2}}{\mathrm{tr}^{1/2}(Q_0 Q_1^{-1})} \quad . \tag{2.5.10}$$

(b) *Suppose that* $\mathrm{int}(\mathcal{E}_2) \neq \emptyset$. *Then the ellipsoid with maximal volume that is contained in the difference* $\mathcal{E}_1 \dot{-} \mathcal{E}_2$ *is of the form* $\mathcal{E}(a_1 - \varepsilon a_0, Q_\varepsilon(-q^*))$, *where* q^* *is defined by (2.5.10).*

Proof. Denote by $\mu_1 \leq \mu_2 \leq \ldots \leq \mu_n$ the relative eigenvalues of the matrices $\mathcal{E}_1$ and $\mathcal{E}_0$. Comparing them with the relative eigenvalues $\lambda_1 \leq \lambda_2 \leq \ldots \leq \lambda_n$ of $\mathcal{E}_1$ and $\mathcal{E}_2$ we obviously have for all $i \in \overline{1,n}$

$$\varepsilon^2 \lambda_i = \mu_i \quad . \tag{2.5.11}$$

We look for the root of equation (2.5.2) where we have to use (2.5.11) and according to the equality $q = p\varepsilon$. Rewriting equation (2.5.2), into the form

$$\sum_{i=1}^{n} \frac{\lambda_i}{\lambda_i + p} = \frac{np}{p+1} \quad , \tag{2.5.12}$$

and carrying out the substitutions, by the analytic dependence of the roots on the parameters, we obtain for the Taylor-series expansion in ε of equation (2.4.20)

$$\sum_{i=1}^{n} \frac{1}{1 + \varepsilon q \mu_i^{-1}} = \frac{n}{1 + \varepsilon q^{-1}} \quad ,$$

and

$$n - \varepsilon q^* \sum_{i=1}^{n} \mu_i^{-1} = n(1 - \varepsilon (q^*)^{-1}) + o(\varepsilon) \quad .$$

From the comparison of the coefficients of ε,

$$(q^*)^2 = \frac{n}{\sum_{i=1}^{n} \mu_i^{-1}} \quad ,$$

and then (a) follows.[17] For part (b) we have to take the negative sign, and the condition of positive definiteness follows in the same way as for Lemma 2.5.2.

Q.E.D.

Lemma 2.5.7 *Assume the relation the relation (2.5.4) to be true.*

(a) Then the ellipsoid with minimal $\mathrm{tr}(Q^2)$ *that contains the sum* $\mathcal{E}_1 + \mathcal{E}_2$ *is of the form* $\mathcal{E}(a_1 + \varepsilon a_0, Q_\varepsilon(q^*))$, *with formula* $Q_\varepsilon(q)$ *given by (2.5.5)*

$$q^* = \frac{\mathrm{tr}^{1/2}(Q_1^2)}{\mathrm{tr}^{1/2}(Q_1 Q_0)} \quad . \tag{2.5.13}$$

[17] In calculating q^* the terms of higher order in ϵ have been omitted.

(b) *Suppose that* $\mathrm{int}(\mathcal{E}_2) \neq \emptyset$. *Then the ellipsoid with maximal* $\mathrm{tr}(Q^2)$ *that is contained in the difference* $\mathcal{E}_2 \dot{-} \mathcal{E}_1$ *is of the form* $\mathcal{E}(a_1 - \varepsilon a_0, Q_\varepsilon(-q^*))$, *where* q^* *is defined by (2.5.13).*

Proof. The notation of equation (2.5.3)

$$\gamma_{ij} = \mathrm{tr}(Q_i Q_j) \quad ,$$

is now used for $i, j = 1$ or 0. Using the above scheme, we look for the root again in the form of $q = p\varepsilon$. Substituting this into equation (2.5.3)

$$\text{(2.5.14)} \qquad \varepsilon\gamma_{00}q^3 + \gamma_{10}q^2 - \varepsilon\gamma_{10}q - \gamma_{11} = 0$$

and from the comparison of coefficients in the Taylor-series expansion in ε for this equation

$$\text{(2.5.15)} \qquad (q^*)^2 = \frac{\mathrm{tr}(Q_1^2)}{\mathrm{tr}(Q_1 Q_0)} \quad .$$

The rest of the proof is analogous. Q.E.D.

Corresponding to the above, we formulate the converse statements concerning the internal estimates of maximal volume of the Minkowski-sum and the external estimates of minimal volume of the difference. Part (a) of the following theorem is given in [73], but this proof is different and appears to be more general. On the other hand, part (b) is a new result, which is proved by the technique used there.

Lemma 2.5.8

(a) *There exists a unique ellipsoid of maximal volume contained in the sum* $\mathcal{E}_1 + \mathcal{E}_2$. *It is of the form* $\mathcal{E}(a_1 + a_2, Q_+^*)$ *where*

$$\text{(2.5.16)} \quad Q_+^* = Q_1 + Q_2 + 2{Q_2}^{1/2}[{Q_2}^{-1/2}Q_1{Q_2}^{-1/2}]^{1/2}{Q_2}^{1/2} \quad .$$

(b) *There exists a unique ellipsoid of minimal volume containing the difference* $\mathcal{E}_1 \dot{-} \mathcal{E}_2$. *It is of the form* $\mathcal{E}(a_1 - a_2, Q_-^*)$ *where*

$$\text{(2.5.17)} \quad Q_-^* = Q_1 + Q_2 - 2{Q_2}^{1/2}[{Q_2}^{-1/2}Q_1{Q_2}^{-1/2}]^{1/2}{Q_2}^{1/2} \quad .$$

Proof. It is not difficult to prove, as in [73], that (a) is valid with

$$Q_+^* = S^{-1}[(SQ_1S')^{1/2} + (SQ_2S')^{1/2}]^2 S'^{-1} \quad ,$$

where S is the matrix diagonalizing both Q_1 and Q_2. It is also possible to observe that although S is not unique, this expression is independent of the choice of S. Let us select

$$S = N{Q_2}^{-1/2} \quad ,$$

where N is orthogonal. Then for a suitable N the matrix S will meet the requirements. A substitution now yields (a).

Let us now consider (b). By the affine invariance of the volume function, we can use again the matrix S to diagonalize Q_1 and Q_2, i.e., to get $SQ_1S' = D_1 = \mathrm{diag}\{r_1, \ldots r_n\}$. and $SQ_2S' = P$, where $P = \mathrm{diag}\{p_1, \ldots p_n\}$ is a partial identity. Our aim is to find a minimal volume ellipsoid $\mathcal{E}(a, D)$ among those with the property:

$$\rho(l|\mathcal{E}(a, D)) \geq \rho(l|\mathcal{E}(0, D_1) \dot{-} \mathcal{E}(0, P))$$

By the argument of the proof of Part 2 in [84], the existence and uniqueness of such an ellipsoid follows, and the same argument implies that $a = 0$ and D is diagonal. This is justified because of the inclusion $\mathrm{int}(\mathcal{E}_1) \supset \mathcal{E}_2$ holds. Substituting the unit coordinate-vectors into the previous inequality we shall require

$$d_i^{1/2} \geq {r_i}^{1/2} - {p_i}^{1/2} \quad i \in \overline{1, n} \ .$$

If we define d_i, $i \in \overline{1, n}$ with $d_i = (r_i^{1/2} - {p_i}^{1/2})^2$, then it can be established by direct calculation that D is an external estimate. The statement is thus true. Q.E.D.

Lemma 2.5.9 *Let us have the relation (2.5.4) for* $\varepsilon > 0$. *Then*

(a) the ellipsoid with maximal volume that is contained in the sum $\mathcal{E}_1 + \mathcal{E}_2$ *is of the form* $\mathcal{E}(a_1 + \varepsilon a_0, Q_{\varepsilon+})$, *where*

$$(2.5.18) \quad Q_{\varepsilon+} = Q_1 + 2\varepsilon(Q_0^{1/2}[{Q_0}^{-1/2} Q_1 Q_0^{-1/2}]^{1/2} Q_0^{1/2}) + o(\varepsilon)I$$

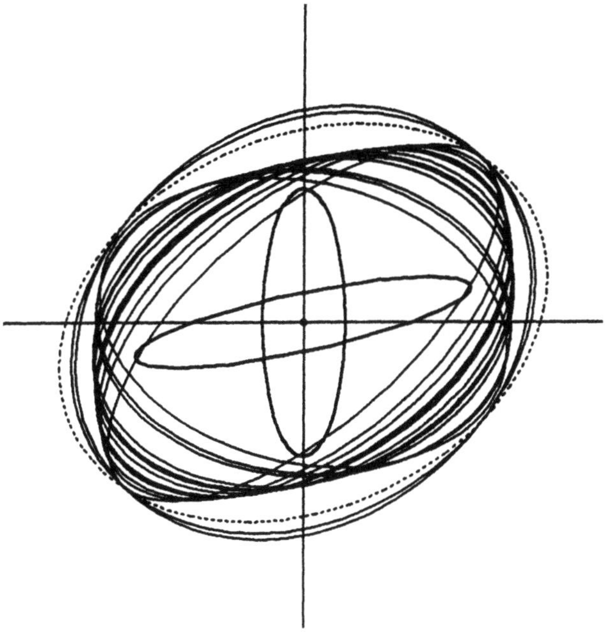

Figure 2.5.1(a).

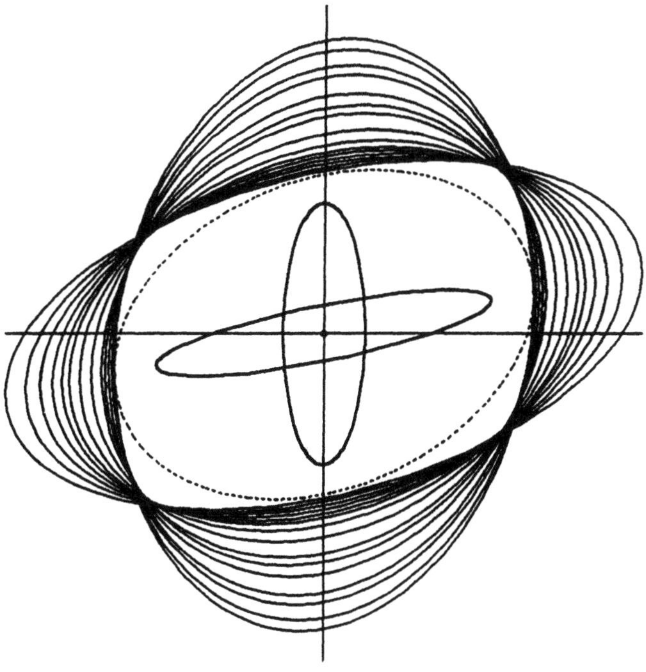

Figure 2.5.1(b).

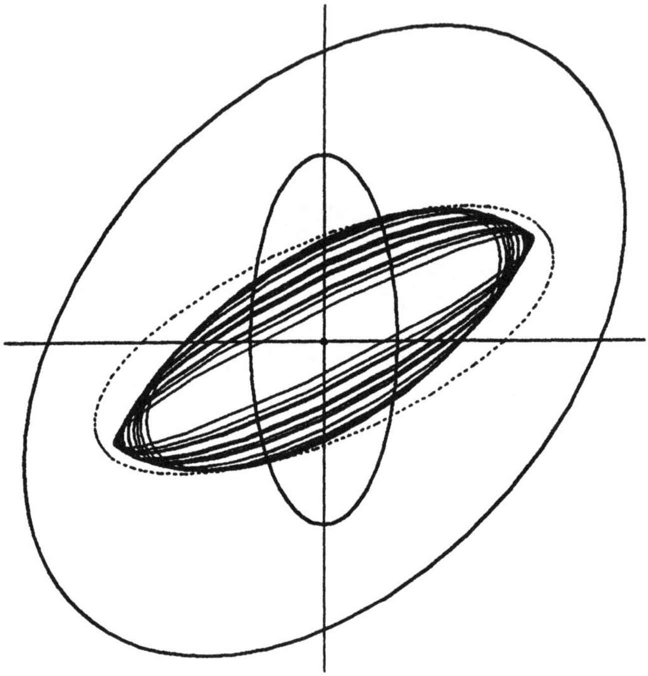

Figure 2.5.2(a).

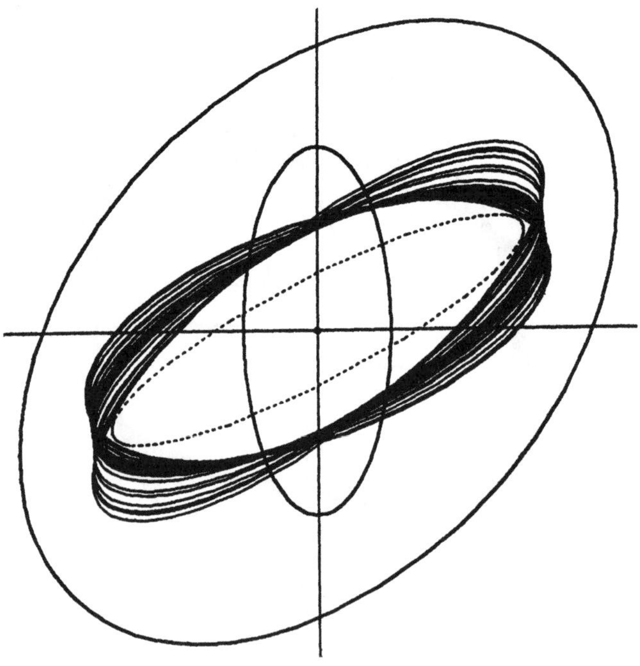

Figure 2.5.2(b).

(b) the ellipsoid with minimal volume that contains the difference $\mathcal{E}_1 \dot{-} \mathcal{E}_2$ *is of the form* $\mathcal{E}(a_1 - \varepsilon a_0, Q_{\varepsilon -})$, *where*

$$(2.5.19)\quad Q_{\varepsilon-} = Q_1 - 2\varepsilon(Q_0^{1/2}[Q_0^{-1/2}Q_1Q_0^{-1/2}]^{1/2}Q_0^{1/2}) + o(\varepsilon)I \ .$$

The proof follows the lines of the above through expansions of the respective relations in ε, and the solution is found within the terms with ε of power 1. The reader may verify this as an exercise.

To illustrate the material of this section, we introduce several figures. Thus Figure 2.5.1(a) is the same as 2.3.1(a), except that in addition to the internal estimates that indicate the sum of two ellipsoids, we also indicate the external estimates that are of minimal trace (tr$Q$), minimal tr(Q^2) (both drawn in continuous lines) and minimal volume (drawn in dotted line). Figure 2.5.1(b) is the same as 2.2.1(a), but in addition to the external estimates that indicate the sum of two ellipsoids, we also indicate the internal ellipsoid of maximal volume (shown with dotted line). Figure 2.5.2(a) is the same as 2.2.1(b), but in addition the dotted line shows the volume-minimal ellipsoid for the Minkowski-difference of two ellipsoids. Finally, Figure 2.5.2(b) is the same as 2.3.1(b), but in addition the dotted line shows the volume-maximal internal ellipsoid for the Minkowski difference.

The next issue is to consider intersections of ellipsoids. The description of this operation is more complicated than what has gone before and does not reach the same degree of detail, being confined mostly to the external ellipsoidal estimates of these intersections.

2.6 Intersections of Ellipsoids

Let us consider m nondegenerate ellipsoids $\mathcal{E}_i = \mathcal{E}(a^{(i)}, Q_i)$, $i = 1, \ldots, m$. Their intersection

$$\bigcap_{i=1}^{m} \mathcal{E}(a^{(i)}, Q_i) = \mathcal{P}[m]$$

consists of all the vectors $x \in \mathbb{R}^m$ that simultaneously satisfy the inequalities

$$(2.6.1)\qquad (x - a^{(i)}, Q_i^{-1}(x - a^{(i)})) \leq 1, \quad (i = 1, \ldots, m) \ .$$

Assuming

$$\mathcal{A} = \{\alpha \in \mathbb{R}^m : \sum \alpha_i = 1,\ \alpha_i \geq 0,\ i = 1, \ldots, m\} \ ,$$

take the inequality

$$\sum_{i=1}^{m} \alpha_i (x - a^{(i)}, Q_i^{-1}(x - a^{(i)})) \leq 1 \ . \tag{2.6.2}$$

The following assertion is obvious

Lemma 2.6.1 *If $x^* \in \mathbb{R}^m$ is a solution to the system (2.6.1), then x^* satisfies (2.6.2) for any $\alpha \in \mathcal{A}$ (and vice versa).*

By direct calculation we may observe that for a given $\alpha \in \mathcal{A}$ the inequality (2.6.2) defines an ellipsoid

$$\mathcal{E}[\alpha] = \{x : (x - a[\alpha],\ Q[\alpha](x - a[\alpha])) \leq 1 - h^2[\alpha]\} \ , \tag{2.6.3}$$

where

$$a[\alpha] = \left(\sum_{i=1}^{m} \alpha_i Q_i^{-1}\right)^{-1} \left(\sum_{i=1}^{m} \alpha_i Q_i^{-1} a^{(i)}\right) \ ,$$

$$Q[\alpha] = \sum_{i=1}^{m} \alpha_i Q_i^{-1} \ ,$$

$$\begin{aligned} h^2[\alpha] &= \sum_{i=1}^{m} \alpha_i \left(a^{(i)}, Q_i^{-1} a^{(i)}\right) \\ &- \left(\sum_{i=1}^{m} \alpha_i Q_i^{-1} a^{(i)}, \left(\sum_{i=1}^{m} \alpha_i Q_i^{-1}\right)^{-1} \left(\sum_{i=1}^{m} \alpha_i Q_i^{-1} a^{(i)}\right)\right) \ . \end{aligned}$$

It is not difficult to check that $h^2[\alpha] \in [0, 1]$. In other terms

$$\mathcal{E}[\alpha] = \mathcal{E}(a[\alpha], (1 - h^2[\alpha])(Q[\alpha])^{-1}) \ . \tag{2.6.4}$$

A direct consequence of Lemma 2.6.1 is

Lemma 2.6.2 *The following assertion is true. The set*

$$\mathcal{P}[m] = \{\cap \mathcal{E}[\alpha] | \alpha \in \mathcal{A}\} \ .$$

Each of the ellipsoids $\mathcal{E}[\alpha]$ is therefore an *external estimate* for $\mathcal{P}[m]$, so that $P_m \subseteq \mathcal{E}[\alpha]$, for any $\alpha \in \mathcal{A}$.

The intersection $\mathcal{P}[m]$ of m ellipsoids $\mathcal{E}_i$ is now presented as an intersection of a parametrized family of ellipsoids $\mathcal{E}[\alpha], \alpha \in \mathcal{A}$. Among these we may select, if necessary, an optimal external ellipsoidal estimate for $\mathcal{P}[m]$.

It is clear that the variety $\mathcal{E}[\alpha]$, $\alpha \in \mathcal{A}$, contains each of the ellipsoids $\mathcal{E}_i$, so that

$$\mathcal{E}_i = \mathcal{E}[e^{(i)}]; \quad i = 1, \ldots, m \quad ,$$

where $e^{(i)} \in \mathbb{R}^m$ is a unit vector (an orth) along the i-th coordinate axis and its coordinates are defined as

$$e_j^{(i)} = \delta_{ij} \quad ;$$

$$\delta_{ij} = 1 \text{ if } i = j, \ \delta_{ij} = 0 \ \text{ if } i \neq j \quad j = 1, \ldots, m \quad .$$

In Figure 2.6.1(a) one may observe an illustration of Lemma 2.6.2. Here numbers 1,2 indicate the intersecting ellipsoids while the unmarked ones are the external estimates $\mathcal{E}[\alpha]$ calculated due to formula (2.6.4). The intersecting ellipsoids 1,2 correspond to values $\alpha_1 = 1, \alpha_2 = 0$ and $\alpha_1 = 0, \alpha_2 = 1$ in (2.6.4). Figure 2.6.1(b) shows the two intersecting ellipsoids and with dotted line the volume-minimal external estimate obtained by a one-dimensional search in $\alpha_1 \in [0, 1]$, $(\alpha_2 \geq 0, \ \alpha_1 + \alpha_2 = 1)$.

However, in general, the optimal external ellipsoidal estimate $\mathcal{E}[\alpha^0]$ for $\mathcal{P}[m]$ (taken, for example, for one of the criteria of the above) may not be among the ellipsoids $\mathcal{E}_i$. One of the questions that arise here is whether the variety $\mathcal{E}[\alpha]$ is sufficiently complete in the sense of the following question: will the optimal external estimate $\mathcal{E}[\alpha^0]$ (with respect to a function $\Psi(\mathcal{E})$) chosen only among the ellipsoids $\mathcal{E}[\alpha], \alpha \in \mathcal{A}$ be the same as the optimal ellipsoid $\mathcal{E}$ (also with respect to $\Psi(\mathcal{E})$) chosen among some other class of external ellipsoids or, particularly, among *all the possible* external estimates?

In the sequel of this section we shall produce some examples (see Examples 2.6.2, 2.6.3) that give an answer to this question.

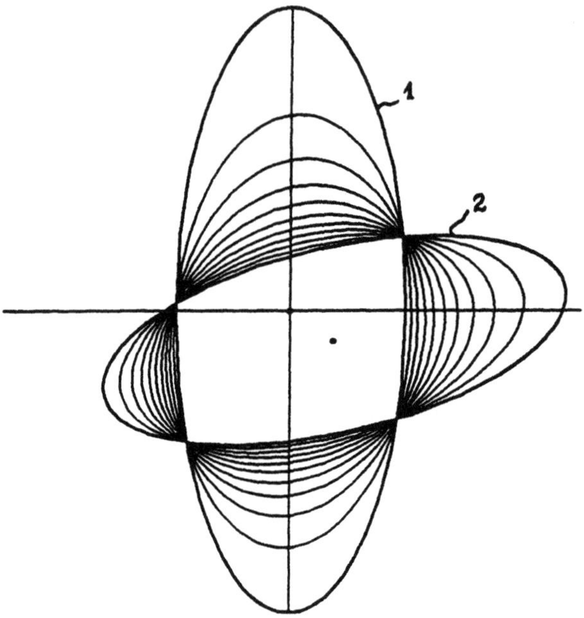

Figure 2.6.1(a).

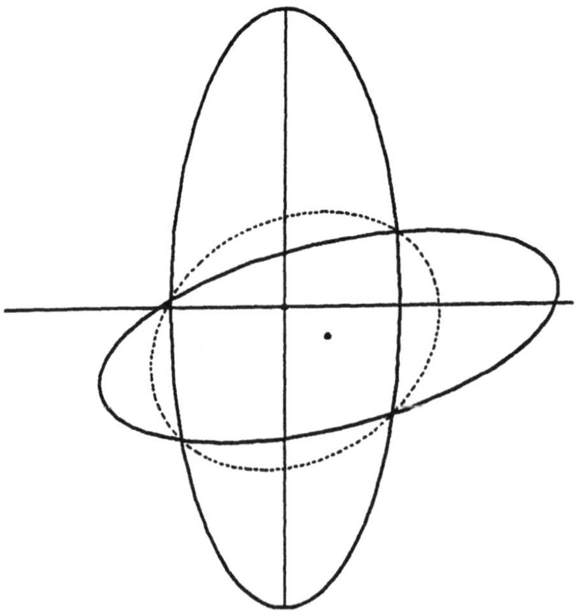

Figure 2.6.1(b).

Meanwhile let us introduce another formula for the intersection

$$\bigcap_{i=1}^{m} \mathcal{E}(a^{(i)}, Q_i) = \mathcal{P}[m]$$

of nondegenerate ellipsoids $\mathcal{E}(a^{(i)}, Q_i)$.

Assumption 2.6.1 *The intersection $\mathcal{P}[m]$ has an interior point: int$\mathcal{P}[m] \neq \emptyset$.*

We shall further proceed under this assumption. Taking the support functions $\rho(\ell|\mathcal{E}(a^{(i)}, Q_i))$ for the ellipsoids $\mathcal{E}(a^{(i)}, Q_i)$, we may apply the convolution formula [265]

$$(2.6.5) \qquad \rho(\ell|\mathcal{Q}) = \{ \inf \sum_{i=1}^{m} \rho(\ell^{(i)}|\mathcal{Q}_i) | \sum_{i=1}^{m} \ell^{(i)} = \ell \}$$

that relates the support function $\rho(\ell|\mathcal{Q})$ of an intersection $\mathcal{Q} = \cap_{i=1}^{m} \mathcal{Q}_i$ with the support functions $\rho(\ell|\mathcal{Q}_i)$ for each set $\mathcal{Q}_i$.

Then, assuming

$$(2.6.6) \qquad \ell^{(i)} = M^{(i)}\ell,$$

where the matrix $M^{(i)} \in \mathbb{R}^{n \times n}$ does exist for any $\ell, \ell^{(i)} \in \mathbb{R}^n$, $\ell \neq \{0\}$, we may substitute (2.6.6) into (2.6.5), coming to

$$(2.6.7)\ \rho(\ell|\mathcal{Q}) = \left\{ \inf \sum_{i=1}^{m} \rho(\ell|M^{(i)\prime}\mathcal{Q}_i) \,|M^{(i)} : \sum_{i=1}^{m}(M^{(i)} - I)\ell = 0 \right\},$$

or to the relations

$$\rho(\ell|\mathcal{Q}) \leq \rho(\ell|\sum_{i=1}^{m} M^{(i)\prime}\mathcal{Q}_i), \quad (\sum_{i=1}^{m} M^{(i)} - I)\ell = 0 \quad ,$$

which should be true for any $\ell \in \mathbb{R}^n$ and any array of matrices $M^{(i)}$ that satisfy the last equality.

Otherwise this means

$$(2.6.8) \qquad \mathcal{Q} \subseteq \sum_{i=1}^{n} M^{(i)\prime}\mathcal{Q}_i \quad ,$$

whatever are the matrices $M^{(i)}$ that satisfy the equality

$$\sum_{i=1}^{m} M^{(i)} = I \ . \tag{2.6.9}$$

Moreover, (2.6.7) implies

$$\mathcal{Q} = \cap \sum_{i=1}^{m} M^{(i)} Q_i$$

over all the matrices $M^{(i)}$ that satisfy (2.6.9) (we may omit the transpose, since $M^{(i)}$ are chosen arbitrarily, provided only that (2.6.9) does hold).

In terms of ellipsoids and in view of the formula

$$M\mathcal{E}(a, Q) = \mathcal{E}(Ma, MQM')$$

this gives

$$\mathcal{P}[m] \subseteq \sum_{i=1}^{m} \mathcal{E}(M^{(i)}a^{(i)}, M^{(i)}Q_i M^{(i)\prime}), \ \sum_{i=1}^{m} M^{(i)} = I \ , \tag{2.6.10}$$

and for the same class of matrices (2.6.9) we have

$$\mathcal{P}[m] = \bigcap \sum_{i=1}^{m} \mathcal{E}(M^{(i)}a^{(i)}, M^{(i)}Q_i M^{(i)\prime}), \ \sum_{i=1}^{m} M^{(i)} = I \ . \tag{2.6.11}$$

We therefore come to the assertion

Lemma 2.6.3 *The intersection $\mathcal{P}[m]$ of m nondegenerate ellipsoids $\mathcal{E}(a^{(i)}, Q_i)$ satisfies the inclusion (2.6.10), whatever are the matrices $M^{(i)}$ of (2.6.9). Moreover, the equality (2.6.11) is true with the intersection taken over all $M^{(i)}$ of (2.6.9).*

Particularly, for $m = 2$ we have

$$\mathcal{P}[2] \subseteq \tag{2.6.12}$$

$$\subseteq \mathcal{E}(Ma^{(1)}, MQ_1M') + \mathcal{E}((I - M)a^{(2)}, (I - M)Q_2(I - M)')$$

for any $n \times n$ matrix $M \in \mathcal{L}(\mathbb{R}^n, \mathbb{R}^n)$.

The intersection $\mathcal{P}[m]$ of n ellipsoids $\mathcal{E}(a^{(i)}, Q_i)$ is therefore approximated from above in (2.6.10) by the sum of m ellipsoids $\mathcal{E}(a^{(i)}, M^{(i)}Q_iM^{(i)\prime})$ restricted only by the equality (2.6.9). As we have seen earlier, the sum of m ellipsoids may, however, be approximated from above by one ellipsoid. Namely,

$$\sum_{i=1}^{m} \mathcal{E}(M^{(i)}a^{(i)}, M^{(i)}Q_iM^{(i)\prime}) \subseteq \mathcal{E}(a[m,\mathcal{M}], Q[m,p,\mathcal{M}]) \ ,$$

where

$$a[m,\mathcal{M}] = \sum_{i=1}^{m} M^{(i)}a^{(i)} \ ,$$

$$Q[m,p,\mathcal{M}] = (\sum_{i=1}^{m} p_i)\sum_{i=1}^{m} p_i^{-1}M^{(i)}Q_iM^{(i)\prime}, \quad p_i \geq 0 \ ,$$

$$\mathcal{M} = \{M^{(1)}, \ldots, M^{(m)}\}, \quad p = \{p_1, \ldots, p_m\} \ .$$

Combining the results of Lemma 2.6.3 and Theorem 2.4.1 (formula 2.4.5), we conclude that the intersection $\mathcal{P}[m]$ may be presented through the inclusion

$$\mathcal{P}[m] \subseteq \mathcal{E}(a[m,\mathcal{M}], Q[m,p,\mathcal{M}]) \ , \tag{2.6.13}$$

which is true for any $\mathcal{M}, p > 0$, provided $\mathcal{M}$ satisfies (2.6.9), or the equality

$$\mathcal{P}[m] = \bigcap_{p}\bigcap_{\mathcal{M}} \mathcal{E}(a[m,\mathcal{M}], Q[m,p,\mathcal{M}]) \tag{2.6.14}$$

with $\mathcal{M}$ of (2.6.9), $p > 0$.

Lemma 2.6.4 *The intersection $\mathcal{P}[m]$ satisfies the inclusion (2.6.13) over all $\mathcal{M}$ of (2.6.9) and the equality (2.6.14) over all $\mathcal{M}$ of (2.6.9).*

Among the ellipsoids $\mathcal{E}(a[m,\mathcal{M}], Q[m,p,\mathcal{M}])$ we may now select those that are optimal relative to some criteria, taking perhaps one of the above, defined at the end of Section 2.1.

Let us first consider two ellipsoids with centers $a^{(1)} = a^{(2)} = 0$ so that

(2.6.15) $\mathcal{P}[2] =$

$$= \mathcal{E}(0, Q_1) \cap \mathcal{E}(0, Q_2) \subseteq \mathcal{E}(0, MQ_1M') + \mathcal{E}(0, (I-M)Q_2(I-M)') \ .$$

The external bounding ellipsoid may be now designed through the following schemes.

Scheme A

For a matrix Q positive symmetrical we may rewrite

$$MQM' = (Q^{1/2}M')'(Q^{1/2}M')$$

and introduce the norm

$$\|MQM'\|^2 = \langle Q^{1/2}M', Q^{1/2}M' \rangle = \mathrm{tr} MQM' \ ,$$

where the scalar product $\langle K, L \rangle$ of two $n \times n$-matrices $K, L \in \mathbb{R}^{n \times n}$ is defined as

$$\langle K, L \rangle = \mathrm{tr} K'L \ .$$

The present scheme is now defined through minimizing

$$\|MQ_1M'\|^2 + \|(I-M)Q_2(I-M)'\|^2 =$$

$$= \langle Q_1^{1/2}M', Q_1^{1/2}M' \rangle + \langle Q_2^{1/2}(I-M)', Q_2^{1/2}(I-M)' \rangle \qquad (2.6.16)$$

over M which leads to the optimal $M = M_0$:

$$M_0 = (Q_1 + Q_2)^{-1} Q_2 \ .$$

Further on, following (2.6.13) we have

$$\mathcal{E}(0, M_0Q_1M_0') + \mathcal{E}(0, (I-M_0)Q_2(I-M_0)') \qquad (2.6.17)$$

$$\subseteq \mathcal{E}(0, (1+p^{-1})M_0Q_1M_0' + (1+p)(I-M_0)Q_2(I-M_0)')$$

whatever is the $p > 0$. The bounding ellipsoid may now be optimalized over p due to one of the criteria of the above (see Section 2.1).

Let us for example select an optimal $p = p_0$, minimizing over p the trace

$$\mathrm{tr}((1+p^{-1})S_1^0 + (1+p)S_2^0) = f_1(p) \ ,$$

where

$$S_1^0 = M_0Q_1M_0', \ S_2^0 = (I-M_0)Q_2(I-M_0)' \ . \qquad (2.6.18)$$

Solving this problem through the equation $f_1'(p) = 0$, (check here that what one gets is precisely a minimum), we observe

$$p_0^2 = \mathrm{tr} S_1^0 / \mathrm{tr} S_2^0 \ .$$

The final calculation gives an upper bound for $\mathcal{P}[m]$, which is

$$\mathcal{P}[m] \subseteq \mathcal{E}(0, (1 + p_0^{-1}) S_1^0 + (1 + p_0) S_2^0) = \mathcal{E}(0, S^0) \ , \tag{2.6.19}$$

where

$$\mathrm{tr} S^0 = ((\mathrm{tr} S_1^0)^{1/2} + (\mathrm{tr} S_2^0)^{1/2})^2 \ . \tag{2.6.20}$$

Consider a specific case

Example 2.6.1

Take the two-dimensional ellipsoids $\mathcal{E}_1 = \mathcal{E}(0, Q_1)$, $\mathcal{E}_2 = \mathcal{E}(0, Q_2)$, where

$$Q_1 = \begin{pmatrix} 1 & 0 \\ 0 & k^2 \end{pmatrix}, \quad Q_2 = \begin{pmatrix} 4k^2 & 0 \\ 0 & 1 \end{pmatrix} \ .$$

Then

$$M_0 = (Q_1 + Q_2)^{-1} Q_2 = \begin{pmatrix} 4k^2(1 + 4k^2)^{-1} & 0, \\ 0 & (1 + k^2)^{-1} \end{pmatrix} \ .$$

$$\mathrm{tr} M^0 Q_1 M^{0'} = 16k^4(1 + 4k^2)^{-2} + k^2(1 + k^2)^{-2} = \alpha^2(k) \ ,$$

$$\mathrm{tr}(I - M^0) Q_2 (I - M^0)' = 4k^2(1 + 4k^2)^{-2} + k^4(1 + k^2)^{-2} = \beta^2(k) \ ,$$

$$p^0 = \alpha(k) \beta^{-1}(k) \ .$$

Following (2.6.19), (2.6.20), we have

$$\mathrm{tr} S^0 = (\alpha(k) + \beta(k))^2 \ ,$$

$$S^0 = (1 + p^0)(p^{0-1} S_1^0 + S_2^0) \ .$$

Scheme B

The next option is not to minimize (2.6.16) first over M, then over p, but to take the bounding ellipsoid $\mathcal{E}(0, S[p, M])$ given by the inclusion

$$\mathcal{E}(0, MQM') + \mathcal{E}(0, (I - M) Q_2 (I - M)') \subseteq$$

$$\subseteq \mathcal{E}(0,(1+p^{-1})MQM' + (1+p)(I-M)Q_2(I-M)') = \mathcal{E}(0,S[p,M])$$

and to minimize $\mathcal{E}(0,S[p,M])$ directly over the pair $p, M (p > 0,\ \ M \in \mathbb{R}^{n\times n})$ having $\mathrm{tr}S[p,M] = \min$ as the criterion. After a minimization of $\mathrm{tr}S[p,M]$ over p, this leads to the problem of minimizing the function

$$f_2(M) = ((\mathrm{tr}MQ_1M')^{1/2} + (\mathrm{tr}(I-M)Q_2(I-M)')^{1/2})^2$$

over M. Since $f_2(M)$ is strictly convex and $f_2(M) \to \infty$ with $M \to \infty$, there exists a unique minimizer M^*.

We also gather that

$$p^* = (\mathrm{tr}M^*Q_1M^*)^{1/2}(\mathrm{tr}(I-M^*)Q_2(I-M^*)')^{-1/2}$$

so that the optimal ellipsoid

$$\mathcal{E}^* = \mathcal{E}(0,S[p^*,M^*]) \ .$$

We have thus indicated two options for the bounding ellipsoid

$$\mathcal{P}[m] \subseteq \mathcal{E}(0,S^0) \ .$$

The one of *Scheme A* is when S^0 is taken due to (2.6.19). The value M^0 for this case is calculated through the minimum of (2.6.16) which is

$$\langle M'Q_1M'\rangle + \langle (I-M)', Q_2(I-M)'\rangle =$$

$$= \mathrm{tr}MQ_1M' + \mathrm{tr}(I-M)Q_2(I-M)'$$

On the other hand, in *Scheme B*, we have

$$\mathcal{P}[m] \subseteq \mathcal{E}(0,S[p^*,M^*])$$

where M^* is calculated by minimizing $f_2[M]$, which is equivalent to the minimization of

$$(\mathrm{tr}MQ_1M')^{1/2} + (\mathrm{tr}(I-M)Q_2(I-M)')^{1/2} \ .$$

We shall now illustrate the given schemes on two-dimensional examples, comparing on them the results given by Schemes A and B. Apart from distinguishing these two cases, we shall also distinguish for each case a minimization over diagonal matrices M only (cases AD and BD, respectively) from a minimization over all possible matrices M (cases AA and BA). In all the consecutive figures the intersecting ellipsoids are marked by numbers 1,2, while the approximating ellipsoids are marked as A (AD,AA) and B (BD,BA).

Consider first the case when the centers of $\mathcal{E}_1$ and $\mathcal{E}_2$ coincide.[18]

Example 2.6.2

(**a**) The ellipsoids 1,2 are centerd at 0. Here both schemes AA and BA give the same external ellipsoid (Figure 2.6.2(a)). However, one may observe, that scheme AD gives a larger one than AA. At the same time, scheme BD does not give any other ellipsoid except 1,2.

(**b**) The ellipsoids 1,2 are centered at 0. Here schemes A and B give different external ellipsoids AA and BA (Figure 2.6.2(b)). At the same time, for each of these schemes the ellipsoids AA,BA are smaller (by inclusion) than AD,BD (which are not shown).

The schemes A,B are now applied to ellipsoids $\mathcal{E}_1, \mathcal{E}_2$ with different centers.

Example 2.6.3

(**a**) Here schemes AA,BA give the same external ellipsoid which clearly is not optimal by either trace or volume. Note that scheme BD gives nothing more than ellipsoids 1,2 (Figure 2.6.3(a)).

(**b**) Here schemes A,B give different external ellipsoids, but AA coincides with AD and BA with BD (Figures 2.6.3(b) and 2.6.3(b1)).

(**c**) Here schemes AA,BA give the same external ellipsoid which is close to optimal by trace or volume (Figure 2.6.3(c)). Note that AD,BD give worse results in both cases.

Scheme C

This one is similar to Scheme 1, but instead of minimizing the trace $f_1(p)$, we minimize

$$f_3(p) = \mathrm{tr}(((1+p^{-1})S_1^0 + (1+p)S_2^0)((1+p^{-1})S_1^0 + (1+p)S_2^0)')$$

[18] Examples 2.6.2 and 2.6.3 were calculated by S. Fefelov.

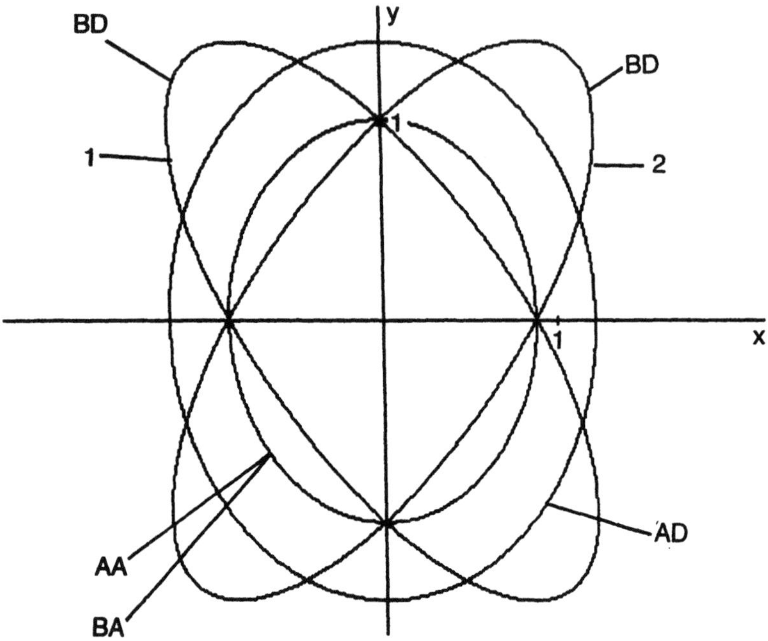

Figure 2.6.2(a).

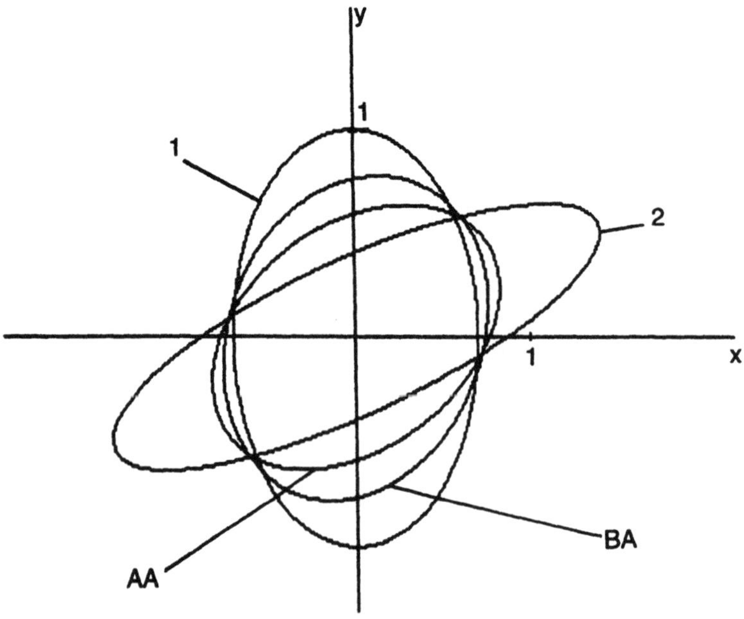

Figure 2.6.2(b).

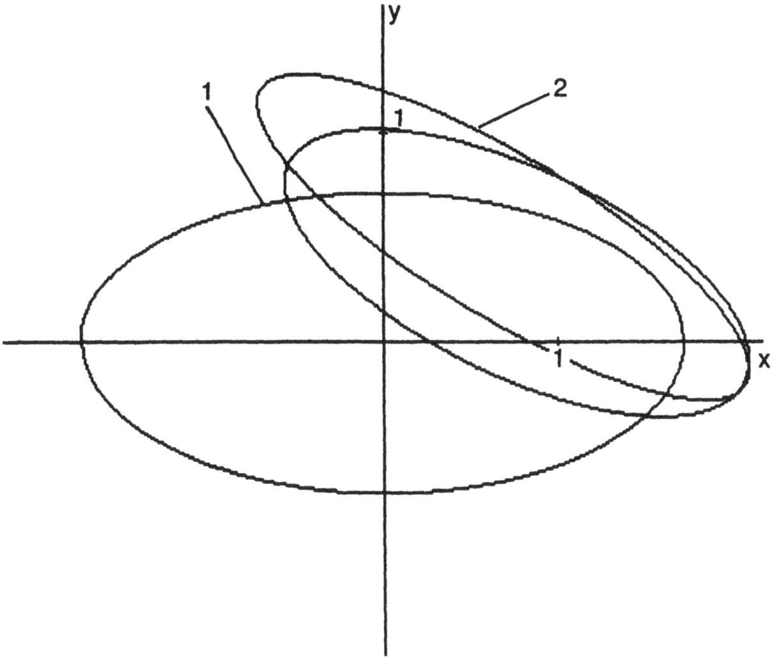

Figure 2.6.3(a).

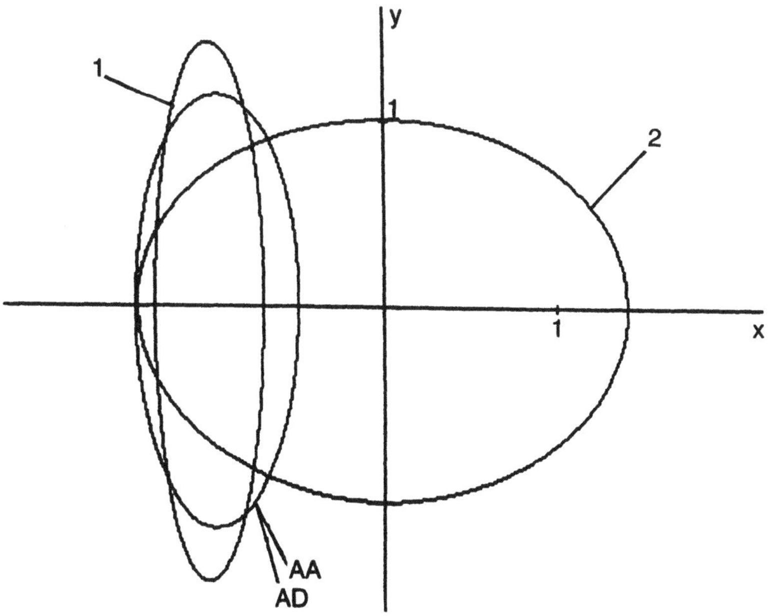

Figure 2.6.3(b).

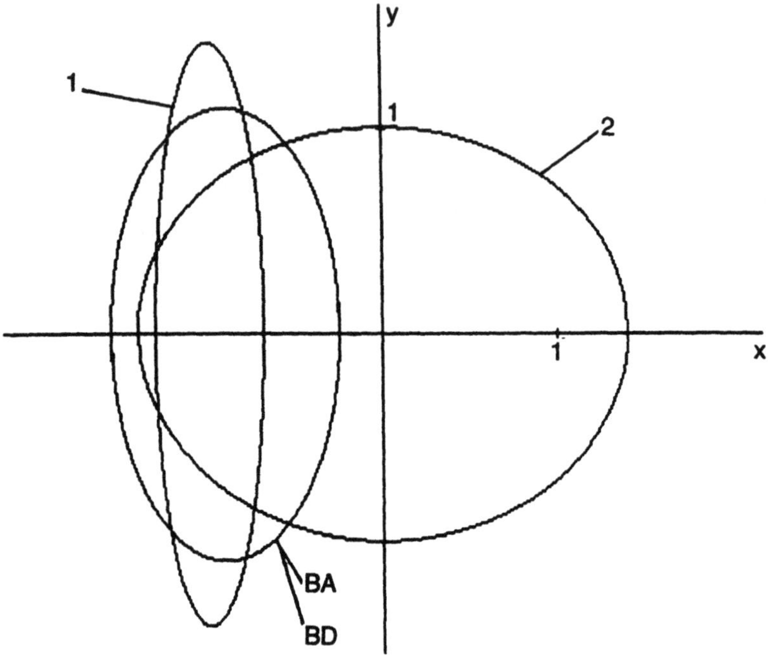

Figure 2.6.3(b1).

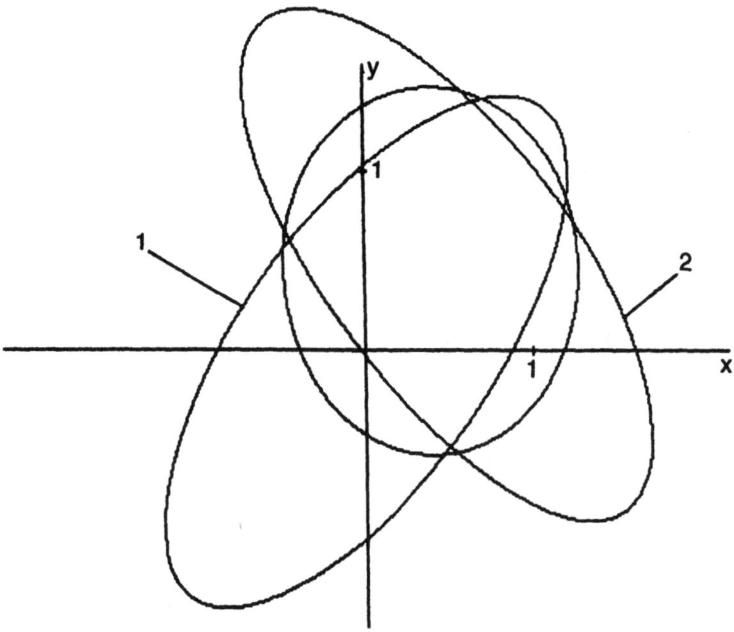

Figure 2.6.3(c).

Equation $f_3'(p) = 0$ now yields a cubic polynomial

$$a_0 p^3 + a_1 p^2 + a_2 p + a_3 = 0 \quad ,$$

where $(S_1^0 = S_1^{0'}, S_2^0 = S_2^{0'})$

$$a_0 = \mathrm{tr} S_2^0 S_2^0, \quad a_1 = -a_2 = \mathrm{tr} S_1^0 S_2^0 \quad ,$$

$$a_3 = -\mathrm{tr} S_1^0 S_1^0 \quad .$$

It may be checked, without difficulty, that the given polynomial has a unique positive root $p = p^* > 0$, which turns out to be the optimalizer and therefore may be substituted into $f_3(p)$ allowing us to write

$$f_3(p^*) = \min_p f_3(p), \quad p > 0 \quad .$$

The optimal circumscribed (external) ellipsoid

$$\mathcal{E}(0, S^0) \supseteq \mathcal{E}(0, S_1^0) + \mathcal{E}(0, S_2^0)$$

is given by

$$S^0 = (1 + p^{*-1}) S_1^0 + (1 + p^*) S_2^0 \quad .$$

Let us now return to the case of an arbitrary finite number m of intersecting ellipsoids and select the external circumscribed ellipsoid as a trace-minimal set. We have

$$\mathrm{tr} Q[m, p] = \sum_{i=1}^{m} \gamma_i c_i^2 = \varphi(p), \tag{2.6.21}$$

where

$$\gamma_i = \left(\sum_{i=1}^{m} p_i \right) p_i^{-1}, \quad c_i^2 = \mathrm{tr} M^{(i)} Q_i M^{(i)'} \quad ,$$

and

$$\sum_{i=1}^{m} M^{(i)} = I \quad . \tag{2.6.22}$$

Minimizing $trQ[m, p]$ over $p = \{p_i, \ldots, p_m\}$ and assuming

$$\sum_{i=1}^{m} p_i \neq 0, \; p_i > 0 \quad ,$$

we come to the system

$$\frac{\partial \varphi}{\partial p^i} = 0, \quad i = 1, \ldots, m \quad ,$$

or otherwise, to the equations

$$\left(\sum_{i=1}^{m} p_i\right)^{-1}\left(-c_i^2\gamma_i^2 + \sum_{i=1}^{m} \gamma_i c_i^2\right) = 0, \quad i = 1, \ldots, m \ ,$$

the solution to which is given by

$$\gamma_i^0 = \left(\sum_{i=1}^{m} c_i\right) c_i^{-1} \ ,$$

and therefore, by $c_i = p_i (i = 1, \ldots, m)$ so that the optimal value

$$\varphi(p^0) = (\sum_{i=1}^{m} c_i)^2 \ .$$

Further on we shall briefly describe a possible approach to the calculation of *internal ellipsoidal approximations* of an *intersection* of two nondegenerate ellipsoids $\mathcal{E}_1 = \mathcal{E}(a^{(1)}, Q_1)$ and $\mathcal{E}_2 = \mathcal{E}(a^{(2)}, Q_2)$. *We assume that this intersection has an interior point* : $intE_1 \cap \mathcal{E}_2 \neq \emptyset$ (Assumption 2.6.1).

Consider the direct product

$$\mathcal{E}_1 \otimes \mathcal{E}_2 = \mathcal{E}(a_*^{(1)}, Q^{(1)}) + \mathcal{E}(a_*^{(2)}, Q^{(2)}) = \mathcal{H} \ ,$$

where

$$a_*^{(1)} = \begin{pmatrix} a^{(1)} \\ 0 \end{pmatrix}, \quad a_*^{(2)} = \begin{pmatrix} 0 \\ a^{(2)} \end{pmatrix}$$

and

$$Q^{(1)} = \begin{pmatrix} Q_1 & 0 \\ 0 & 0 \end{pmatrix}, \quad Q^{(2)} = \begin{pmatrix} 0 & 0 \\ 0 & Q_2 \end{pmatrix} \ .$$

Clearly, $a_*^{(1)}, a_*^{(2)} \in \mathbb{R}^{2n}, Q^{(1)}, Q^{(2)} \in \mathcal{L}(\mathbb{R}^{2n}, \mathbb{R}^{2n})$.

The set $\mathcal{H}$ is the sum of two degenerate ellipsoids in $\mathbb{R}^{2n}$. Nevertheless, since $\mathcal{E}_1, \mathcal{E}_2$ are nondegenerate in $\mathbb{R}^n$ and the set $\mathcal{H}$ is assumed to have an interior point in $\mathbb{R}^{2n}$, it may still be approximated internally according to formula (2.3.3) and Corollary 2.3.1 (where one just has to take the closure of the approximating variety). We may therefore write

$$\mathcal{H} \supset \mathcal{E}(a_*^{(1)} + a_*^{(2)}, Q[S]) \ ,$$

where

$$Q[S] = S^{-1}[(SQ^{(1)}S')^{1/2} + (SQ^{(2)}S')^{1/2}]^2(S')^{-1} \quad ,$$

and S is a symmetrical matrix of dimension $2n \times 2n$.

Let us denote $a = a_*^{(1)} + a_*^{(2)}$ and

$$Q^{-1}[S] = \begin{pmatrix} Q_{11}^- & Q_{12}^- \\ Q_{21}^- & Q_{22}^- \end{pmatrix}, \quad z = \begin{pmatrix} x^{(1)} \\ x^{(2)} \end{pmatrix} \quad ,$$

where $x^{(i)} \in \mathbb{R}^n, Q_{i,j}^- \in \mathcal{L}(\mathbb{R}^n, \mathbb{R}^n), i,j = 1,2$.

Then

(2.6.23) $$\mathcal{E}(a, Q[S]) = \{z : (z-a, Q^{-1}[S](z-a)) \leq 1\} =$$
$$\{z : \Sigma_{i,j}(x^{(i)} - a^{(i)}, Q_{ij}^-(x^{(i)} - a^{(j)}))|i,j = 1,2\} \quad .$$

Let us now *intersect* sets $\mathcal{H}$ and $\mathcal{E}(a, Q[S])$ with the hyperplane $\{x^{(1)} \equiv x^{(2)} = \mathcal{L}\}$. Then

(2.6.24) $$\mathcal{H} \cap \mathcal{L} \supset \mathcal{E}(a, Q[S]) \cap \mathcal{L}, \ \ \forall S \in \Sigma.$$

(Here Σ is the set of all symmetrical matrices in $\mathbb{R}^{2n}$). Moreover, due to an extension of Corollary 2.3.1, we will have

(2.6.25) $$\mathcal{H} \cap \mathcal{L} = \overline{\cup\{\mathcal{E}(a, Q[S]) \cap \mathcal{L} | S \in \Sigma\}} \quad .$$

The obtained relations may be now rewritten in $\mathbb{R}^n$. Namely, taking $x^{(1)} = x^{(2)} = x, x \in \mathbb{R}^n$, we may observe that then

$$\mathcal{H} \cap \mathcal{L} = \mathcal{E}_1 \cap \mathcal{E}_2$$

and

$$\mathcal{E}(a, Q[S]) \cap \mathcal{L} = \{x : \Sigma_{i,j}(x - a^{(i)}, Q_{ij}^-(x - a^{(j)})) \leq 1\}$$

We may now rearrange the previous relation and rewrite (3.6.23) as

(2.6.26) $$\mathcal{E}_1 \cap \mathcal{E}_2 \supset \mathcal{E}(q[S], (1 - h^2[S])Q^*[S]) \quad ,$$

where

$$Q^*[S] = (\Sigma_{i,j=1}^2 Q_{ij}^-)^{-1}, \ \ h^2[S] = \Sigma_{i,j=1}^2 (a^{(i)}, Q_{ij}^- a^{(j)}) -$$
$$-(q[S], (Q^*[S])^{-1} q[S]) \quad ,$$

and

$$q[S] = \frac{1}{2} Q^*[S] \cdot b[S], \quad b[S] = \Sigma_{i,j=1}^2 (Q_{i,j}^{-\prime} a^{(i)} + Q_{i,j}^- a^{(j)}) \quad .$$

The previous reasoning results in

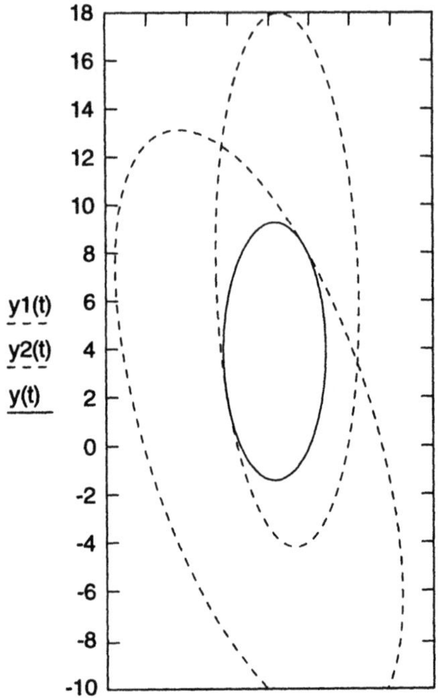

Figure 2.6.4.(a).

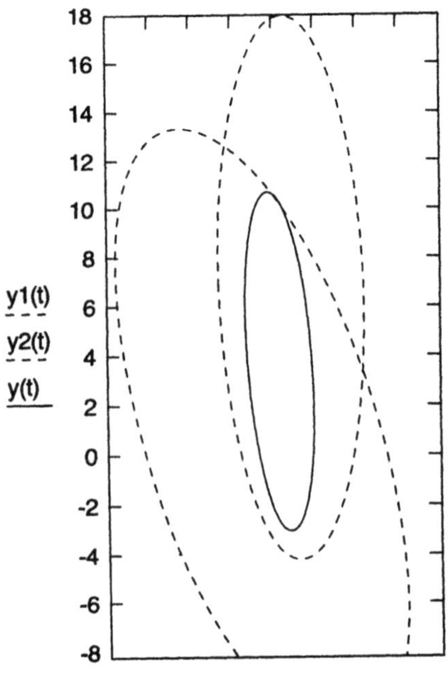

Figure 2.6.4.(b).

Lemma 2.6.5 *Suppose Assumption 2.6.1 holds: the intersection $\mathcal{E}_1 \cap \mathcal{E}_2$ of two nondegenerate ellipsoids has an interior point ($\mathrm{int}\mathcal{E}_1 \cap \mathcal{E}_2 \neq \emptyset$). Then the internal ellipsoidal approximation of $\mathcal{E}_1 \cap \mathcal{E}_2$ may be described due to formula (2.6.26), where S is any symmetrical matrix in $\mathbb{R}^{2n}$.*

The following relation is true

$$(2.6.27) \qquad \mathcal{E}_1 \cap \mathcal{E}_2 = \overline{\cup\{\mathcal{E}(q[S], (1 - h^2[S])Q^*[S]) | S \in \Sigma\}} \ .$$

The last relation follows from (2.6.25).

Remark 2.6.1 *Under nondegeneracy conditions similar to those of Lemma 2.6.5 relations analogous to (2.6.26), (2.6.27) are true for intersections of a finite number n of ellipsoids.*

An interesting exercise here would be to specify some types of optimal or extremal internal ellipsoids and also to describe some minimal variety of internals that would nevertheless wipe out the set $\mathcal{E}_1 \cap \mathcal{E}_2$ from inside. We leave this to the interested reader. However we shall finalize this section with yet another illustration.

Example 2.6.4

Here we demonstrate some internal ellipsoids for an intersection $\mathcal{E}_1 \cap \mathcal{E}_2$ of two ellipsoids where these are marked by numbers 1,2, as before. The internal ellipsoids, calculated due to relations (2.6.25), are unmarked (Figures 2.6.4(a) and 2.6.4(b)).[19]

Exercise 2.6.1. Apply the scheme used in Lemma 2.6.5 to external ellipsoidal approximations of the intersection $\mathcal{E}_1 \bigcap \mathcal{E}_2$.

2.7 Finite Sums and Integrals: External Approximations

Consider m nondegenerate ellipsoids $\mathcal{E}_i = \mathcal{E}(q_i, Q_i)$, $q_i \in \mathbb{R}^n$, $Q_i \in \mathcal{L}(\mathbb{R}^n, \mathbb{R}^n)$, $Q_i > 0$, $i = 1, \ldots, m$. Let us find the external estimates of

[19] This example was calculated by D. Potapov.

their Minkowski sum

(2.7.1) $$\mathcal{S}[m] = \sum_{i=1}^{m} \mathcal{E}_i$$

which is, by definition,

$$\mathcal{E} = \bigcup_{p^{(1)} \in \mathcal{E}_1} \dots \bigcup_{p^{(m)} \in \mathcal{E}_m} \left\{ p^{(1)} + \dots + p^{(m)} \right\} .$$

We shall try to get a hint at the type of formula required. Let us first take three ellipsoids:

$$\mathcal{E}_1 = \mathcal{E}(0, Q_1),\ \mathcal{E}_2 = \mathcal{E}(0, Q_2),\ \mathcal{E}_3 = \mathcal{E}(0, Q_3) .$$

Applying formula (2.2.1) first to $\mathcal{E}_1 + \mathcal{E}_2$, one comes to

$$\mathcal{E}_1 + \mathcal{E}_2 \subseteq \mathcal{E}(p[2]) = \mathcal{E}(0, Q(p[2])) ,$$

where

$$Q(p[2]) = (p_1 + p_2)(p_1^{-1} Q_1 + p_2^{-1} Q_2) ,$$

and parameter $p = p[2]$ of (2.2.1) is presented in the form $p = p_1/p_2$, $p_1 > 0$, $p_2 > 0$. Applying (2.2.1) once more (now to $\mathcal{E}(p[2])$ and $\mathcal{E}_3$), one obtains

$$\mathcal{E}(p[3]) = \mathcal{E}(0, Q(p[3])) ,$$

where

$$Q(p[3]) = (1 + p^{-1}) Q(p[2]) + (1 + p) Q_3 ,$$

with parameter $p > 0$ taken as

$$p = \frac{p_1 + p_2}{p_3} \quad p_3 > 0 .$$

This gives

(2.7.2) $$Q(p[3]) = (p_1 + p_2 + p_3)(p_1^{-1} Q_1 + p_2^{-1} Q_2 + p_3^{-1} Q_3) .$$

Now the general assertion is as follows:

Theorem 2.7.1 *The external estimate*

(2.7.3) $$\mathcal{E}(p[m])$$

of the Minkowski sum $\mathcal{S}[m] = \sum_{i=1}^{m} \mathcal{E}_i$ *of* m *nondegenerate ellipsoids* $\mathcal{E}_i = \mathcal{E}(q_i, Q_i)$ *is given by*

$$\mathcal{E}(p[m]) = \mathcal{E}(q^*[m], Q(p[m])) \tag{2.7.4}$$

where

$$q^*[m] = \sum_{i=1}^{m} q_i \tag{2.7.5}$$

and

$$Q(p[m]) = \sum_{i=1}^{m} p_i \cdot \left(\sum_{i=1}^{m} p_i^{-1} Q_i \right) \tag{2.7.6}$$

for each set of $p_i > 0$, $i = 1, \ldots, m$.

Proof. The proof is given by induction. For $m = 2$ the statement clearly follows from (2.2.1). Assuming (2.7.4) - (2.7.6) to be true for given m and applying (2.2.1) to

$$\mathcal{E}(p[m]) + \mathcal{E}_{m+1}$$

one comes to

$$q^*[m+1] = q^*[m] + q_{m+1}, \tag{2.7.7}$$

$$Q(p[m+1]) = (1 + p^{-1})Q[m] + (1 + p)Q_{m+1} \ . \tag{2.7.8}$$

After taking $p > 0$ as

$$p = \frac{\sum_{i=1}^{m} p_i}{p_{m+1}}$$

this gives

$$Q(p[m+1]) = \left(\sum_{i=1}^{m+1} p_i \right) \sum_{i=1}^{m+1} p_i^{-1} Q_i \ .$$

Q.E.D.

In the form of recurrence relations, one has

$$Q(p[k+1]) = (1 + p_{k+1} p^{-1}[k])Q[k] + (1 + p_{k+1}^{-1} p[k])Q_{k+1}, \tag{2.7.9}$$

$$p[k+1] = p[k] + p_{k+1}, \quad p_k > 0, \ k = 1, \ldots, m \ . \tag{2.7.10}$$

Direct calculations yield the following

Lemma 2.7.1 *If the parameter* $p[m] = \{p_1, \ldots, p_m\}$ *of (2.7.10) is selected as*

$$p_i = (Q_i\ell^*, \ell^*)^{\frac{1}{2}}, \quad i = 1, \ldots, m \tag{2.7.11}$$

with $\ell \in \mathbb{R}^n$, $(\ell^*, \ell^*) = 1$ *fixed, then*

$$\rho(\ell^* | \mathcal{E}(q^*[m], Q(p[m])) = \rho(\ell^* | \mathcal{S}[m]) \ . \tag{2.7.12}$$

Formula (2.7.12) implies

Lemma 2.7.2 *The following relation is true*

$$\mathcal{S}[m] = \bigcap \{ \mathcal{E}(q^*[m], Q(p[m])) \,|\, p[m] \in \mathbb{R}^m \} \ . \tag{2.7.13}$$

As in the case of two ellipsoids, the finite sum $\mathcal{S}[m]$ may be presented as an intersection of ellipsoids, which now belong to the parametrized variety $\mathcal{E}(q^*[m], Q(p[m])$.

Although the equality (2.7.13) is true, this does not mean that the variety $\mathcal{E}(q^*[m], Q(p[m])$ contains all the inclusion-minimal ellipsoids circumscribed around $\mathcal{S}[m]$.

The following example illustrates that in the case of adding three (or in general more than two) ellipsoids, the family

$$\{\mathcal{E}(0, Q[p_1, p_2, p_3]) \ : \ p_i > 0, \ i = 1, 2, 3 \}, \quad Q[p_1, p_2, p_3] = Q(p[3]) \ ,$$

does not contain the covering ellipsoid of minimal volume.

Example 2.7.1

Consider the segments $\mathcal{F}_i = [A_i, B_i] \subset \mathbb{R}^2$, $i = 1, 2, 3$ *where*

$$\begin{aligned} A_1 &= (-1, 0), & B_1 &= (1, 0), \\ A_2 &= \left(-\frac{1}{2}, \frac{\sqrt{3}}{2}\right), & B_2 &= \left(\frac{1}{2}, -\frac{\sqrt{3}}{2}\right), \\ A_3 &= \left(\frac{1}{2}, \frac{\sqrt{3}}{2}\right), & B_3 &= \left(-\frac{1}{2}, -\frac{\sqrt{3}}{2}\right) \ . \end{aligned} \tag{2.7.14}$$

The Minkowski sum

$$\mathcal{F} = \sum_{i=1}^{3} \mathcal{F}_i$$

is the regular hexagon, that is covered by the ball of radius 2 around the origin, $\mathcal{S}(0,2) \subset \mathbb{R}^2$, *with*

$$\frac{1}{\pi^2} \cdot \mathrm{Vol}^2(\mathrm{S}(0,2)) = 16 \ .$$

On the other hand

$$\min\left\{\frac{1}{\pi^2} \cdot \mathrm{Vol}^2 \mathcal{E}(0, Q[p_1,p_2,p_3]) \ : \ p_i > 0, \ i = 1,2,3\right\} = \frac{81}{4} \ .$$

Proof. We have

$$\mathcal{F}_i = \mathcal{E}(0, Q_i) \qquad i = 1,2,3$$

with

$$Q_1 = \begin{bmatrix} 1 & 0 \\ 0 & 0 \end{bmatrix} ,$$

$$Q_2 = \begin{bmatrix} \frac{1}{4} & -\frac{\sqrt{3}}{4} \\ -\frac{\sqrt{3}}{4} & \frac{3}{4} \end{bmatrix} ,$$

$$Q_3 = \begin{bmatrix} \frac{1}{4} & \frac{\sqrt{3}}{4} \\ \frac{\sqrt{3}}{4} & \frac{3}{4} \end{bmatrix} .$$

Consider the matrix

$$Q[p_1,p_2,p_3] = (p_1 + p_2 + p_3) \begin{bmatrix} \frac{1}{p_1} + \frac{1}{4p_2} + \frac{1}{4p_3} & \frac{\sqrt{3}}{4}\left(\frac{1}{p_2} - \frac{1}{p_3}\right) \\ \frac{\sqrt{3}}{4}\left(\frac{1}{p_2} - \frac{1}{p_3}\right) & \frac{3}{4}\left(\frac{1}{p_2} + \frac{1}{p_3}\right) \end{bmatrix} .$$

Calculating the determinant, we obtain:

$$\det(Q[p_1,p_2,p_3]) = \frac{1}{\pi^2} \cdot \mathrm{Vol}^2(\mathcal{E}(0, Q[p_1,p_2,p_3])) = \frac{3}{4} \cdot \left(\frac{p_1 + p_2 + p_3}{(p_1 p_2 p_3)^{\frac{1}{3}}}\right)^3 .$$

The well-known inequality between the arithmetic and geometric mean completes the proof. Q.E.D.

Exercise 2.7.1. Consider the variety

$$\mathbf{E}[m] = \mathcal{E}(q^*[m], Q(p[m]))$$

by vectors $p[m] > 0$. Select an optimal ellipsoid among those of the form $\mathcal{E} \in \mathbf{E}[m]$ relative to the criterion

$$\psi[Q(p[m])] = \min$$

where the function ψ is one of those given in Section 2.1.1.

A further step is to approximate set-valued integrals. Assume an ellipsoidal valued function

$$\mathcal{P}(t) = \mathcal{E}(q(t), Q(t)), \quad t \in [t_0, t_1]$$

with the functions $q : [t_0, t_1] \longrightarrow \mathbb{R}^n$, $Q : [t_0, t_1] \longrightarrow \mathcal{L}(\mathbb{R}^n, \mathbb{R}^n)$ continuous and $Q(t) > 0$ for all $t \in [t_0, t_1]$ given. What would be its set-valued integral

$$I[t_0, t_1] = \int_{t_0}^{t_1} \mathcal{E}(q(t), Q(t))dt \quad ?$$

Since the sum of a finite number of ellipsoids is not obliged to be an ellipsoid, this, obviously is all the more true for the integral of an ellipsoidal valued function $\mathcal{P}(\cdot)$.

With the functions $q(\cdot)$, $Q(\cdot)$ continuous, the integral $I[t_0, t_1]$ can be treated as a set-valued Riemann-integral with integral sums

$$I(\Sigma_N) = \sum_{i=1}^{N} \mathcal{E}(q(\tau_i), Q(\tau_i))\sigma_i \tag{2.7.15}$$

with

$$\Sigma_N = \{\tau_0 = t_0, \tau_i = \tau_{i-1} + \sigma_{i-1},\ \sigma_i = 0, i = 1, \ldots, N\}$$

and

$$\sigma(N) = \max\{\sigma_i : i = 1, \ldots, N\}$$

converging to $I[t_0, t_1]$ in the Hausdorff-metric h for any subdivision Σ_N

$$\lim_{\sigma_N \to 0\ N \to \infty} h(I(\Sigma_N), I[t_0, t_1]) = 0 \ . \tag{2.7.16}$$

In the sequel assume $\sigma_i = (t_1 - t_0)/N = \sigma(N)$ for $i = 0, \ldots, N-1$. Applying Theorem 2.7.1 to (2.7.15) we have

$$I(\Sigma_N) \subseteq \mathcal{E}(q^*(\Sigma_N), Q(\Sigma_N)) \ , \tag{2.7.17}$$

where

$$q^*(\Sigma_N) = \sum_{i=1}^{N} q(\tau_i)\sigma_i \ ,$$

and

$$Q(\Sigma_N) = \left(\sum_{i=1}^{N} p_i^*(N)\right) \sum_{i=1}^{N} \sigma_i^2 (p_i^*)^{-1}(N) Q(\tau_i) \ ,$$

with $p_i^* > 0$.

After substitution $p_i(N) = \sigma^{-1}(N) p_i^*(N)$ the last equality transforms into

$$Q(\Sigma_N) = \left(\sum_{i=1}^{N} p_i(N)\sigma(N)\right) \sum_{i=1}^{N} \sigma_i p_i^{-1}(N) Q(\tau_i) \ .$$

Assuming $p : [t_0, t_1] \longrightarrow \mathbb{R}$ to be a continuous function with positive values, taking

$$p_i(N) = p(t_0 + \sigma(N))$$

and having in view the continuity of Q we observe

(2.7.18)

$$\lim_{N\to\infty} Q(\Sigma_N) = \left(\int_{t_0}^{t_1} p(\tau)d\tau\right)\left(\int_{t_0}^{t_1} p^{-1}(\tau)Q(\tau)d\tau\right) = \mathcal{Q}^+(t_1|p(\cdot))$$

while

(2.7.19) $$\lim_{N\to\infty} q^*(\Sigma_N) = \int_{t_0}^{t_1} q(\tau)d\tau = q_{t_0}(t_1) \quad .$$

Making a limit transition in (2.7.17) in view of (2.7.16), (2.7.18), and (2.7.19), we arrive at the inclusion

(2.7.20) $$I[t_0, t_1] \subseteq \mathcal{E}(q_{t_0}(t_1), \mathcal{Q}^+(t_1|p(\cdot)))$$

whatever is the function $p(\cdot) > 0$. The last argument allows us to formulate

Theorem 2.7.2 *An external ellipsoidal estimate for the integral $I[t_0, t_1]$ is given by relation (2.7.20). Moreover, the following equality holds*

(2.7.21) $$I[t_0, t_1] = \bigcap \{\mathcal{E}(q_{t_0}(t_1), \mathcal{Q}^+(t_1|p(\cdot)))|p(\cdot) \in C_+[t_0, t_1]\}$$

where $C_+[t_0, t_1]$ denotes the open cone of continuous, positive valued functions over the interval $[t_0, t_1]$.

Equality (2.7.21) follows from propositions similar to Lemmas 2.7.1, 2.7.13, namely, from

Lemma 2.7.3 *If the function $p(\cdot) \in C_+[t_0, t_1]$ of (2.7.20) is selected as*

$$p(t) = (Q(t)\ell^*, \ell^*)^{\frac{1}{2}}, \quad t \in [t_0, t_1] \ ,$$

with $\ell^* \in \mathbb{R}^n$, $(\ell^*, \ell^*) = 1$, fixed, then the respective support function verify the equality:

$$\rho(\ell^* | I[t_0, t_1]) = \rho(\ell^* | \mathcal{E}(q_{t_0}(t_1), \mathcal{Q}^+(t_1 | p(\cdot)))) \ . \tag{2.7.22}$$

Proof. The proof follows from direct substitution. Q.E.D.

Let us finally indicate some differential relations for $q_{t_0}(t)$ and $\mathcal{Q}(t) = \mathcal{Q}^+(t|p(\cdot))$, taking $p(\cdot)$ to be fixed. Recalling (2.7) we have the representation

$$\mathcal{Q}^+(t) = \left(\int_{t_0}^t p(\tau)d\tau\right)\left(\int_{t_0}^t p^{-1}(\tau)Q(\tau)d\tau\right) \ ,$$

or, after differentiating both sides by t and introducing the notation

$$\pi(t) = p(t)\left(\int_{t_0}^t p(\tau)d\tau\right)^{-1} \ , \tag{2.7.23}$$

the differential equation

$$\dot{\mathcal{Q}}^+(t) = \pi(t)\mathcal{Q}^+(t) + \pi^{-1}(t)Q(t) \tag{2.7.24}$$

$$\mathcal{Q}^+(t_0) = 0 \tag{2.7.25}$$

complemented by

$$\dot{q_{t_0}}(t) = q(t) \tag{2.7.26}$$

$$q_{t_0}(t_0) = q_0 \ . \tag{2.7.27}$$

Exercise 2.7.2. Prove that for the sum

$$\mathcal{E}(q^0, Q^0) + \int_{t_0}^{t_1} \mathcal{E}(q(t), Q(t))dt \subseteq \mathcal{E}(q_{t_0}(t_1), \mathcal{Q}^+(t_1)) \tag{2.7.28}$$

the external ellipsoidal representation is still given by equations (2.7.24), and (2.7.26), the change appearing only in the initial conditions (2.7.25) and (2.7.27), so that Q^0 and q^0 have to be added on the respective right hand side.

Before ending this section, let us single out some individual external ellipsoids. We shall discuss two ways of selecting these. Integrating relation (2.7.4), in view of (2.7.25), we have

$$\mathcal{Q}^+(\tau) = \int_{t_0}^{\tau} \mathcal{F}(\pi(t), \mathcal{Q}^+(t), \mathcal{Q}(t))dt \ ,$$

where

$$\mathcal{F}(\pi(t), \mathcal{Q}^+(t), \mathcal{Q}(t)) = \pi(t)\mathcal{Q}^+(t) + \pi^{-1}(t)\mathcal{Q}(t) \ . \tag{2.7.29}$$

Let us now minimize the matrix $\mathcal{F}[t] = \mathcal{F}(\pi, \mathcal{Q}^+(t), \mathcal{Q}(t))$ over π (at each instant $t \in [t_0, \tau]$), taking, for example, the following local optimality criteria (see Section 2.1).

(a)

$$\psi[\mathcal{F}[t]] = \mathrm{tr}(\mathcal{F}[t]) \ ,$$

(b)

$$\psi[\mathcal{F}[t]] = \mathrm{tr}(\mathcal{F}^2[t]) \ ,$$

(c)

$$\psi[\mathcal{F}[t]] = \det \mathcal{F}[t] \ .$$

Through calculations similar to those of Section 2.5 one may observe that the respective optimalizers are

$$(a) \quad (\pi(t))^{-1} = \frac{\mathrm{tr}^{1/2}(\mathcal{Q}^+(t))}{\mathrm{tr}^{1/2}(\mathcal{Q}(t))} \ , \tag{2.7.30}$$

$$(b) \quad (\pi(t))^{-2}(t) = \frac{\mathrm{tr}^{1/2}((\mathcal{Q}^+(t))^2)}{\mathrm{tr}^{1/2}((\mathcal{Q}(t))^2)} \ , \tag{2.7.31}$$

$$(c) \quad (\pi(t))^{-1} = \frac{n^{1/2}}{\mathrm{tr}^{1/2}(\mathcal{Q}(t)(\mathcal{Q}^+(t))^{-1})} \ . \tag{2.7.32}$$

Summarizing these results, we come to

Lemma 2.7.4 *(a)The parameters $\mathcal{Q}^+(\tau)$ of the external ellipsoids*

$$\mathcal{E}(q_{t_0}(\tau, \mathcal{Q}^+(\tau)) = \mathcal{E}^+[\tau]$$

singled out through the local optimality criteria (a),(b),(c) taken for each $t \in [t_0, \tau]$, may be calculated due to equation (2.7.24), where the function $\pi(t), t \in [t_0, \tau]$ has to be selected due to equalities (2.7.30)–(2.7.32), respectively.

(b) Each of the ellipsoidal tubes $\mathcal{E}^+[t], t_0 \leq t \leq \tau$, generated by equations (2.7.24)–(2.7.27) is nondominated with respect to inclusion. (In the sense that for each t the respective set $\mathcal{E}^+[t]$ is an inclusion-minimal external ellipsoidal estimate of $I[t_0, t]$).

One may observe that in equation (2.7.24) *the functions $\pi(t)$ may treated as (positive-valued) controls.* The problem of selecting optimal ellipsoids may then be reduced to an open-loop *terminal control* problem, where the *nonlocal optimality criteria* to be minimized over $\pi(t), t \in [t_0, \tau]$, could be

(2.7.33) $$\mathrm{tr}\mathcal{Q}^+(\tau), \ \ \mathrm{tr}(\mathcal{Q}^+(\tau))^2, \ \ \det(\mathcal{Q}^+(\tau))$$

accordingly.[20]

Exercise 2.7.3. Compare the solutions of the optimal terminal control problem for system (2.7.24) with control $\pi(t)$, due to optimality criteria (2.7.33), with the solutions obtained due to local criteria (a),(b),(c), as specified in Lemma 2.7.4.

We shall also calculate the internal ellipsoidal approximations for finite sums of ellipsoids and for integrals of ellipsoidal-valued functions.

2.8 Finite Sums and Integrals: Internal Approximations

Consider again the sum

$$\mathcal{S}[m] = \sum_{i=1}^{m} \mathcal{E}_i$$

[20] One should be aware, in view of Example 2.7.1, that these criteria would be minimized only in the class of ellipsoids described by formula (2.7.24).

of m nondegenerate ellipsoids $\mathcal{E}_i = \mathcal{E}(q_i, Q_i)$. We shall introduce the internal ellipsoidal approximation of these, assuming again , without loss of generality, that $q_i = 0, i = 1, ..., m$.

Applying formula (2.4.2) to $\mathcal{E}_0, \mathcal{E}_1$, we have,

$$\mathcal{E}_0 \ + \ \mathcal{E}_1 \ = \mathcal{E}(0, Q_0) + \mathcal{E}(0, Q_1) \ \supseteq \mathcal{E}(0, Q[S_1]) \ ,$$

where $S[1] = S_1$ and

$$(2.8.1) \qquad Q(S[1]) = S_1^{-1}[(S_1 Q_0 S_1')^{\frac{1}{2}} \ + \ (S_1 Q_1 S_1')^{1/2}]^2 S'^{-1} \ .$$

Moreover , the representation of Theorem 2.4.1 yields

$$(2.8.2) \qquad \mathcal{E}(0, Q_0) \ + \ \mathcal{E}(0, Q_1) = \cup\{\mathcal{E}(0, Q(S[1])) | S[1] \in \Sigma\} \ .$$

Continuing this procedure, we have, due to the same representations

$$(2.8.3) \qquad \mathcal{S}[2] = \mathcal{E}_0 + \mathcal{E}_1 + \mathcal{E}_2 \supseteq \mathcal{E}(0, Q(S[2])) \ ,$$

$$\forall S[2] = \{S_1, S_2\}, \ S_i \in \Sigma \ ,$$

where

$$Q(S[2]) = S_2^{-1}[(S_2 Q(S[1]) S_2')^{\frac{1}{2}} + (S_2 Q_2 S_2')^{\frac{1}{2}}]^2 S_2'^{-1} \ .$$

Further on, assuming that the last relations are true for $S[m-1]$, we have,
(2.8.4)

$$\mathcal{S}[m] = \sum_{i=1}^{m} \mathcal{E}_i \supseteq \mathcal{E}(0, Q(S[m-1])) \ + \ \mathcal{E}(0, Q_m) \supseteq \mathcal{E}(0, Q(S[m])) \ ,$$

$$\forall S[m] = \{S_1, ..., S_m\} \ ,$$

where

$$(2.8.5) Q(S[k]) = S_k^{-1}[(S_k Q(S[k-1]) S_k')^{\frac{1}{2}} \ + \ (S_k Q_k S_k')^{\frac{1}{2}}]^2 S_k'^{-1} \ ,$$

$$S[k] = \{S_1, ..., S_k\}, \ Q(S[0]) = Q_0 \ .$$

Applying the representation of Theorem 2.4.1 to (2.8.3), we come to

$$\mathcal{E}_0 + \mathcal{E}_1 + \mathcal{E}_2 \ = \ \cup\{\mathcal{E}(0, Q(S[1])) | S_1\} \ + \ \mathcal{E}(0, Q_2) \ ,$$

$$\mathcal{E}(0, Q(S[1])) + \mathcal{E}(0, Q_2) = \cup\{\mathcal{E}(0, Q(S[2]))|S_2\} ,$$

which gives

$$\mathcal{S}[2] = \cup\cup\{\mathcal{E}(0, Q(S[2]))|S_1, S_2\} .$$

Similarly, by induction,

(2.8.6) $$\mathcal{S}[m] = \cup\{\mathcal{E}(0, Q(S[m-1]))|S[m-1]\} + \mathcal{E}(0, Q_m) =$$

$$= \cup\{\mathcal{E}(0, Q(S[m]))|S[m]\} .$$

Concluding the discussion, we are now able to formulate

Theorem 2.8.1 *The internal ellipsoidal estimate*

$$\mathcal{E}^-[m] \subseteq \sum_{i=0}^{m} \mathcal{E}(0, Q_i) = \mathcal{S}[m] ,$$

for the sum $\mathcal{S}[m]$ *of* $m+1$ *nondegenerate ellipsoids* $\mathcal{E}(0, Q_i)$ *is given by the inclusion (2.8.4) with exact representation (2.8.6), where the union is taken over all the sequences* $S[m]$ *of symmetrical matrices* $S_i \in \Sigma, i = 1, ..., m$.

The general case, with

$$\mathcal{S}[m] = \sum_{i=0}^{m} \mathcal{E}(q_i, Q_i) ,$$

is treated similarly. This allows

Corollary 2.8.1 *The inclusion*

(2.8.7) $$\mathcal{S}[m] \supseteq \mathcal{E}^-[m] = \mathcal{E}(q[m], Q(S[m])) ,$$

$$q[m] = \sum_{i=1}^{m} q_i ,$$

holds for any sequence $S[m]$. *The following representation is true*

(2.8.8) $$\mathcal{S}[m] = \cup\{\mathcal{E}(q[m], Q[m])|S[m]\} .$$

Remark 2.8.1 *The last assertions were proved for the sum $\mathcal{S}[m]$ of $m+1$ nondegenerate ellipsoids $\mathcal{E}_i, i = 0, ..., m$. The basic relations turn out to be also true if these are degenerate. However, the union in the right-hand side of (2.8.8) has to be substituted by its closure.*

Exercise 2.8.1. Prove the assertion of the previous remark.

Let us now pass to the internal approximation of the set-valued integral

$$I[t_0, t_1] = \int_{t_0}^{t_1} \mathcal{E}(q(t), Q(t))dt \ .$$

Its Riemannean integral sum is the one given in (2.7.15) with convergence property (2.7.16). Applying Theorem 2.8.1 to (2.7.15), we observe

$$(2.8.9) \quad I(\Sigma_k) \supseteq \mathcal{E}(\sum_{i=1}^{k-1} \sigma_i q_i, Q_\sigma(S[k-1])) + \mathcal{E}(\sigma_k q_k, \sigma^2 Q_k) \supseteq$$

$$\supseteq \mathcal{E}(\sum_{i=1}^{k} \sigma_i q_i, Q_\sigma(S[k])) \ ,$$

and

$$(2.8.10) \qquad I(\Sigma_k) = \bigcup\{\mathcal{E}(\sum_{i=1}^{k} \sigma_i q_i, Q_\sigma(S[k]))|S[k]\} \ ,$$

where $S[k]$ is such that $S_i \in \Sigma, i = 1, ..., k$, and

$$Q_\sigma(S[k]) = S_k^{-1}[(S_k Q_\sigma(S[k-1])S_k')^{\frac{1}{2}} + \sigma_k(S_k Q_k S_k')^{\frac{1}{2}}]^2 S_k'^{-1} \ .$$

The last relations are equivalent to

$$Q_\sigma(S[k]) - Q_\sigma(S[k-1]) =$$

$$= \sigma_k S_k^{-1}((S_k Q_\sigma(S[k-1])S_k')^{\frac{1}{2}}(S_k Q_k S_k')^{\frac{1}{2}} +$$

$$+ (S_k Q_k S_k')^{\frac{1}{2}}(S_k Q_\sigma(S[k-1])S_k')^{\frac{1}{2}})S_k^{-1} + \sigma_k^2 Q_k \ .$$

Denoting

$$\tau_k = t, \Sigma_k = \{\tau_0 = t_0, \tau_i = \tau_{i-1} + \sigma_{i-1}, \ \sigma_i > 0, \ i = 1, ..., k\}$$

and

$$S[k] = S[t], \ S[i] = S[\tau_i] = \{S(\tau_j); j = 0, ..., i\}, S(\tau_j) = S_j \ ,$$

$$q(i) = q(\tau_i), q_\sigma(t) \; = \; \sum_{i=1}^{k} q(i)\sigma_i$$

we observe, that the previous relations may be rewritten as

(2.8.11)

$$Q_\sigma(S[t]) \; - \; Q_\sigma(S[t-\sigma]) \; =$$
$$= \; \sigma_k S^{-1}[t]((S[t]Q_\sigma(S[t-\sigma_k])S'[t])^{\frac{1}{2}}(S[t]Q(t)S'[t])^{\frac{1}{2}} \; +$$
$$+ \; (S[t]Q(t)S'[t])^{\frac{1}{2}}(S[t]Q_\sigma(S[t-\sigma_k])S'[t])^{\frac{1}{2}})S^{-1}[t] \; + \; \sigma_k^2 Q(t) \quad ,$$

(2.8.12) $$q_\sigma(t) \; - \; q_\sigma(t-\sigma_k) \; = \; q(t)\sigma_k \quad .$$

Let us assume that the values $S[\tau_i]$ in the above are generated by a measurable, matrix-valued function $S[\tau], \tau \in [t_0, t]$ with values in Σ.

Passing to the limit in (2.8.11), (2.8.12) with

(2.8.13) $$\max\{\sigma_i, i = 1, ..., k\} \to 0, \; k \to \infty \quad ,$$

(for an arbitrary t) and denoting

$$\lim q_\sigma(t) = q_0(t), \; \lim Q_\sigma(S[t]) \; = \; \mathcal{Q}^-(t)$$

we arrive at the differential equations

(2.8.14) $$d\mathcal{Q}^-(t)/dt \; = \; S^{-1}[t]((S[t]\mathcal{Q}^-(t)S'[t])^{\frac{1}{2}}(S[t]Q(t)S'[t])^{\frac{1}{2}} +$$
$$+ \; (S[t]Q(t)S'[t])^{\frac{1}{2}}(S[t]\mathcal{Q}^-(t)S'[t])^{\frac{1}{2}})S'^{-1}[t] \quad ,$$

(2.8.15) $$dq_0(t)/dt \; = \; q(t),$$
$$\mathcal{Q}^-(t_0) = 0 \; , \; q_0(t_0) \; = \; 0 \quad .$$

The inclusion (2.8.9) and the relation

(2.8.16) $$\lim_{k\to\infty} I[\Sigma_k] \; = \; I[t_0, t]$$

imply

Lemma 2.8.1 *The inclusion*

(2.8.17) $$I[t_0, t] \; \supseteq \mathcal{E}(q_0(t), \mathcal{Q}^-(t)) \quad ,$$

is true, whatever is the measurable function $S[t]$ *with values in* Σ.

Further, since (2.8.10) is true for any value of k and since (2.8.16) is true with (2.8.13), the limit transition in (2.8.10), (2.8.16) yields

$$I[t_0, t] = \overline{\{\cup\{\mathcal{E}(q_0(t), Q^-(t)) | S[t]\}\}} \tag{2.8.18}$$

over all measurable functions $S[\cdot]$ of the type considered above. The result may be summarized in

Theorem 2.8.2 *The integral $I[t_0, t]$ allows an internal approximation (2.8.17) where $q_0(t), Q^-(t)$ satisfy the differential equations (2.8.14),(2.8.15) with zero initial conditions.*

The representation (2.8.18) is true, where the union is taken over all measurable functions $S[t]$ with values in Σ.

An obvious consequence of this Theorem is

Corollary 2.8.2 *The sum*

$$\mathcal{E}(q^0, Q^0) + \int_{t_0}^{t_1} \mathcal{E}(q(t), Q(t))dt = I[t_0, t_1]$$

allows an internal approximation (2.8.17) and a representation of type (2.8.18), where $q_0(t), Q(t)$ are the solutions to the differential equations (2.8.14), (2.8.15) with initial conditions

$$q_0(t_0) = q^0, \;\; Q(t_0) = Q^0 \; . \tag{2.8.19}$$

We finally offer the reader to formulate and solve a problem similar to Exercise 2.7.3, but taken for internal ellipsoids.

This section finalizes Part II. We shall now apply the results of this part to the problems of Part I.

Part III. ELLIPSOIDAL DYNAMICS: EVOLUTION and CONTROL SYNTHESIS

Introduction

In this part we apply the calculus of Part II to the problems of Part I. We start from systems with no uncertainty, constructing external and internal ellipsoidal- valued approximations of the attainability (reachability) domains and tubes. In order to achieve these results we introduce two corresponding types of evolution funnel equations with *ellipsoidal-valued solutions.* Each of these evolution equations generates a respective variety of ellipsoidal-valued tubes that approximate the original attainability tubes externally or internally and finally yield, through their intersections or unions, an exact representation of the approximated tube. This result is similar to those achieved for static situations in Sections 2.2–2.4, but is now given for a dynamic problem (Sections 3.2 and 3.3). The main point, however, is that the time-varying coefficients of the approximating ellipsoidal tubes are further described *through ordinary differential equations* with right-hand sides depending on parameters. The same result is given in backward time (Section 3.4). This gives us the internal approximations for synthesizing the control strategies in the target control problem. It is shown that the scheme of Section 1.4 remains true except that the *the solvability tube of Definition 1.4.3 is substituted for its internal ellipsoidal approximation*, and the control strategy is constructed accordingly (Section 3.6).

The specific advantage of such solutions is that the strategies are given (relative to the solution of a simple algebraic equation) in the form of an *analytical design.*

One should realize, however, that attainability domains for linear systems are among the relatively simpler constructions in control theory. The problem is substantially more difficult if the system is under the action of uncertain (unknown but bounded) inputs. The approximation of the domains of *attainability under counteraction* or of the *solvability domains for uncertain systems* requires, in its general setting, the incorporation of both internal and external approximations of sums or geometrical (Minkowski) differences of ellipsoids. The external and internal ellipsoidal approximations of the solvability tubes for uncertain systems are derived in Section 3.5 (under conventional nondegeneracy conditions). The important point is that these ellipsoidal approximations that reflect *the evolution dynamics of uncertain or conflict-control systems* are again described through the solutions of *ordinary differential equations.* Once the internal approximation of the solvability tubes are known, it is again possible (now following the schemes of Section 1.8), to implement an ellipsoidal control synthesis in the form of an analytical design (relative to the solution of an algebraic equation). Moreover, the ellipsoidal solvability tubes constructed here are such that they retain the property of being "Krasovski bridges". Namely, once the starting position is in a specific internal ellipsoidal solvability tube, there exists an analytical control design that keeps the trajectory within this tube despite the unknown disturbances.

We should emphasize the key elements that allow us to use the ellipsoidal tubes introduced here for designing synthesizing control strategies (both with and without uncertainty). These are, first, that the approximating (internal) ellipsoidal tubes are *nondominated* with respect to inclusion, their crossections being *inclusion-maximal* at each instant of time and – second – that the respective ellipsoidal-valued mappings satisfy *a semi-group property* which we call the lower and upper semigroup property – for internal and external tubes accordingly. It is these two elements that allow the internal ellipsoidal approximations to retain the property of being bridges, specifically, to be the ellipsoidal-valued bridges.

The techniques of this part are illustrated in Sections 3.7 and 3.9, where one may observe some examples on solvability tubes and ellipsoidal control synthesis for 4-dimensional systems animated through *computer windows.*

3.1 Ellipsoidal-Valued Constraints

Let us again consider system (1.1.1) and pass to its transformed version (1.14.1), where $A(t) \equiv 0$. Namely, taking

$$\dot{x} = u + f(t) \tag{3.1.1}$$

$$x(t_0) = x^0, \;\; t_0 \leq t \leq t_1 \;\; ,$$

we shall further presume the constraints on u, f, x^0 to be ellipsoidal-valued:

(3.1.2) $$(u - p(t), P^{-1}(t)(u - p(t)) \leq 1 \;\; ,$$

(3.1.3) $$(f - q(t), Q^{-1}(t)(f - q(t)) \leq 1 \;\; ,$$

(3.1.4) $$(x^0 - x^*, X_0^{-1}(x^0 - x^*)) \leq 1 \;\; ,$$

where the continuous functions $p(t), q(t)$ and the vector x^* are given together with continuous matrix functions $P(t) > 0, Q(t) > 0$ and matrix $X_0 > 0$.

In terms of inclusions we have,

(3.1.5) $$u \in \mathcal{E}(p(t), P(t)) \;\; ,$$

(3.1.6) $$f \in \mathcal{E}(q(t), Q(t)) \;\; ,$$

(3.1.7) $$x^0 \in \mathcal{E}(x^*, X_0)$$

or, in terms of support functions, the inequalities

(3.1.8) $$(l, u) \leq (l, p(t)) + (l, P(t)l)^{\frac{1}{2}} \;\; ,$$

(3.1.9) $$(l, f) \leq (l, q(t)) + (l, Q(t)l)^{\frac{1}{2}} \;\; ,$$

(3.1.10) $$(l, x^0) \leq (l, x^*) + (l, X_0 l)^{\frac{1}{2}} \;\; ,$$

$$\forall l \in \mathbb{R}^n \;\; .$$

With $f(t)$ given, the attainability domain $\mathcal{X}[t] = \mathcal{X}(t, t_0, \mathcal{E}(x^*, X_0))$ is defined by the set-valued integral (1.3.1), which is now

(3.1.11) $$\mathcal{X}[t] = \mathcal{E}(x^*, X_0) + \int_{t_0}^{t} \mathcal{E}(p(t) + f(t), P(t))dt \;\; .$$

With $f(t)$ continuous, the set-valued function $\mathcal{X}[t]$ satisfies the evolution equation, based on (1.3.3) or (1.14.6)

$$(3.1.12)\ \lim_{\sigma\to 0} \sigma^{-1} h(\mathcal{X}[t+\sigma], \mathcal{X}[t] + \sigma\mathcal{E}(p(t)+f(t), P(t))) = 0 \ ,$$

with boundary condition

$$\mathcal{X}[t_0] = \mathcal{E}(x^*, X_0)$$

for the *attainability tube* $\mathcal{X}[\cdot]$. On the other hand, with terminal set $\mathcal{M}$ being an ellipsoid,

$$(3.1.13)\ \ \mathcal{M} = \mathcal{E}(m, M),\ m \in \mathbb{R}^n, M \in \mathcal{L}(\mathbb{R}^n, \mathbb{R}^n), M > 0 \ ,$$

we have an evolution equation

$$(3.1.14) \lim_{\sigma\to 0} \sigma^{-1} h(W[t-\sigma], W[t] - \sigma\mathcal{E}(p(t)+f(t), P(t))) = 0 \ ,$$

$$W[t_1] = \mathcal{E}(m, M) \ ,$$

for the *solvability tube.*

Passing to an *uncertain system* with $f(t)$ measurable, bounded by restriction (3.1.3), we come to the equation for the *solvability tube under uncertainty*, which is[21]

$$\lim_{\sigma\to 0} \sigma^{-1} h_+(W[t-\sigma] + \sigma\mathcal{E}(q(t), Q(t)), W[t] - \sigma\mathcal{E}(p(t), P(t))) = 0 \ ,$$

$$(3.1.15) \qquad W[t_1] = \mathcal{E}(m, M) \ .$$

After the introduction of an additional ellipsoidal-valued state constraint

$$(3.1.16) \qquad G(t)x(t) \in \mathcal{E}(y(t), K(t))$$

or

$$x(t) \in \mathcal{Y}(t) = \{x : G(t)x \in y(t) + \mathcal{E}(0, K(t))\},$$

[21] We recall that this equation was introduced under nondegeneracy Assumptions 1.7.1 or 1.7.2 which imply that the tube $W[t], t \in [t_0, t_1]$ contains an ***internal tube*** of type $\beta(t)\mathcal{S} + \tilde{x}(t), \beta(t) > 0$.

with $K(t) \in \mathcal{L}(\mathbb{R}^k, \mathbb{R}^k), K'(t) = K(t) > 0, y(t) \in \mathbb{R}^k$, $y(t), K(t)$ continuous, the equation for the solvability tube *under state constraints* is as follows

$$\lim_{\sigma \to 0} \sigma^{-1} h_+(\mathcal{W}[t-\sigma], \mathcal{W}[t] \cap \mathcal{Y}(t) - \sigma\mathcal{E}(p(t), P(t) + f(t))) = 0$$

$$\mathcal{W}[t_1] = \mathcal{E}(m, M) \cap \mathcal{Y}(t) \ . \tag{3.1.17}$$

(The solvability tube is the *maximal solution* to (3.1.17).)

If the function $u(t)$ is given, and the constraint (3.1.16) is due to a measurement equation with observed values $y(t)$ (assuming $u(t), y(t)$ to be continuous), then the attainability domain $\mathcal{X}[t]$ for system (3.1.1), (3.1.6), (3.1.7), (3.1.16) is the corresponding *information domain* that satisfies the evolution equation

$$\lim_{\sigma \to 0} \sigma^{-1} h_+(\mathcal{X}[t+\sigma], \mathcal{X}[t] \cap \mathcal{Y}(t) + \tag{3.1.18}$$

$$+ \sigma\mathcal{E}(q(t) + u(t), Q(t))) = 0 \ ,$$

$$\mathcal{X}[t_0] = \mathcal{E}(x^*, X_0) \ .$$

($\mathcal{X}[t]$ is the *maximal solution* to this equation.) The set $\mathcal{X}[t]$ gives a *guaranteed estimate* of the state space vector $x(t)$ of system (3.1.1) ($u(t)$ given), under unknown but bounded disturbances $f(t) \in \mathcal{E}(q(t), Q(t))$, through the measurement of vector

$$y(t) \in G(t)x(t) + \mathcal{E}(0, K(t)) \ . \tag{3.1.19}$$

As we have observed in Part II, the sets $\mathcal{X}[t], \mathcal{W}[t]$ generated by the solutions to the evolution equations of this section, are not obliged to be ellipsoids. We shall therefore introduce external and internal ellipsoidal approximations of these within a scheme that would generalize the results of Part II, propagating them to continuous-time dynamic processes. Our further subject is therefore the one of *ellipsoidal-valued dynamics*. Following the sequence of topics of Part I, we start from the simplest attainability problem.

3.2 Attainability Sets and Attainability Tubes: The External and Internal Approximations

Our first subject is to consider the differential inclusion

$$\dot{x} \in \mathcal{E}(p(t), P(t)) + f(t), \quad t_0 \leq t \leq t_1 \ , \tag{3.2.1}$$

$$x(t_0) = x^0, \quad x^0 \in \mathcal{E}(x^*, X_0) \ ,$$

and to approximate its attainability domain $\mathcal{X}[t] = X(t, t_0, \mathcal{E}(x^*, X_0))$, where

$$\mathcal{X}[t] = \mathcal{E}(x^*, X_0) + \int_{t_0}^{t} \mathcal{E}(p(s) + f(s), P(s))ds \ . \tag{3.2.2}$$

The external ellipsoidal approximation for such a sum has been indicated in Section 2.7, particularly, through relations (2.7.20), (2.7.24), and (2.7.22). Applying these relations to the present situation and changing the notations to those of (3.2.2), we have

$$\mathcal{X}[t] \subseteq \mathcal{E}(x^*(t), X^+(t)) \ , \tag{3.2.3}$$

where

$$\dot{x}^*(t) = p(t) + f(t) \ , \tag{3.2.4}$$

$$\dot{X}^+(t) = \pi(t)X^+(t) + \pi^{-1}(t)P(t), \quad \pi(t) > 0 \ , \tag{3.2.5}$$

$$x^*(t_0) = x^*, X^+(t_0) = X_0 \ . \tag{3.2.6}$$

Here $X^+(t)$ actually depends on $\pi(\cdot)$, so that if necessary, we shall also use the notation

$$X^+(t) = X^+(t|\pi(\cdot)) \ .$$

It follows from Theorem 2.7.2 and the substitution (2.7.23) that the inclusion

$$\mathcal{X}[t] \subseteq \mathcal{E}(x^*(t), X^+(t|\pi(\cdot))) \tag{3.2.7}$$

is true, whatever is the function $\pi(t) > 0$ that allows representation (2.7.20), (2.7.23) with $\pi(t) > 0$. Moreover, the equality

$$\mathcal{X}[t] = \cap\{\mathcal{E}(x^*(t), X^+(t|\pi(\cdot))|\pi(\cdot)\} \tag{3.2.8}$$

is true if the intersection is taken over all the functions $\pi(\cdot)$ of the type indicated above. We leave it to the reader to observe that (3.2.7) remains true if the intersection is taken over all piece-wise continuous or even continuous functions $\pi(t) > 0$. This finally leads to the proof of the following assertion

Theorem 3.2.1 *The external ellipsoidal approximation to the attainability domain $\mathcal{X}[t] = \mathcal{X}(t, t_0, \mathcal{E}(x^*, X_0))$ of the differential inclusion (3.2.1) is given by the inclusion (3.2.7) with exact representation (3.2.8), where the intersection may be taken over all piecewise continuous (or even continuous) functions $\pi(t) > 0$.*

Let us now return to the last Theorem, approaching it through another scheme – the technique of funnel equations. Following Sections 1.4 and 3.1, we observe that the tube $\mathcal{X}[t]$ satisfies the funnel equation (3.1.12). This allows us to write

$$\mathcal{X}[t+\sigma] \subseteq \mathcal{X}[t] + \sigma\mathcal{E}(p(t), P(t)) + o(\sigma)\mathcal{S} ,$$

where $\sigma^{-1}o(\sigma) \to 0$ if $\sigma \to 0$, and $\mathcal{S}$ is a unit ball in $\mathbb{R}^n$, as before.

With $\mathcal{X}[t]$ being an ellipsoid of type $\mathcal{X}[t] = \mathcal{E}(x^*(t), X^+(t))$, we may apply the expansion (2.5.6), so that the external approximation to $\mathcal{X}[t+\sigma]$ would be

$$\mathcal{X}[t+\sigma] \subseteq \mathcal{E}(x^*(t+\sigma), X^+(t+\sigma)) , \tag{3.2.9}$$

where

$$X^+(t+\sigma) = X^+(t) + \sigma\pi^{-1}(t)X^+(t) + \sigma\pi(t)P(t) + \sigma^2 P(t) , \tag{3.2.10}$$

with $\pi(t) > 0$ continuous. Relations (3.2.9) , (3.2.10) are true for any $\sigma > 0$ and any $\pi(t)$ of the indicated type.

Dividing the interval $[t_0, t]$ into subintervals with subdivision

$$\sum = \{\sigma_1, ..., \sigma_s\} ,$$

$$\tau_0 = t_0,\ \tau_s = t_0 + \sum_{i=0}^{s} \sigma_i,\ t = \tau_s ,$$

where

$$\sigma_i > 0,\ \sum_{i=0}^{s} \sigma_i = t - t_0 ,$$

we have:

$$\mathcal{X}(\tau_1) = \mathcal{E}(x_*, X^+(t_0)) + \sigma_1\mathcal{E}(p(t_0) + f(t_0), P(t_0)) \subseteq$$

$$\subseteq \mathcal{E}(x^*(\tau_1), X^+(\tau_1)) = \mathcal{E}^+[\tau_1] ,$$

where

(3.2.11) $$x^*(\tau_1) = x^* + \sigma_1(p(t_0) + f(t_0))$$

(3.2.12) $$X^+(\tau_1) = (1 + \sigma_1\pi^{-1}(t_0))X^+(t_0) + \sigma_1\pi(t_0)P(t_0) + \sigma_1^2 P(t_0) .$$

We further have:

$$\mathcal{X}^+(\tau_k) \subseteq \mathcal{E}(x^*(\tau_{k-1}), X^+(\tau_{k-1})) + \sigma_k\mathcal{E}(p(\tau_{k-1}) + f(\tau_{k-1}), P(\tau_{k-1})) \subseteq$$

$$\subseteq \mathcal{E}(x^*(\tau_k), X^+(\tau_k)), \quad (k = 1, ..., s) ,$$

where

(3.2.13) $$x^*(\tau_k) = x^*(\tau_{k-1}) + \sigma_k(p(\tau_{k-1}) + f(\tau_{k-1})) ,$$

(3.2.14) $$X^+(\tau_k) = (1 + \sigma_k\pi^{-1}(\tau_{k-1}))X^+(\tau_{k-1}) + \sigma_k\pi(\tau_{k-1})P(\tau_{k-1}) .$$

Dividing relations (3.2.13), (3.2.14) by σ_k and passing to the limit, with

$$\max\{\sigma_k | k = 1, ..., s\} \to 0 , \; s \to \infty,$$

and t being fixed as the end-point of the interval $[t_0, t)$, whatever is the subdivision $\sum$ and the integer s, we again come to equations (3.2.4) and (3.2.5) with initial condition (3.2.6). This gives an alternative proof for the relation (3.2.7) of Theorem 3.2.1.

Let us now assume $A(t) \not\equiv 0$. Then Theorem 3.2.1 transforms into

Corollary 3.2.1 *For every* $t \in [t_0, t_1]$ *the following equality is fulfilled*

$$\mathcal{X}(t) = \cap\{\mathcal{E}(x(t), X^+(t|\pi(\cdot)))|\pi(\cdot)\} ,$$

where $X^+(t) = X^+(t|\pi(\cdot))$, *are the solutions of the following differential equations*

$$\dot{x} = A(t)x + p(t); x(t_0) = x^* ,$$

$$\dot{X}^+ = A(t)X^+ + X^+A'(t) + \pi^{-1}(t)X^+ + \pi(t)P(t); \quad X^+(t_0) = X_0 .$$

We shall now indicate that with a certain modification this result remains true for the special case when $\mathcal{E}(p(t), P(t))$ is *a product of ellipsoids* and therefore does not fall under the requirements of Section 3.1. Let us start from a generalization of Lemma 2.2.1.

Lemma 3.2.1 *Suppose* $E_1 = \mathcal{E}(q_1; Q_1), E_2 = \mathcal{E}(q_2; Q_2)$ *where*

$$Q_1 = \begin{pmatrix} A_1 & 0 \\ 0 & 0 \end{pmatrix}, \qquad Q_2 = \begin{pmatrix} 0 & 0 \\ 0 & A_2 \end{pmatrix} ,$$

$A_1(A_2)$ *is a symmetric positively defined* $m \times m$*- (respectively,* $k \times k-$*) matrix,* $m + k = n$*. Then*

$$E_1 + E_2 = \cap\{\mathcal{E}(q_1 + q_2, Q(p)) | p > 0\} ,$$

$$Q(p) = \begin{pmatrix} (1+p^{-1})A_1 & 0 \\ 0 & (1+p)A_2 \end{pmatrix} .$$

Proof. The upper estimate

$$E_1 + E_2 \subseteq \mathcal{E}(q_1 + q_2; Q(p)), \ \ p > 0 ,$$

can be obtained along the lines of the proof of Lemma 2.3.1.

Consider now an arbitrary vector $v = \{l, b\} \in \mathbb{R}^n$, $l \in \mathbb{R}^m$, $b \in \mathbb{R}^k$ such that $l \neq 0$, $b \neq 0$. It is not difficult to demonstrate that

$$\rho(v|E_1 + E_2) = v'(q_1 + q_2) + (l'A_1 l)^{\frac{1}{2}} + (b'A_2 b)^{\frac{1}{2}} =$$

$$= v'(q_1 + q_2) + (v'Q(p)v)^{\frac{1}{2}}$$

for $p = (l'A_1 l)^{\frac{1}{2}}/(b'A_2 b)^{\frac{1}{2}}$.

This yields

$$\rho(v|E_1 + E_2) = \rho(v|\mathcal{E}(q_1 + q_2, Q(p))) ,$$

for every direction $v = \{l, b\}$ with $l \neq 0$, $b \neq 0$. From the continuity of the support functions of the convex compact sets $E_1 + E_2$ and of the set $\cap\{\mathcal{E}(q_1 + q_2, Q(p)) | p > 0\}$ we conclude that equality

$$\rho(v|E_1 + E_2) = \rho(v| \cap \{\mathcal{E}(q_1 + q_2, Q(p)) | p > 0\})$$

is true for all $v \in \mathbb{R}^n$. The last relation implies the assertion of this Lemma.

Q.E.D.

Denote symbol $\Pi^+[t_0, t_1]$ to stand for the set of all continuous positive functions from $[t_0, t_1]$ to $\mathbb{R}$.

Combining the last Lemma with Corollary 3.2.1, we come to the conclusion

Corollary 3.2.2 *Consider the differential inclusion*

$$\dot{x} \in A(t)x + \mathcal{P}(t) \quad ,$$

$$x(t_0) \in \mathcal{E}(x^*, X_0), t_0 \leq t \leq t_1 \quad ,$$

with

$$\mathcal{P}(t) = \mathcal{E}_k(s(t), S(t)) \times \mathcal{E}_m(q(t), Q(t)) \quad ,$$

where $\mathcal{E}_k(s(t), S(t)) \subset \mathbb{R}^k, \mathcal{E}_m(q(t), Q(t)) \subset \mathbb{R}^m, k + m = n.$

For every $t \in [t_0, t_1]$ *the following equality is true*

$$X(t) = \cap\{\mathcal{E}(z(t), Z(t|\pi(\cdot), \chi(\cdot))|\pi(\cdot), \chi(\cdot)\}$$

where $\{\pi(\cdot), \chi(\cdot)\} \in \Pi^+[t_0, t_1]$, *and*

$$z : [t_0, t_1] \to \mathbb{R}^n, Z : [t_0, t_1] \to \mathcal{L}(\mathbb{R}^n, \mathbb{R}^n)$$

are the solutions to differential equations

$$\dot{z} = Az + v(t), \quad v(t) = \{s(t), q(t)\}, \quad z(t_0) = x^* \quad .$$

$$\dot{Z} = A(t)Z + ZA'(t) + \chi^{-1}(t)Z + \chi(t)Q[t] \quad ,$$

$$Q[t] = Q(t, \pi(t)) = \begin{pmatrix} (1 + \pi^{-1}(t))S(t) & 0 \\ 0 & (1 + \pi(t))Q(t) \end{pmatrix}$$

and $Z(t_0) = X_0$.

In order to deal with internal approximations, we will follow the last scheme, dealing now with funnel equation (3.1.12). This time we have

$$(3.2.15) \qquad \mathcal{X}[t] + \sigma\mathcal{E}(p(t), P(t)) \subseteq \mathcal{X}[t+\sigma] + o_1(\sigma)\mathbf{S} \ ,$$

where

$$\sigma^{-1}o_1(\sigma) \to 0 \ , \sigma \to 0.$$

With $\mathcal{X}[t]$ being an ellipsoid of type

$$\mathcal{X}[t] = \mathcal{E}(x^*(t), X^-(t)) \ ,$$

we may apply formula (2.3.3) with (2.3.1). Changing the respective notations, namely, taking

$$Q_1 = X^-(t) \ , \ Q_2 = \sigma^2 P(t) \ , \ S = H(t) \ ,$$

we have, in view of (3.2.15)

$$\mathcal{X}[t+\sigma] \supseteq \mathcal{E}(x^*(t+\sigma), X^-(t+\sigma)) \ ,$$

where

$$\begin{aligned} X^-(t+\sigma) = \ & H^{-1}(t)[H(t)X^-(t)H'(t) + \\ & + \sigma(H(t)X^-(t)H'(t))^{\frac{1}{2}}(H(t)P(t)H'(t))^{\frac{1}{2}} + \\ & + \sigma(H(t)P(t)H'(t))^{\frac{1}{2}}(H(t)X^-(t)H'(t))^{\frac{1}{2}} + \\ & + \sigma^2 H(t)P(t)H'(t)]H'^{-1}(t) \ , \end{aligned}$$

and

$$x^*(t+\sigma) = x^*(t) + \sigma(p(t) + f(t)) \ .$$

After a discretization and a limit transition in the last equations, similar to the one for the external approximations of the above, we come to ordinary differential equations which are equation (3.2.4) and

$$\begin{aligned} dX^-(t)/dt = \ & H^{-1}((H(t)X^-(t)H'(t))^{\frac{1}{2}}(H(t)P(t)H'(t))^{\frac{1}{2}} + \\ (3.2.16) \qquad & + (H(t)P(t)H'(t))^{\frac{1}{2}}(H(t)X^-(t)H(t))^{\frac{1}{2}})H'^{-1} \ , \end{aligned}$$

with initial conditions

$$x^*(t_0) = x^0 \ , \ X^-(t_0) = X_0 \ .$$

What follows from here is the inclusion

$$\mathcal{X}[t] \supseteq \mathcal{E}(x^*(t), X^-(t)) \ , \tag{3.2.17}$$

where $x^*(t), X(t)$ satisfy (3.2.4), (3.2.16) and $H(t)$ is a continuous function of t with values in Σ – the variety of symmetric matrices. A detailed proof of the same inclusion follows from Theorem 2.9.1, where one just has to change notations $S(t), q_0(t), Q^-(t)$ to $H(t), x^*(t), X^-(t)$, respectively. The given reasoning allows us to formulate

Theorem 3.2.2 *The internal approximation of the attainability domain $\mathcal{X}[t] = X(t, t_0, \mathcal{E}(x^*, X_0))$ of the differential inclusion (3.2.1) is given by the inclusion (3.2.17) ,where $x^*(t), X^-(t)$ satisfy the equations (3.2.4) and (3.2.16). Moreover, the following representation is true*

$$\mathcal{X}[t] = \overline{\cup\{\mathcal{E}(x^*(t), X^-(t)) | H(\cdot)\}}, \tag{3.2.18}$$

where the union is taken over all measurable matrix-valued functions with values in Σ.

Relation (3.2.18) is a direct consequence of Corollary 2.8.1.

One may remark, of course, that all the earlier conclusions of this section were made under the assumptions that ellipsoids $\mathcal{E}(p(t), P(t)), \mathcal{E}(x^*, X_0)$ are nondegenerate. However, the given relations still *remain true under relaxed assumptions that allow degeneracy* in the following sense.

Consider system

$$\dot{x} \in A(t)x + B(t)\mathcal{E}(p(t), P(t)) \ , \tag{3.2.19}$$

$$x(t_0) \ = \ x^0, \ \ x^0 \in \mathcal{E}(x^*, X_0) \ ,$$

where $B(t)$ is continuous, $p(t) \in \mathbb{R}^m, P(t) \in \mathcal{L}(\mathbb{R}^m, \mathbb{R}^m), m < n$.

The parameters of this sytem allow to generate the set-valued integral

$$\mathcal{X}^*[t] = \int_{t_0}^{t} S(\tau, t) B(\tau) \mathcal{E}(0, P(\tau)) d\tau, \tag{3.2.20}$$

where matrix $S(\tau, t)$ is defined in Section 1.1, see (1.1.6).

Assumption 3.2.1 *There exists a continuous scalar function* $\beta(t) > 0, t > t_0$ *such that the support function*

$$\rho(l|\mathcal{X}^*[t]) \geq \beta(t)(l,l)^{1/2} \ ,$$

for all $t \geq t_0$.

This assumption implies that the attainability domain $\mathcal{X}[t]$ of system (3.2.19) has a nonempty interior ($int\mathcal{X}[t] \neq \emptyset$). It is actually equivalent to the requirement that system (3.2.19) with unrestricted control $u(t)$ would be *completely controllable* [147], [212], on every finite time interval $[t_0, t]$.

Under the last Assumption the analogies of Theorems 3.1, 3.2 for system (3.2.19) still remain true. Namely, taking equations

(3.2.21)
$$\dot{X}^+(t) = A(t)X^+(t) + X^+(t)A'(t) +$$
$$+\pi(t)X^+(t) + \pi^{-1}(t)B(t)P(t)B'(t) \ ,$$

(3.2.22)
$$\dot{x}^*(t) = A(t)x^* + B(t)p(t) + f(t) \ ,$$

we have the following assertion.

Lemma 3.2.2 *Suppose Assumption 3.2.1 for system (3.2.19) is true. Then the results of Theorems 3.2.1 and Corollary 3.2.1 (for the attainability domain* $\mathcal{X}[t]$ *of this system) remain true with equations (3.2.5) and (3.2.4) substituted by (3.2.22) and (3.2.21).*

The details of the proof follow the lines of Section 2.7 and the reasoning of the present section.

Remark 3.2.1 *An assertion similar to Lemma 3.2.2 is also true for internal ellipsoidal approximations (under the same Assumption 3.2.1). System (3.2.16) is then substituted by*

(3.2.23)
$$dX^-(t)/dt =$$
$$= A(t)X^-(t) + X^-(t)A'(t) + H^{-1}((H(t)X^-(t)H'(t))^{\frac{1}{2}}(H(t)P(t)H'(t))^{\frac{1}{2}} +$$
$$+ (H(t)P(t)H'(t))^{\frac{1}{2}}(H(t)X^-(t)H(t))^{\frac{1}{2}})H'^{-1} \ ,$$

Exercise 3.2.1. Prove the statement of Remark 3.2.1.

Remark 3.2.2 *It is now possible to single out individual ellipsoidal tubes that approximate $\mathcal{X}[t]$ externally or internally. This, particularly, may be done as described in Lemma 2.7.4 (due to a local optimality condition) or due to nonlocal criteria, of types (2.7.33), for example (see Exercise 2.7.3).*

We emphasize once again that that functions $\pi(t), H(t)$ in equations (3.2.5) and (3.2.16) may be interpreted as *controls* which, for example, may be selected on an interval $[t_0, \tau]$ so as to optimalize the terminal ellipsoid $\mathcal{E}(x^*(\tau), X^+(\tau))$ or $\mathcal{E}(x^*(\tau), X^-(\tau))$ in the classes of ellipsoids determined by equations (3.2.21)–(3.2.23).

The next natural step would be to introduce ellipsoidal approximations for *solvability tubes.* Prior to that, however, we shall introduce some evolution equations for ellipsoidal-valued mappings.

3.3 Evolution Equations with Ellipsoidal-Valued Solutions

Having found the external and internal ellipsoidal approximations for the attainability domains $\mathcal{X}[t]$ and recalling that $\mathcal{X}[t]$ satisfies an evolution funnel equation, we come to what seems to be a natural question: do the ellipsoidal mappings that approximate $\mathcal{X}[t]$ satisfy, in their turn, some evolution equations with ellipsoidal-valued solutions? Let us investigate this issue.

Writing down the evolution equation (3.1.12) for $\mathcal{X}[t]$ with the ellipsoidal data of Section 3.1, we have

$$(3.3.1)\quad \lim_{\sigma\to 0} \sigma^{-1} h(\mathcal{X}[t+\sigma], \mathcal{X}[t] \;+\; \sigma\mathcal{E}(p(t)+f(t), P(t))) = 0 \quad ,$$

$$\mathcal{X}[t_0] = \mathcal{E}(x^*, X_0) \quad .$$

As indicated in the above, it should be clear that in general the solution to (3.3.1) *is not ellipsoidal-valued.*

Let us now introduce another equation, namely,

$$(3.3.2) \qquad \lim_{\sigma \to 0} h_-(\mathcal{E}[t+\sigma], \mathcal{E}[t] + \sigma\mathcal{E}(p(t)+f(t), P(t))) = 0 \ ,$$

$$\mathcal{E}(t_0) \ = \ \mathcal{E}(x^*, X_0) \ .$$

Definition 3.3.1 *A function $\mathcal{E}^+[t]$ is said to be a solution to the evolution equation (3.3.2) if*

(i) $\mathcal{E}^+[t]$ satisfies (3.3.2) almost everywhere,

(ii) $\mathcal{E}^+[t]$ is ellipsoidal-valued,

(iii) $\mathcal{E}^+[t]$ is the minimal solution to (3.3.2) with respect to inclusion.

From the definition of the semidistance h_- and of the solution $\mathcal{E}^+[t]$ (points it (i),(ii)), it follows that always

$$\mathcal{E}^+[t] \supseteq \mathcal{X}[t] \ ,$$

so that $\mathcal{E}^+[t]$ is an *external approximation* of $\mathcal{X}[t]$.

Lemma 3.3.1 *The external approximation*

$$\mathcal{E}^+[t] = \mathcal{E}(x^*(t), X^+(t|\pi(\cdot)))$$

is a solution to the equation (3.3.2) in the sense of Definition 3.3.1, provided $\pi(\tau) > 0$ is selected as

$$(3.3.3) \pi(\tau) = (l^*(\tau), P(\tau) l^*(\tau))^{\frac{1}{2}} \ , \ (l^*(\tau), l^*(\tau)) = 1, \ t_0 \le \tau \le t \ ,$$

where $l^*(\tau)$ is a measurable function of t.

This follows, due to (3.3.1), (3.2.7), from the inclusion

$$\mathcal{E}^+(x^*(t+\sigma), X^+(t+\sigma|\pi(\cdot))) \ + \ o(\sigma)\mathcal{S} \supseteq \mathcal{E}(x^*(t), X^+(t|\pi(\cdot))) +$$

$$+ \ \sigma\mathcal{E}(p(t), P(t)) \ ,$$

that ensures the ellipsoidal-valued function $\mathcal{E}[t]$ to satisfy (3.3.2) and from the equalities

$$\rho(l|\mathcal{E}^+[t]) = \rho(l|I_0[t_0,t_1]) ,$$

$$I_0[t_0,t_1] = \mathcal{E}(x^*,X_0) + \int_{t_0}^{t_1} \mathcal{E}(p(t)+f(t),P(t))dt ,$$

(taken for $\pi(t)$ selected due to (3.3.3)) that ensure the minimality property (iii) of Definition 3.3.1.

The last Lemma indicates that the solution to equation (3.3.1) *is not unique.* This is all the more true due to

Corollary 3.3.1 *The ellipsoidal function* $\mathcal{E}^+[t] = \mathcal{E}(x^*(t), X^+(t|\pi(\cdot)))$ *is a solution to (3.3.2) , whatever is the measurable function* $\pi(\tau) > 0, t_0 \leq \tau \leq t_1$ *selected due to the inequalities*

$$\min\{(l,P(\tau)l)|(l,l)=1\} \leq \pi(\tau) \leq \max\{(l,P(\tau)l)|(l,l)=1\} .$$

The proof of this corollary follows from the results of Sections 2.3 and 2.7.

For a given function $\pi(t)$ and given initial pair x^*, X^*, we shall also denote

$$E^+(t,\tau,\mathcal{E}(x^*,X^*)) = \mathcal{E}(x^*(t,\tau,x^*), X^+(t,\tau,X^*)) ,$$

where $x^*(t,\tau,x^*), X^+(t,\tau,X^*)$ satisfy (3.2.4),(3.2.16) with

$$x^*(\tau,\tau,x^*) = x^*,\ X^+(\tau,\tau,X^*) = X^* .$$

Then, obviously

$$E^+(t,\tau,\mathcal{E}(x^*,X^*)) \supseteq \mathcal{E}(x^*,X^*) + \int_{\tau}^{t} \mathcal{E}(p(s)+f(s),P(s))ds$$

and a direct substitution leads to

Lemma 3.3.2 *The following relation is true*

$$E^+(t,\tau,E^+(\tau,t_0,\mathcal{E}(x^0,X^0))) = E^+(t,t_0,\mathcal{E}(x^0,X^0)) , \quad (3.3.4)$$

$$t_0 \leq \tau \leq t .$$

Relations (3.3.4) describe the dynamics of the ***external ellipsoidal estimates*** $E^+(t, \tau, \mathcal{E}(x^*, X^*))$. They thus define an upper semigroup property of the respective mappings. The sets $E^+(t, t_0, \mathcal{E}(x^0, X^0))$ are sometimes referred to as *supperattainability domains.*

Together with (3.3.2) consider equation

$$(3.3.5) \quad \lim_{\sigma \to 0} \sigma^{-1} h_+(\mathcal{E}[t+\sigma], \mathcal{E}[t] + \sigma \mathcal{E}(p(t) + f(t), P(t))) = 0 \ ,$$

$$\mathcal{E}[t_0] = \mathcal{E}(x^*, X_0) \ .$$

Definition 3.3.2 *A function $\mathcal{E}^-[t]$ is said to be a solution to the evolution equation (3.3.5) if*

(i) $\mathcal{E}^-[t]$ satisfies (3.3.5) almost everywhere,

(ii) $\mathcal{E}^-[t]$ is ellipsoidal-valued,

(iii) $\mathcal{E}^-[t]$ is the maximal solution to (3.3.5) with respect to inclusion.

From the definition of the semidistance h_+ and of the solution $\mathcal{E}^-[t]$ (points *(i),(ii)*) it follows that always

$$\mathcal{E}^-[t] \subseteq \mathcal{X}[t] \ .$$

Thus, we have

Lemma 3.3.3 *Any solution $\mathcal{E}^-[t]$ to (3.2.5) that satisfies points (i),(ii) of Definition 3.3.2 is an internal approximation for $\mathcal{X}[t]$.*

Moreover, representation (3.2.18) yields the fulfillment of the requirement of point (iii) of Definition 3.3.3 for any function $\mathcal{E}_-[t] = \mathcal{E}(x^*(t), X^-(t))$ generated by the solutions $x^*(\cdot), X^-(\cdot)$ to equations (3.2.4), (3.2.16). This leads to

Theorem 3.3.1 *The internal approximation*

$$\mathcal{E}^-[t] = \mathcal{E}(x^*(t), X^-(t))$$

is a solution to the evolution equation (3.3.5), whenever $x^*(t), X^-(t)$ are the solutions to differential equations (3.2.4), (3.2.16).

For a given function $H(t)$ and a given initial pair x^*, X^* denote

$$E_-(t,\tau,\mathcal{E}(x^*,X^*)) = \mathcal{E}(x^*(t), X^-(t)) \ ,$$

where $x^*(t), X^-(t)$ are the solutions to (3.2.4), (3.2.16) with initial conditions

$$x^*(\tau) = x^*, X^-(\tau) = X^* \ .$$

Then, clearly,

$$E_-(t,\tau,\mathcal{E}(x^*,X^*)) \subseteq \mathcal{E}(x^*,X^*) + \int_\tau^t \mathcal{E}(p(s)+f(s), P(s))ds,$$

and a direct substitution leads to

Lemma 3.3.4 *The following relation is true*

$$(3.3.6) \quad E^-(t,\tau,E^-(\tau,t_0,\mathcal{E}(x^0,X^0))) = E^-(t,t_0,\mathcal{E}(x^0,X^0)), \ t_0 \leq \tau \leq t \ .$$

The last relation describes the dynamics of the *internal estimates* $E_-(t,\tau,\mathcal{E}(x^*,X^*))$ for the attainability domains $\mathcal{X}[t]$, defining thus a lower semigroup property for the respective mappings. The sets $E_-(t,t_0,\mathcal{E}(x^0,X^0))$ are sometimes referred to as *subattainability domains.* A similar type of description may now be introduced for *solvability tubes.*

3.4 Solvability in Absence of Uncertainty

We shall now pass to the treatment of *solvability tubes* for the simplest case of systems without uncertainty and state constraints. Our aim is to approximate these tubes by *ellipsoidal-valued functions.*

Returning to relation (1.4.4), we recall that in our case

$$(3.4.1) \qquad W[t] = \mathcal{E}(m,M) \ - \ \int_t^{t_1} \mathcal{E}(p(t)+f(t), P(t))dt \ .$$

Then, following the approximation schemes of Sections 2.8, 2.9, 3.2, and 3.3, with obvious changes of signs, we come to the differential equations

$$(3.4.2) \qquad \dot{x} \ = \ p(t) \ + \ f(t) \ ,$$

(3.4.3) $$\dot{X}_+(t) = -\pi(t)X_+(t) - \pi^{-1}(t)P(t) \ ,$$

(3.4.4) $$\dot{X}_-(t) = -H^{-1}(t)[(H(t)X_-(t)H'(t))^{\frac{1}{2}}(H(t)P(t)H'(t))^{\frac{1}{2}} + $$
$$+ (H(t)P(t)H'(t))^{\frac{1}{2}}(H(t)X_-(t)H(t))^{\frac{1}{2}}]H'^{-1}(t) \ ,$$

with boundary conditions

(3.4.5) $$x(t_1) = m \ , \ X_+(t_1) = M, \ X_-(t_1) = M \ .$$

Denote the solutions to (3.4.2)–(3.4.4) with boundary conditions (3.4.5) as $x(t)$, $X_+(t)$, $X_-(t)$, respectively. Similarly to (3.1.3), (3.1.17), we then come to

Theorem 3.4.1 *The following inclusions are true*

(3.4.6) $$\mathcal{E}_-(x(t), X_-(t)) \subseteq W[t] \subseteq \mathcal{E}_+(x(t), X_+(t)) \ ,$$

whatever are the solutions to differential equations (3.4.2)–(3.4.5) with $\pi(t) > 0, H(t) \in \Sigma$.

As in the previous Sections 3.2 and 3.3, the last assertion develops into exact representations.

Theorem 3.4.2 *(i) The following external representation is true*

(3.4.7) $$W[t] = \cap\{\mathcal{E}_+(x(t), X_+(t))|\pi(\cdot)\} \ ,$$

where the intersection is taken over all measurable functions $\pi(t)$ *that satisfy the inequalities*

(3.4.8) $$\min\{(l, P(t)l)|(l,l) = 1\} \leq \pi(t) \leq \max\{(l, P(t)l)|(l,l) = 1\} \ .$$

(ii) The following internal representation is true

(3.4.9) $$W[t] = \overline{\cup\{\mathcal{E}_-(x(t), X_-(t))|H(\cdot)\}} \ ,$$

where the union is taken over all measurable functions $H(t)$ *with values in* Σ.

The next issue is to write down an evolution equation with ellipsoidal-valued solutions for each of the approximating functions $\mathcal{E}_-(t), \mathcal{E}_+(t)$.

Consider equations

(3.4.10) $\lim_{\sigma \to 0} \sigma^{-1} h_-(W[t-\sigma], W[t] - \sigma\mathcal{E}(p(t) + f(t), P(t))) = 0$,

(3.4.11) $\lim_{\sigma \to 0} \sigma^{-1} h_+(W[t-\sigma], W[t] - \sigma\mathcal{E}(p(t) + f(t), P(t))) = 0$

with boundary condition

(3.4.12) $$W[t_1] = \mathcal{E}(m, M) \ .$$

The solution to these equations are not obliged to be ellipsoidal-valued. Therefore, in analogy with (3.3.2), we introduce another pair of equations, namely,

(3.4.13) $\lim_{\sigma \to 0} h_-(\mathcal{E}[t-\sigma], \mathcal{E}[t] - \sigma\mathcal{E}(p(t) + f(t), P(t))) = 0$,

(3.4.14) $\lim_{\sigma \to 0} h_+(\mathcal{E}[t-\sigma], \mathcal{E}[t] - \sigma\mathcal{E}(p(t) + f(t), P(t))) = 0$,

with the same boundary condition

(3.4.15) $$\mathcal{E}[t_1] = \mathcal{E}(m, M) \ .$$

Definition 3.4.1 *A function $\mathcal{E}_+[t]$ (respectively $\mathcal{E}_-[t]$), is said to be a solution to the evolution equation (3.4.13) (respectively 3.4.14), if*

(i) $\mathcal{E}_-[t]$ (respectively $\mathcal{E}_+[t]$) satisfies (3.4.13) (respectively 3.4.14) almost everywhere,

(ii) $\mathcal{E}_+[t]$ (respectively $\mathcal{E}_-[t]$) is ellipsoidal-valued,

(iii) $\mathcal{E}_+[t]$ (respectively $\mathcal{E}_-[t]$) is the minimal (respectively maximal) solution to (3.4.13) (respectively 3.4.14) with respect to inclusion.

From the definitions of the semidistances h_-, h_+ and of the solutions $\mathcal{E}_+[t], \mathcal{E}_-[t]$, (properties *(i),(ii)*) it follows that always

(3.4.16) $$\mathcal{E}_-[t] \subseteq W[t] \subseteq \mathcal{E}_+[t] \ .$$

It also follows that the last relations are true for any functions $\mathcal{E}_+[t], \mathcal{E}_-[t]$, that satisfy properties *(i),(ii)* of Definition 3.4.1.

The minimality and maximality properties of the respective solutions are described similarly to Sections 3.2 and 3.3. This gives

Theorem 3.4.3 *(i) The ellipsoidal-valued function $\mathcal{E}_+[t] = \mathcal{E}(x(t), X_+(t))$ generated by the solutions $x(t), X_+(t)$ to the differential equations (3.4.2), (3.4.3), (3.4.5) is a solution to the evolution equation (3.4.13), whatever is the measurable function $\pi(\cdot)$ selected due to the inequalities (3.4.8).*

(ii) The ellipsoidal-valued function $\mathcal{E}_-[t] = \mathcal{E}(x(t), X_-(t))$ generated by the solutions to the differential equations (3.4.2), (3.4.4), (3.4.5) is a solution to the evolution equation (3.4.14), whatever is the measurable function $H(t)$ with values in Σ.

For a given pair of functions $\pi(\cdot), H(\cdot)$ and a given pair of boundary values m^*, M^* we shall denote

(3.4.17) $$E_+(t, \tau, \mathcal{E}(m^*, M^*)) = \mathcal{E}(x(t, \tau, m^*), X_+(t, \tau, M^*)) \ ,$$

(3.4.18) $$E_-(t, \tau, \mathcal{E}(m^*, M^*)) = \mathcal{E}(x(t, \tau, m^*), X_-(t, \tau, M^*)) \ ,$$

where $x(t, \tau, m)$, $X_+(t, \tau, M^*)$, $X_-(t, \tau, M^*)$ satisfy (3.4.2), (3.4.3), (3.4.4), with boundary condition

$$x(\tau, \tau, m^*) = m^*, \ X_+(\tau, \tau, M^*) = X_-(\tau, \tau, M^*) = M^* \ .$$

Then, obviously,

$$E_-(t, \tau, \mathcal{E}(m^*, M^*)) \subseteq \mathcal{E}(x^*, M^*)) -$$

$$- \int_t^\tau \mathcal{E}(p(s), P(s)) ds \subseteq E_+(t, \tau, \mathcal{E}(m^*, M^*)) \ ,$$

and a direct substitution leads to

Lemma 3.4.1 *The following relations are true with $t \leq \tau \leq t_1$:*

(3.4.19) $$E_-(t, \tau, E(\tau, t_1, \mathcal{E}(m, M))) = E_-(t, t_1, \mathcal{E}(m, M)) \ ,$$

(3.4.20) $$E_+(t, t_1, \mathcal{E}(m, M)) = E_+(t, \tau, E_+(\tau, t_1, \mathcal{E}(m, M))) \ .$$

Relations (3.4.20), (3.4.21) describe the dynamics of the ellipsoidal estimates for $W[t]$, respectively defining, (now, in backward time, however) the lower and upper semigroup properties of the corresponding mappings.

Exercise 3.4.1. Assume that the original system (3.1.1) is given in the form (1.1.1), that is with $A(t) \neq 0$.

By direct calculations prove that equations (3.4.2)–(3.4.4) are then substituted by

$$\dot{x} = A(t)x + p(t) + f(t) \ , \tag{3.4.21}$$

$$\dot{X}_+(t) = A(t)X_+(t) + X_+(t)A'(t) - \\ -\pi(t)X_+(t) - \pi^{-1}(t)P(t) \ , \tag{3.4.22}$$

$$\dot{X}_-(t) = A(t)X_-(t) + X_-(t)A'(t) - \\ H^{-1}(t)[(H(t)X_-(t)H'(t))^{\frac{1}{2}}(H(t)P(t)H'(t))^{\frac{1}{2}} - \\ + (H(t)P(t)H'(t))^{\frac{1}{2}}(H(t)X_-(t)H(t))^{\frac{1}{2}}]H'^{-1}(t) \ , \tag{3.4.23}$$

with same boundary condition (3.4.5) as before.

The next step is to proceed with the approximations of solvability tubes for *systems with uncertainty.*

3.5 Solvability Under Uncertainty

In this section we discuss solvability tubes for uncertain systems with unknown input disturbances. Taking equation (3.1.15) for the solvability tube of such a system, we again observe that in general its set-valued solution $W[t]$ *is not ellipsoidal-valued.* How should we construct the ellipsoidal approximations for $W[t]$ now, that $f(t)$ is unknown but bounded?

Since we do not expect ellipsoidal-valued functions to be produced by solving (3.1.15), we will try, as in the previous section, to introduce other evolution equations than (3.1.15), constructing them such that their solutions, on the one hand, would be ellipsoidal-valued and, on the other, would form an appropriate variety of external and internal ellipsoidal approximations to the solution $W[t]$ of (3.1.15). Being interested in the solvability set (under uncertainty), we shall further presume that $W[t]$

is *inclusion-maximal*, namely, as shown in Sections 1.6 and 1.7, the one that gives precisely the solvability set.

Indeed, relation (3.1.15), $t \in [t_0, t_1]$, yields[22]

$$(3.5.1)\; W[t-\sigma] + \sigma\mathcal{E}(q(t), Q(t)) \subseteq W[t] - \sigma\mathcal{E}(p(t), P(t)) + o(\sigma)\mathcal{S} \; .$$

We shall now look for an ellipsoidal-valued function $\mathcal{E}(x(t), X(t)) = \mathcal{E}[t]$ that would ensure an *internal approximation* for the *left-hand side* of (3.5.1) and an *external approximation* for the *right-hand side.* Due to (2.4.1) and (2.4.2), this would give

$$(3.5.2) \qquad W[t-\sigma] + \sigma\mathcal{E}(q(t), Q(t)) \supseteq$$

$$\supseteq \mathcal{E}(x(t-\sigma), X(t-\sigma)) + \sigma\mathcal{E}(q(t), Q(t)) \; ,$$

$$(3.5.3) \qquad \mathcal{E}(x(t-\sigma), X(t-\sigma)) + \sigma\mathcal{E}(q(t), Q(t)) \supseteq$$

$$\supseteq \mathcal{E}(x(t-\sigma) + \sigma q(t), H^{-1}[(H(t)X(t-\sigma)H'(t))^{\frac{1}{2}} +$$

$$+ \sigma(H(t)Q(t)H'(t))^{\frac{1}{2}}]^2 H'^{-1}) \; ,$$

and

$$(3.5.4) \; W[t] - \sigma\mathcal{E}(p(t), P(t)) \subseteq \mathcal{E}(x(t), X(t)) - \sigma\mathcal{E}(p(t), P(t)) \; ,$$

$$(3.5.5) \qquad \mathcal{E}(x(t), X(t)) - \sigma\mathcal{E}(p(t), P(t)) \subseteq$$

$$\subseteq \mathcal{E}(x(t) - \sigma p(t), (1 + \sigma\pi(t))X(t) + \sigma^2(1 + (\sigma\pi(t))^{-1})P(t)) \; .$$

Combining (3.5.3), (3.5.5) and requiring that the right-hand parts of these inclusions are equal (within the terms of first order in σ), we require the equality

$$(3.5.6) \qquad \mathcal{E}(x(t-\sigma) + \sigma q(t), X[t-\sigma] +$$

$$+ \sigma H^{-1}(t)[(H(t)X(t-\sigma)H'(t))^{\frac{1}{2}}(H(t)Q(t)H'(t))^{\frac{1}{2}} +$$

$$+ (H(t)Q(t)H'(t))^{\frac{1}{2}}(H(t)X(t-\sigma)H'(t))^{\frac{1}{2}}]H'^{-1}(t)) =$$

$$= \mathcal{E}(x(t) - \sigma p(t), (1 + \sigma\pi(t))X(t) + \sigma^2(1 + (\sigma\pi(t))^{-1})P(t)) \; ,$$

which is ensured if $x(t), X(t)$ satisfy the following equalities

$$(3.5.7) \qquad x(t-\sigma) + \sigma q(t) = x(t) \; - \; \sigma p(t) \; ,$$

[22] This equation is treated under nondegeneracy Assumptions 1.7.1, 1.7.2, see footnote for formula (3.1.15).

and

$$X(t-\sigma)+ \tag{3.5.8}$$
$$\sigma H^{-1}(t)[(H(t)X(t-\sigma)H'(t))^{\frac{1}{2}}(H(t)Q(t)H'(t))^{\frac{1}{2}}+$$
$$+(H(t)Q(t)H'(t))^{\frac{1}{2}}(H(t)X(t-\sigma)(H'(t))^{\frac{1}{2}}]H'^{-1}(t)=$$
$$=(1+\sigma\pi(t))X(t)-\sigma^2(1+(\sigma\pi(t))^{-1})P(t) \ .$$

Dividing both parts by σ and passing to the limit ($\sigma \to 0$), we come to the differential equations (with further notation $X = X_+$)

$$\dot{x} \ = \ p(t) \ + \ q(t) \ , \tag{3.5.9}$$

and

$$\dot{X}_+(t) = -\pi(t)X_+(t) - \pi^{-1}(t)P(t) + \tag{3.5.10}$$
$$+H^{-1}(t)[(H(t)Q(t)H'(t))^{\frac{1}{2}}(H(t)X_+(t)H'(t))^{\frac{1}{2}}+$$
$$+(H(t)X_+(t)H'(t))^{\frac{1}{2}}(H(t)Q(t)H'(t))^{\frac{1}{2}}]H'^{-1}(t) \ ,$$

which have to be taken with boundary conditions

$$x(t_1) = m, \ X_+(t_1) = M \ . \tag{3.5.11}$$

Let us introduce an evolution equation

$$\lim_{\sigma\to 0} \sigma^{-1}h_-(\mathcal{E}[t-\sigma]+ \tag{3.5.12}$$
$$+\sigma\mathcal{E}(q(t),Q(t)), \mathcal{E}[t] - \sigma\mathcal{E}(p(t),P(t))) = 0 \ ,$$

with boundary condition

$$\mathcal{E}[t_1] = \mathcal{E}(m, M) \ . \tag{3.5.13}$$

Definition 3.5.1 *A solution to (3.5.12), (3.5.13) will be defined as an ellipsoidal-valued function $\mathcal{E}[t]$ that satisfies (3.5.12) almost everywhere together with boundary condition (3.5.13).*

A solution to the evolution equation (3.5.12) obviously satisfies the inclusion

$$\mathcal{E}[t-\sigma] + \sigma\mathcal{E}(q(t),Q(t)) + o(\sigma)\mathcal{S} \supseteq \tag{3.5.14}$$
$$\supseteq \mathcal{E}[t] - \sigma\mathcal{E}(p(t),P(t))$$

Lemma 3.5.1 *The ellipsoid $\mathcal{E}(x(t), X_+(t|\pi(\cdot), H(\cdot))) = \mathcal{E}_+[t]$ given by equations (3.5.9)–(3.5.11) satisfies the inclusion (3.5.14).*

Introducing the support function $\rho(l|\mathcal{E}_+[t])$ and calculating its derivative in t, we have

$$\partial\rho(l|\mathcal{E}_+[t])/\partial t = \frac{1}{2}(\dot{X}_+(t)l, l)(X_+(t)l, l))^{-\frac{1}{2}} + (l, p+q) \ .$$

By equation (3.5.10) this implies

$$\partial\rho(l|\mathcal{E}_+[t])/\partial t = (l, p+q) - \tag{3.5.15}$$

$$-\frac{1}{2}(P(t)l, l)^{\frac{1}{2}}((X_+(t)l, l)^{\frac{1}{2}}\pi(t)(P(t)l, l)^{-\frac{1}{2}} +$$

$$+\pi^{-1}(t)(P(t)l, l)^{\frac{1}{2}}(X_+(t)l, l)^{-\frac{1}{2}}) +$$

$$+(X_+(t)l, l)^{-\frac{1}{2}}((H(t)Q(t)H(t))^{\frac{1}{2}}H^{-1}l, (H(t)X_+(t)H(t))^{\frac{1}{2}}H^{-1}l) \ .$$

From inequality $a + a^{-1} \geq 2$ and the inequality of Cauchy-Buniakowski it then follows (for all $l \in \mathbb{R}^n$)

$$\partial\rho(l|\mathcal{E}_+[t])/\partial t \leq (l, p+q) \ - \ (l, P(t)l)^{\frac{1}{2}} + (l, Q(t)l)^{\frac{1}{2}} \ . \tag{3.5.16}$$

Integrating this inequality within the interval $[t-\sigma, t]$ and having in view the continuity of $P(t), Q(t)$, we come to (3.5.14) and therefore, to the proof of the Lemma.

Lemma 3.5.2 *The ellipsoid $\mathcal{E}_+[t]$ is an external estimate for $W[t]$ – the solvability set under uncertainty, whatever are the parametrizing functions $\pi(t) > 0, H(t) \in \Sigma$.*

Following the scheme of Section 1.6 and incorporating relation (3.5.1), we have

$$-\partial\rho(l|W[t])/\partial t \leq \rho(-l|\mathcal{E}(p(t), P(t))) - \rho(l|\mathcal{E}(q(t), Q(t))) \ , \tag{3.5.17}$$

Together with (3.5.16), this gives

$$-\partial(\rho(l|W[t]) - \rho(l|\mathcal{E}_+[t]))/\partial t \leq 0 \ , \tag{3.5.18}$$

and since the boundary conditions are $W[t_1] = \mathcal{E}(m, M) = \mathcal{E}_+[t_1]$, this yields relation

(3.5.19) $\rho(l|W[t]) \leq \rho(l|\mathcal{E}_+[t]) = \rho(l|\mathcal{E}(x(t), X_+(t|\pi(\cdot), H(\cdot))), \ \forall l$,

or, in other terms,

$$W[t] \subseteq \mathcal{E}_+[t] = \mathcal{E}(x(t), X_+(t|\pi(\cdot), H(\cdot)) \ ,$$

for $t_1 - \sigma \leq t \leq t_1$ and consequently, the same inclusion for all $t \in [t_0, t_1]$, whatever are the respective functions $\pi(\cdot), H(\cdot)$.

We shall now indicate the inclusion-minimal solutions to (3.5.12). Indeed, for a given vector l, take

$$\pi(t) = (X_+(t)l, l)^{-\frac{1}{2}}(P(t)l, l)^{\frac{1}{2}} \ , \tag{3.5.20}$$

and take $H(t)$ due to a relation similar to (2.4.4), so as to ensure

$$(X_+(t)l, l)^{-\frac{1}{2}}((H(t)Q(t)H(t))^{\frac{1}{2}}H^{-1}l, (H(t)X_+(t)H(t))^{\frac{1}{2}}H^{-1}l) =$$

$$- (l, Q(t)l)^{\frac{1}{2}} \ . \tag{3.5.21}$$

For a given vector l, with π, H selected due to (3.5.20), (3.5.21), the value of the respective derivative

$$\partial\rho(l|\mathcal{E}_+[t])/\partial t = \rho(l|\mathcal{E}(q(t), Q(t))) - \rho(-l|\mathcal{E}(p(t), P(t)))$$

is clearly the largest among all feasible π, H:

$$\partial\rho(l|\mathcal{E}_+[t])/\partial t \geq \partial\rho(l|\mathcal{E}_+^*[t])/\partial t \ ,$$

where $\mathcal{E}^*$ is any other external estimate. Integrating the last inequality from t to t_1 and having in view that $\mathcal{E}_+[t_1] = \mathcal{E}_+^*[t_1] = \mathcal{E}(m, M)$, we come to

$$\rho(l|\mathcal{E}_+[t]) \leq \rho(l|\mathcal{E}^*[t]) \ .$$

This means that along the direction l there is no other external ellipsoid governed by equations (3.5.9), (3.5.10) that could be squeezed between $W[t]$ and $\mathcal{E}_+[t]$ if the last one is chosen due to (3.5.20) and (3.5.21). This implies

Lemma 3.5.3 *With $\pi[t], H[t]$ selected due to (3.5.20), (3.5.21), $l \in \mathbb{R}^n$, the ellipsoid $\mathcal{E}_+[t] = \mathcal{E}(x(t), X_+(t|\pi(\cdot), H(\cdot)))$ is inclusion-minimal in the class of all external ellipsoids governed by equations (3.5.9) and (3.5.10).*

The selected functions $\pi(t), H(t)$ may be treated as *feedback controls* selected, as we shall see, so as to ensure a tightest external bound for $W[t]$.[23]

We now note that the maximal solution $W[t]$ to (3.1.15) ensures an equality in (3.5.19), if for each l the ellipsoid $\mathcal{E}_+[t] = \mathcal{E}(x(t), X_+(t|\pi(\cdot), H(\cdot)))$ is selected to be inclusion-minimal, due to (3.5.20) and (3.5.21). Indeed, we observe this after integrating (3.5.18) and arriving at (3.5.19), where *for each l there is its own pair of functions π, H.*

Combining this fact with the previous assertions, we now observe that the inclusion-minimal external ellipsoids $\mathcal{E}^+[t]$ ensure that the approximation of $W[t]$ is as tight as possible.

Namely, in view of the indicated relations, we come to the conclusion that for every vector l there exists a pair $\pi(\cdot), H(\cdot)$, such that

$$(3.5.22) \qquad \rho(l|W[t]) \leq \rho(l|\mathcal{E}(x(t), X_+(t|\pi(\cdot), H(\cdot)))) \ ,$$

and also

$$\rho(l|\mathcal{E}(x(t), X_+(t|\pi(\cdot), H(\cdot)))) \leq \rho(l|W[t]) + o(\sigma, l) \ ,$$

for

$$t_1 - \sigma \leq t \leq t_1 \ ,$$

Therefore, particularly, for the indicated values of t one has

$$\rho(l|W[t]) = \inf\{\rho(l, |\mathcal{E}(x(t), X_+(t|\pi(\cdot), H(\cdot))))\} + o(\sigma) \ ,$$

and moreover, for every vector l we have

$$(3.5.23) \qquad \rho(l|W[t]) \ = \ \rho(l|\mathcal{E}(x(t), X_+(t|\pi(\cdot), H(\cdot))))$$

for some $\pi(\cdot), H(\cdot)$. This yields

$$(3.5.24) \ W[t] = \cap\{\mathcal{E}(x(t), X_+(t|\pi(\cdot), H(\cdot))|\pi(\cdot), H(\cdot)\} + o^*(\sigma)\mathcal{S}$$

[23] The explicit relations for (3.5.20) and (3.5.21) indicate that $\pi(\cdot), H(\cdot)$ are continuous or at least measurable in t. With $l = l(t) \neq 0$ in (3.5.20) and (3.5.21), the respective functions $\pi(\cdot), H(\cdot)$ are still measurable, continuous or piece-wise continuous depending on the properties of $l(t)$.

We may consequently repeat this procedure indicating (see 1.7.6) for

$$W(t_1-\sigma_1-\sigma_2, t_1-\sigma_1, W[t_1-\sigma_1]) =$$

$$= \left(W[t_1-\sigma_1] + \int_{t_1-\sigma_1-\sigma_2}^{t_1-\sigma_1} \mathcal{E}(p(t),P(t))dt\right) \dot{-} \int_{t_1-\sigma_1-\sigma_2}^{t_1-\sigma_1} \mathcal{E}(q(t),Q(t))dt$$

and for a given vector l the existence of a pair $\pi(\cdot), H(\cdot)$ that again yield (3.5.24), now for

$$t \in [t_1-\sigma_1-\sigma_2, t_1-\sigma_1] \ .$$

Continuing the procedure yet further , now for the sets defined by (1.7.7) and passing, under Assumption 1.7.1, to the respective limit transition of Lemma 1.7.2 and (1.7.8), we may observe that with $\sigma \to 0$ the equalities (3.5.23), (3.5.24), ($o^*(\sigma) = 0$), are true for all $t \in [t_0, t_1]$. Relations (3.5.9)–(3.5.11) thus define an array of *external ellipsoidal estimates* for the solvability set $W[t]$. Summarizing the results of the above, we have

Theorem 3.5.1 *Under Assumption 1.7.1 there exists, for every vector $l \in \mathbb{R}^n$, a pair $\pi(\cdot), H(\cdot)$ of measurable functions that ensure the following:*

(i) The support function

$$(3.5.25) \quad \rho(l|W[t]) = \rho(l|\mathcal{E}(x(t), X_+(t|\pi(\cdot), H(\cdot)))),\ t \in [t_0, t_1].$$

(ii) The relation (3.5.24) ($o^*(\sigma) \equiv 0$) is true for $t \in [t_0, t_1]$.

(iii) The external estimates $\mathcal{E}_+[t] = \mathcal{E}(x(t), X_+(t|\pi(\cdot), H(\cdot)))$ for the solutions $W[t]$ to the evolution equation (3.1.15) that are generated by solutions $x(t), X_+(t)$ to the differential equations (3.5.9), (3.5.10) satisfy the evolution equation (3.5.12), (3.5.13) and are minimal with respect to inclusion among all solutions to (3.5.12), (3.5.13).

The next stage is to arrive at a similar theorem for *internal estimates.* Returning to (3.1.15), and considering the relation opposite to (3.5.1), we shall look for an ellipsoidal function $\mathcal{E}(x(t), X(t)) = \mathcal{E}[t]$ that ensures an *external approximation* for its *left-hand side* and an *internal approximation* for its *right-hand side.*

Due to (2.4.1) and (2.4.2) this would give

$$W[t-\sigma] + \sigma\mathcal{E}(q(t), Q(t)) \subseteq \tag{3.5.26}$$

$$\subseteq \mathcal{E}(x(t-\sigma), X(t-\sigma)) + \sigma\mathcal{E}(q(t), Q(t)) \ ,$$

and

$$\mathcal{E}(x(t-\sigma), X(t-\sigma) + \sigma\mathcal{E}(q(t), Q(t)))) \subseteq \tag{3.5.27}$$

$$\subseteq \mathcal{E}(x(t-\sigma) + \sigma q(t), (1+\sigma\pi(t))X(t-\sigma) + \sigma^2(1+(\sigma\pi(t))^{-1})Q(t))) \ ,$$

together with

$$\mathcal{E}(x(t), X(t)) - \sigma\mathcal{E}(p(t), P(t)) \subseteq W[t] - \sigma\mathcal{E}(p(t), P(t)) \ , \tag{3.5.28}$$

$$\mathcal{E}(x(t) - \sigma p(t), H^{-1}(t)[(H(t)X(t)H'(t))^{\frac{1}{2}} +$$

$$+ \sigma(H(t)P(t)H'(t))^{\frac{1}{2}}]^2 H'^{-1}(t)) \subseteq$$

$$\subseteq \mathcal{E}(x(t), X(t)) - \sigma\mathcal{E}(p(t), P(t)) \ . \tag{3.5.29}$$

Equalizing the right-hand side of (3.5.28) with the left-hand side of (3.5.29), dividing both parts by σ and passing to the limit with $\sigma \to 0$ (similarly to (3.5.6)–(3.5.8)), we come (with further notation $X = X_-$) to equations (3.5.9) and

$$\dot{X}_-(t) = \pi(t)X_-(t) + \pi^{-1}(t)Q(t) - \tag{3.5.30}$$

$$-H^{-1}(t)[(H(t)P(t)H'(t))^{\frac{1}{2}}(H(t)X_-(t)H'(t))^{\frac{1}{2}} +$$

$$+(H(t)X_-(t)H'(t))^{\frac{1}{2}}(H(t)P(t)H'(t))^{\frac{1}{2}}]H'^{-1}(t) \ ,$$

which have to be taken with same boundary conditions as before, namely,

$$x(t_1) = m, X_-(t_1) = M \ . \tag{3.5.31}$$

Let us introduce an evolution equation

$$\lim_{\sigma\to 0} \sigma^{-1}h_+(\mathcal{E}[t-\sigma] + \tag{3.5.32}$$

$$+\sigma\mathcal{E}(q(t), Q(t)) \ ,$$

$$\mathcal{E}[t] - \sigma\mathcal{E}(p(t), P(t))) = 0$$

with boundary condition

$$\mathcal{E}[t_1] = \mathcal{E}(m, M) \ . \tag{3.5.33}$$

A solution to the last equation is to be considered due to Definition 3.5.1, where (3.5.32), (3.5.33) are to be taken instead of (3.5.12), (3.5.13). It obviously satisfies the inclusion

$$(3.5.34) \qquad \mathcal{E}[t-\sigma] + \sigma\mathcal{E}(q(t), Q(t)) + \subseteq$$

$$\subseteq \mathcal{E}[t] - \sigma\mathcal{E}(p(t), P(t)) + o(\sigma)\mathcal{S} \ .$$

Similarly to Lemma 3.5.1 one may prove the following:

Lemma 3.5.4 *The ellipsoid $\mathcal{E}(x(t), X_-(t|\pi(\cdot), H(\cdot))) = \mathcal{E}_-[t]$ given by equations (3.5.9), (3.5.30), (3.5.31) is a solution to the evolution equation (3.5.32)-(3.5.33).*

A further reasoning is similar to the one that preceded Theorem 3.5.1, except that the ellipsoid $\mathcal{E}_-[t]$ is now an *internal estimate* for the *maximal solution* $W[t]$ to the evolution equation (3.1.15) and that the respective representations are true relative to closures of the corresponding sets (as in Theorem 2.5.1). We leave the details (which are not too trivial, though) to the reader, confining ourselves to the formulation of

Theorem 3.5.2 *Under Assumption 1.7.1, for every vector $l \in \mathbb{R}^n$, the following relations are true:*

(i)

$$(3.5.35) \quad \rho(l|W[t]) = \sup\{\rho(l|\mathcal{E}(x(t), X_-(t|\pi(\cdot), H(\cdot))))|\pi(\cdot), H(\cdot)\} \ ,$$

(ii)

$$(3.5.36) \qquad W[t] = \overline{\cup\{\mathcal{E}(x(t), X_-(t|\pi(\cdot), H(\cdot)))|\pi(\cdot), H(\cdot)\}} \ ,$$

where $\pi(t) > 0, H(t)$ are measurable functions.

The internal estimates $\mathcal{E}_-[t] = \mathcal{E}(x(t), X_-(t, |\pi(\cdot), H(\cdot)))$ for the maximal solutions $W[t]$ of the evolution equation (3.1.15) that are generated by the solutions $x(t), X_-(t)$ to the differential equations (3.5.9) and (3.5.30), are also the maximal solutions to the evolution equation (3.5.32) and (3.5.33).

Maximality is treated here with repect to inclusion, as in the above.

Exercise 3.5.1. Under a nondegeneracy Assumption 1.7.2 (or 1.7.1) prove that equations (3.5.9), (3.5.10), and (3.5.30) may also be derived through the results of Part II (Theorems 2.4.1 and 2.4.2), when the evolution equation (3.5.1) is by definition substituted by

$$(3.5.37)\ W[t-\sigma] \subseteq (W[t] - \sigma\mathcal{E}(p(t),P(t)))\dot{-}\sigma\mathcal{E}(q(t),Q(t)) + o(\sigma)\mathcal{S}$$

and the ellipsoidal estimates (external and internal) are taken accordingly. The nondegeneracy assumption implies in this case that there exist numbers $\epsilon > 0, \delta > 0$ such that

$$(W[t] - \sigma\mathcal{E}(p(t),P(t)))\dot{-}\sigma\mathcal{E}(q(t),Q(t)) \supseteq \epsilon\mathcal{S},\quad \sigma \leq \delta,\quad t \in [t_0,t_1]\ .$$

Exercise 3.5.2. By direct calculation prove that with $A(t) \neq 0$, that is with system (3.1.1) given in the form (1.1.1), equations (3.5.9), (3.5.10), (3.5.30) will be substituted by

$$(3.5.38)\qquad \dot{x} = A(t)x + p(t) + q(t)\ ,$$

and

$$(3.5.39)\qquad \dot{X}_+(t) = A(t)X_+(t) +$$
$$+X_+(t)A'(t) - \pi(t)X_+(t) - \pi^{-1}(t)P(t) +$$
$$+H^{-1}(t)[(H(t)Q(t)H'(t))^{\frac{1}{2}}(H(t)X_+(t)H'(t))^{\frac{1}{2}} +$$
$$+(H(t)X_+(t)H'(t))^{\frac{1}{2}}(H(t)Q(t)H'(t))^{\frac{1}{2}}]H'^{-1}(t)\ ,$$

$$(3.5.40)\qquad \dot{X}_-(t) = A(t)X_-(t) + X_-(t)A'(t) +$$
$$+\pi(t)X_-(t) + \pi^{-1}(t)Q(t) -$$
$$-H^{-1}(t)[(H(t)P(t)H'(t))^{\frac{1}{2}}(H(t)X_-(t)H'(t))^{\frac{1}{2}} +$$
$$+(H(t)X_-(t)H'(t))^{\frac{1}{2}}(H(t)P(t)H'(t))^{\frac{1}{2}}]H'^{-1}(t)\ ,$$

with same boundary conditions (3.5.11), (3.5.31) as in the above.

Remark 3.5.1 *The external and internal ellipsoidal approximations* $\mathcal{E}_+[t]$ *and* $\mathcal{E}_-[t]$ *of the solvability tubes* $W[t]$ *(under uncertainty) may be interpreted as approximations of Krasovski's bridges or Pontryagin's alternated integrals.*

Remark 3.5.2 *Relations (3.5.39), (3.5.40) will be simpler and reduce to those of type (3.4.22), (3.4.23) if $\mathcal{E}(p(t),P(t)),\mathcal{E}(q(t),Q(t))$ satisfy the matching condition of Remark 1.6.2 which means*

$$\mathcal{E}(p(t),P(t))\dot{-}\mathcal{E}(q(t),Q(t)) = \mathcal{E}(p(t)-q(t)),\gamma P(t))),0 \leq \gamma < 1 \ .$$

We finally mention an important property of the estimates $\mathcal{E}_-[t],\mathcal{E}_+[t]$.

Using the notations of (3.4.17), (3.4.18), where now $x(t,\tau,m^*)$, $X_+(t,\tau,M^*)$, $X_-(t,\tau,M^*)$ satisfy (3.5.9)–(3.5.10), (3.5.30) with boundary conditions (3.4.19), one may verify that the external and internal ellipsoidal approximation mappings $E_+(\cdot),E_-(\cdot)$ satisfy *the external and internal semigroup properties*, respectively.

Lemma 3.5.5 *The external and internal approximation mappings (for the solvability tube under counteraction $W[t]$), are defined through relations (3.4.17)–(3.4.19), (3.5.9), (3.5.10), (3.5.30) and satisfy the upper and lower semigroup properties of type (3.4.19) and (3.4.20).*

Remark 3.5.3 *The ellipsoids $\mathcal{E}_+[t],\mathcal{E}_-[t]$ are nondominated (inclusion-minimal and maximal, respectively). Due to this and to the semigroup property of Lemma 3.5.5, the sets $\mathcal{E}_-[t]$ turn to be ellipsoidal-valued bridges as will be indicated in Section 3.8.*

We are now prepared to deal with problems of control synthesis with the aim of using the described relations as a basis for constructive techniques in analytical controller design.

3.6 Control Synthesis Through Ellipsoidal Techniques

In this section we shall apply the results of the previous paragraphs to the analytical design of synthesizing control strategies through ellipsoidal techniques developed in the previous sections.

Let us return to the Control Synthesis problem 1.1.4 of Section 1.4. There the idea of constructing the synthesizing strategy $\mathcal{U}(t,x)$ for this problem was that $\mathcal{U}(t,x)$ should ensure that all the solutions $x[t] = x(t,\tau,x_\tau)$ to equation

$$\dot{x}(t) \in \mathcal{U}(t,x(t)) + f(t), \qquad \tau \le t \le t_1,$$

with initial state $x[\tau] = x_\tau \in \mathcal{W}[\tau]$, ($\mathcal{W}[t]$ is the respective *solvability set* described on the same section) would satisfy the inclusion

$$x[t] \in \mathcal{W}[t], \quad \tau \le t \le t_1$$

and would therefore ensure $x[t_1] \in \mathcal{M}$. This exact solution requires us, as we have seen, to calculate the tube $\mathcal{W}[t]$ and then, for each instant of time t, to solve an extremal problem of type (1.4.9) whose solution finally yields the desired strategy (1.4.12) $\mathcal{U}(t,x)$. This strategy is thus actually defined *as an algorithm.*

In order to obtain a simpler scheme, we will now substitute $\mathcal{W}[t]$ by one of its *internal approximations* $\mathcal{E}_-[t] = \mathcal{E}(x^*, X(t))$. The conjecture is that once $\mathcal{W}[t]$ is substituted by $\mathcal{E}_-[t]$, we should just copy the scheme of Section 1.4, constructing a strategy $\mathcal{U}_-(t,x)$ such that for every solution $x[t] = x(t,\tau,x_\tau)$ that satisfies equation

$$\dot{x}[t] = \mathcal{U}_-(t,x[t]) + f(t), \quad \tau \le t \le t_1, \quad x[\tau] = x_\tau, \quad x_\tau \in \mathcal{E}_-[\tau]\ , \tag{3.6.1}$$

the following inclusion would be true

$$x[t] \in \mathcal{E}_-[t], \quad \tau \le t \le t_1, \tag{3.6.2}$$

and therefore

$$x[t_1] \in \mathcal{E}(m,M) = \mathcal{M} = \mathcal{E}(x^*(t_1), X_-(t_1))\ .$$

It will be proven that once the approximation $\mathcal{E}_-[t]$ is selected appropriately, the desired strategy $\mathcal{U}_-(t,x)$ may be constructed again according to the scheme of Section 1.4, except that $\mathcal{W}[t]$ will now be substituted by $\mathcal{E}_-[t]$, namely,

$$\mathcal{U}_-(t,x) = \begin{cases} \mathcal{E}(p(t), P(t)) & \text{if } x \in \mathcal{E}_-[t] \\ p(t) - P(t)l^0(l^0, P(t)l^0)^{-1/2} & \text{if } x \notin \mathcal{E}_-[t], \end{cases} \tag{3.6.3}$$

where $l^0 = l^0(t,x)$ is the unit vector that solves the problem

(3.6.4) $d[t,x] = (l^0, x) - \rho(l^0|\mathcal{E}_-[t]) = \max\{(l,x) - \rho(l|\mathcal{E}_-[t])| \|l\| \leq 1\}$

and

(3.6.5) $$d[t,x] = h_+(x, \mathcal{E}_-[t]) = \min\{\|x - s\| | s \in \mathcal{E}_-[t]\} \ .$$

One may readily observe that relations (3.6.4), (3.6.3) coincide with (1.4.9), (1.4.12) if set $\mathcal{W}[t]$ is substituted for for $\mathcal{E}_-[t]$ and $\mathcal{P}(t)$ for $\mathcal{E}(p(t), P(t)$.

Indeed, let us start with the maximization problem of (3.6.4). It may be solved in more detail than its analogue (1.4.9) in Section 1.4 (since $\mathcal{E}_-[t]$ is an ellipsoid).

If s^0 is the solution to the minimization problem

(3.6.6) $$s^0 = \arg\min\{\|(x - s)\| | s \in \mathcal{E}_-[t], \ x = x(t)\},$$

then we can take

(3.6.7) $$l^0 = k(x(t) - s^0), \quad k > 0,$$

in (3.6.4), so that l^0 will be *the gradient* of the distance $d(x, \mathcal{E}_-[t])$ with t fixed. (This can be verified by differentiating either (3.6.4) or (3.6.5) in x.)

Lemma 3.6.1 *Consider a nondegenerate ellipsoid $\mathcal{E} = \mathcal{E}(a,Q)$ and a vector $x \notin \mathcal{E}(a,Q)$. Then the gradient*

$$l^0 = \partial d(x, \mathcal{E}(a,Q))/\partial x$$

may be expressed through $l^0 = (x - s^0)/\|x - s^0\|$,

(3.6.8) $$s^0 = (I + \lambda Q^{-1})^{-1}(x - a) + a,$$

where $\lambda > 0$ is the unique root of the equation $h(\lambda) = 0$, with

$$h(\lambda) = \Big((I + \lambda Q^{-1})^{-1}(x - a), Q^{-1}(I + \lambda Q^{-1})^{-1}(x - a)\Big) - 1 \ .$$

Proof. Assume $a = 0$. Then the necessary conditions of optimality for the minization problem

$$\|x - s\| = \min, \quad (s, Q^{-1}s) \leq 1$$

are reduced to the equation

$$-x + s + \lambda Q^{-1}s = 0$$

where λ is to be calculated as the root of the equation $h(\lambda) = 0, (a = 0)$. Since it is assumed that $x \notin \mathcal{E}(0, Q)$, we have $h(0) > 0$. With $\lambda \to \infty$ we also have

$$\left((I + \lambda Q^{-1})^{-1}x, Q^{-1}(I + \lambda Q^{-1})^{-1}x\right) \to 0$$

This yields $h(\lambda) < 0$, $\lambda \geq \lambda_*$ for some $\lambda_* > 0$. The equation $h(\lambda) = 0$ therefore has a root $\lambda^0 > 0$. The root λ^0 is unique since direct calculation gives $h'(\lambda) < 0$ with $\lambda > 0$. The case $a \neq 0$ can now be given through a direct shift $x \to x - a$. Q.E.D.

Corollary 3.6.1 *With parameters a, Q given and x varying, the multiplier λ may be uniquely expressed as a function*

$$\lambda = \lambda(x) \ .$$

Let us now look at relation (1.4.12). In the present case we have $\mathcal{P}(t) = \mathcal{E}(p(t), P(t))$ and problem (1.4.12) therefore reduces to

(3.6.9) $$\arg\max\{(-l^0, u)|u \in \mathcal{E}(p(t), P(t))\} = \mathcal{U}_-(t, x) \ .$$

Relation (3.6.3) now follows from the following assertion:

Lemma 3.6.2 *Given ellipsoid $\mathcal{E}(p, P)$, the maximizer u^* for the problem*

$$\max\{(l, u)|u \in \mathcal{E}(p, P)\} = (l, u^*) \ , l \neq 0,$$

is the vector

$$u^* = p + Pl(l, Pl)^{-\frac{1}{2}} \ .$$

This Lemma is an obvious consequence of the formula for the support function of an ellipsoid, namely,

$$\rho(l|\mathcal{E}(p, P)) = (l, p) + (l, Pl)^{-\frac{1}{2}} \ .$$

We will now prove that the *ellipsoidal valued strategy* $\mathcal{U}_-(t,x)$ of (3.6.3) does solve the problem of control synthesis, provided we start from a point $x_\tau = x[\tau] \in \mathcal{E}_-[\tau]$. Indeed, assume $x_\tau \in \mathcal{E}_-[\tau]$ with $x[t] = x(t,\tau,x_\tau)$, $\tau \le t \le t_1$ being the respective trajectory. We will demonstrate that once $x[t]$ is a solution to equation (3.6.1), we will always have the inclusion (3.6.2). (With isolated trajectory $x[t]$ given, it is clearly driven by a unique control $u[t] = \dot{x}[t] - f(t)$, a.e. such that $u[t] \in \mathcal{E}(p(t),P(t))$.)

Calculating

$$d[t] = d[t,x[t]] = \max\{(l,x[t]) - \rho(l \mid \mathcal{E}_-[t]) \mid \|l\| \le 1\},$$

for $d[t] > 0$, we observe

$$\frac{d}{dt}d[t] = \frac{d}{dt}[(l^0,x[t]) - \rho(l^0|\mathcal{E}_-[t])] \quad ,$$

and since $l^0 \ne 0$ is a unique maximizer,

$$\frac{d}{dt}d[t] = (l^0,\dot{x}[t]) - \frac{\partial}{\partial t}\rho(l^0|\mathcal{E}_-[t]) = \tag{3.6.10}$$

$$= (l^0,u[t]) - \frac{d}{dt}[(l^0,x(t)) + (l^0,X_-(t)l^0)^{1/2}]$$

where $\mathcal{E}_-[t] = \mathcal{E}(x(t),X_-(t))$. For a fixed function $H(\cdot)$ we have $\mathcal{E}_-[t] = \mathcal{E}(x(t),X_-(t))$, where $x(t),X_-(t)$ satisfy the system (3.4.2) and (3.4.4).

Substituting these relations into (3.6.10) and remembering the rule for differentiating a maximum over a variety of functions, we have

$$\frac{d}{dt}d[t] = (l^0,u[t]) - (l^0,p(t)) + \frac{1}{2}(l^0,X_-(t)l^0)^{-1/2}\cdot$$

$$\cdot(l^0,H^{-1}(t)([H(t)X_-(t)H(t)]^{1/2}[H(t)P(t)H(t)]^{1/2}+$$

$$+[H(t)P(t)H(t)]^{1/2}[H(t)X_-(t)H(t)]^{1/2})H^{-1}(t)l^0) \quad ,$$

or, due to the Bunyakovsky-Schwartz inequality,

$$\frac{d}{dt}d[t] \le -(l^0,p(t)) + (l^0,P(t)l^0)^{1/2} + (l^0,u[t]), \tag{3.6.11}$$

where

$$u[t] \in \mathcal{U}_-(t,x) \subseteq \mathcal{E}(p(t),P(t)),$$

with inequality

$$\frac{d}{dt}d[t] \leq 0$$

attained if $u[t] \in \mathcal{U}_-(t,x)$. Integrating $dd^2[t]/dt = dd^2[t,x[t]]/dt$ from τ to t_1 (see notations of Section 1.4), we come to the equality

$$d^2[\tau, x[\tau]] = d^2[t_1, x[t_1]] = h_+^2(x[t_1], \mathcal{M}) = 0$$

which means $x[t_1] \in \mathcal{M}$, provided $x[\tau] \in X_-(\tau)$.

What follows is the assertion

Theorem 3.6.1 *Define an internal approximation $\mathcal{E}_-[t] = \mathcal{E}_-(x(t), X_-(t))$ with given parametrization $H(t)$ of (3.4.4). Once $x[\tau] \in \mathcal{E}_-[\tau]$ and the synthesizing strategy is $\mathcal{U}_-(t,x)$ of (3.6.3) , the following inclusion is true:*

$$x[t] \in \mathcal{E}_-[t], \quad \tau \leq t \leq t_1,$$

and therefore

$$x[t_1] \in \mathcal{E}(m, M) \ .$$

The ellipsoidal synthesis described in this section thus gives a solution strategy $\mathcal{U}_-(t,x)$ for any internal approximation $\mathcal{E}_-[t] = \mathcal{E}_-(x(t), X_-(t))$. With $x \notin \mathcal{E}_-[t]$, the function $\mathcal{U}_-(t,x)$ is single-valued, while with $x \in \mathcal{E}_-[t]$ it is multivalued ($\mathcal{U}_-(t,x) = \mathcal{E}_-[t]$), being uppersemicontinuous in x, measurable in t, and ensuring the existence of a solution to the differential inclusion (3.6.1).

Remark 3.6.1 *(i) Due to Theorem 3.4.2 (see (1.4.8)), each element $x \in intW[t]$ belongs to a certain ellipsoid $\mathcal{E}_-[t]$ and may therefore be steered to the terminal set $\mathcal{M}$ by means of a certain ellipsoidal-based strategy $\mathcal{U}_-(t,x)$. (The assumptions of Section 3.1 imply $intW[t] \neq \emptyset$.)*

(ii) Relations (3.6.3), (3.6.7), (3.6.8) indicate that strategy $\mathcal{U}_-(t,x)$ is given explicitly, with the only unknown being the multiplier λ of Lemma 3.6.1 which can be calculated as the only root of equation $h(\lambda) = 0$. But in view of Corollary 3.6.1 the function $\lambda = \lambda(t,x)$ may be calculated **in advance**, *depending on the parameters of the internal approximation $\mathcal{E}_-[t]$ (which may also be calculated in advance). With this specificity, the suggested strategy $\mathcal{U}_-(t,x)$ may be considered as an* **analytical design**.

(iii) The internal ellipsoids $\mathcal{E}_-[t]$ satisfy the evolution equation (3.5.32) and therefore the equation (1.7.9) which implies Theorem 1.8.1 and its ellipsoidal version, Theorem 2.6.1. The given facts are particularly due to the lower semigroup property of the respective mappings (see Lemma 2.5.3) and the inclusion-maximal property of the ellipsoids $\mathcal{E}_-[t]$.

We shall now proceed with numerical examples that demonstrate the constructive nature of the solutions obtained above.

3.7 Control Synthesis: Numerical Examples

Let us take system (1.1.1), (3.1.2) to be 4-dimensional, and study it throughout the time interval $[t_s, t_e], t_s = 0, t_e = 5$. We will seek for graphical representations of the solutions. And as the ellipsoids appearing in this problem are *four dimensional*, we we shall present them through their *two dimensional projections.* The figures below are therefore divided into four *windows*, and each shows projections of the original ellipsoids onto the planes spanned by the first and second (x_1, x_2), third and fourth (x_3, x_4), first and third (x_1, x_3), and second and fourth (x_2, x_4) coordinate axes, in a clockwise order, starting from bottom left. The drawn segments of coordinate axes corresponding to state variables range from -10 to 10 according to the above scheme. In some of the figures, where we show the graph of solutions and of solvability set, the third, skew axis corresponds to time and ranges from 0 to 5. Let the initial position $\{0, x_0\}$ be given by

$$x_0 = \begin{pmatrix} 4 \\ 1 \\ 0 \\ 0 \end{pmatrix},$$

the target set $\mathcal{M} = \mathcal{E}(m, M)$ by

$$m = \begin{pmatrix} 0 \\ 5 \\ 5 \\ 0 \end{pmatrix}$$

and

$$M = \begin{pmatrix} 1 & 0 & 0 & 0 \\ 0 & 1 & 0 & 0 \\ 0 & 0 & 1 & 0 \\ 0 & 0 & 0 & 1 \end{pmatrix}$$

at the final instant $t_1 = 5$. We consider a case when the right hand side is constant:

$$A(t) \equiv \begin{pmatrix} 0 & 1 & 0 & 0 \\ -1 & 0 & 0 & 0 \\ 0 & 0 & 0 & 1 \\ 0 & 0 & -4 & 0 \end{pmatrix},$$

describing the position and velocity of two independent oscillators. The restriction $u(t) \in \mathcal{E}(p(t), P(t))$ on the control u, is also defined by time independent constraints:

$$p(t) \equiv \begin{pmatrix} 0 \\ 0 \\ 0 \\ 0 \end{pmatrix},$$

$$P(t) \equiv \begin{pmatrix} 1 & 0 & 0 & 0 \\ 0 & 1 & 0 & 0 \\ 0 & 0 & 1 & 0 \\ 0 & 0 & 0 & 1 \end{pmatrix},$$

so that the controls do couple the system. Therefore, the class of feasible strategies is such that

$$U_{\mathcal{P}}^{c} = \{\mathcal{U}(t,x)\}, \;\; \mathcal{U}(t,x) \subseteq \mathcal{P} = \mathcal{E}(p(t), P(t)) \;\; .$$

The results to be presented here we obtain by way of discretization. We divide the interval $[0,5]$ into 100 subintervals of equal lengths, and use the discretized version of (3.4.22) and (3.4.23) implemented through a standard first-order scheme (see, for example, [63], [272], [273] for technical details). Instead of the set valued control strategy (3.6.3) we apply a *single valued* selection:

$$(3.7.1) \quad u(t,x) = \begin{cases} p(t) & \text{if } x \in \mathcal{E}_-[t] \\ p(t) - P(t)l^0(l^0, P(t)l^0)^{-1/2} & \text{if } x \notin \mathcal{E}_-[t] \end{cases},$$

again in its discrete version. The use of a single-valued strategy in the discrete version does not affect the existence of solutions to the respective recurrence equations.

We shall specify the internal ellipsoid $\mathcal{E}_-[t] = \mathcal{E}(x(t), X_-(t))$ of (3.4.22) and (3.4.23) to be used here by selecting

$$H(t) = P^{-1/2}(t), \quad 0 \leq t \leq 5$$

in (3.4.18). The calculations give the following internal ellipsoidal estimate $\mathcal{E}_-[0] = \mathcal{E}(x(0), X_-(0))$ of the solvability set $\mathcal{W}[0] = \mathcal{W}(0,5,\mathcal{M})$:

$$x(0) = \begin{pmatrix} 4.2371 \\ 1.2342 \\ -2.6043 \\ -3.1370 \end{pmatrix},$$

and

$$X_-(0) = \begin{pmatrix} 31.1385 & 0 & 0 & 0 \\ 0 & 31.1385 & 0 & 0 \\ 0 & 0 & 12.1845 & 2.3611 \\ 0 & 0 & 2.3611 & 44.1236 \end{pmatrix}.$$

Now, as it is easy to check, $x_0 \in \mathcal{E}_-[0]$ and therefore we may apply Theorem 3.4.1 with an implication that the control strategy $\mathcal{U}_-(t,x)$ of (3.6.3) should steer every solution of

(3.7.2) $$\dot{x} = A(t)x + \mathcal{U}_-(t,x) + f(t) \;,$$

$x_0 = x(0)$, into $\mathcal{M}$. For the discrete version this produces

$$x[5] = \begin{pmatrix} 0.0264 \\ 4.9512 \\ 4.0457 \\ -0.0830 \end{pmatrix}$$

as a final state. Figure 3.7.1 shows the graph of the ellipsoidal valued map $\mathcal{E}_-[t], t \in [0,5]$ and of the solution of

(3.7.3) $$x(t_{k+1}) - x(t_k) = \sigma^{-1}(A(t_k)x(t_k) \;+\; u(t_k, x(t_k))) \;,$$

$$t_s = t_0 = 0 \leq t \leq 5 = t_{100} = t_e;$$

$$x[0] = x_0 : \sigma = t_{k+1} - t_k > 0, \;\; k = 0, ..., 100,$$

where we use $u(t,x)$ of (3.7.1).

Equation (3.7.3) serves as a discrete-time version of the differential equation

(3.7.4) $$\dot{x}[t] = A(t)x[t] \;+\; u(t, x[t]) \;.$$

However, the last equation has a single-valued but discontinuous right-hand side which leads to additional questions on the existence of solutions to this equation. There is actually no such problem, however, for the discrete-time system (3.7.3). We will therefore avoid the single-valued equation (3.7.4), but will interpret the limit ($\sigma \to \infty$) of solutions to (3.7.3) as a solution to the differential inclusion (3.7.2).

Figure 3.7.2 shows the target set $\mathcal{M} = \mathcal{E}(m, M)$, (projections appearing as circles), the solvability set $\mathcal{E}_-[0] = \mathcal{E}(x(0), X_-(0))$ at the initial instant $t = 0$, and the trajectory of the solution of (3.7.2), which, within the accuracy of the computation, may be treated as a solution of (3.7.2) constructed for the same tube $\mathcal{E}_-[t]$ as in (3.7.3), (3.7.1).

In the next example we show *by way of numerical evidence*, what can happen if the initial state $x_0 \notin \mathcal{E}_-[0]$. Leaving the rest of the data to be the same, we change the initial state x_0 in such a way that the inclusion

$$x_0 \in \mathcal{E}_-[0]$$

is hurt, but not very much, taking

$$x_0 = \begin{pmatrix} 4 \\ 1 \\ 0 \\ 2 \end{pmatrix} .$$

Though Theorem 3.6.1 cannot be used, let us still apply formulae (3.7.1) and (3.7.3). Analogously to Figure 3.7.2, Figure 3.7.3 shows the phase portrait of the result. The trajectory of the solution to (3.7.3) is drawn with a thick line, as long as it is outside of the respective ellipsoidal solvability set, and with a thin line if it is inside. The drawn projections of the initial state are inside, except one (upper left window). As the illustration shows, at one point in time the trajectory enters the tube $\mathcal{E}_-[t]$, the line changing into thin. After this happens, Theorem 3.6.1 does take effect, and the trajectory remains inside for the rest of the time interval. In this way we obtain

$$x[5] = \begin{pmatrix} 0.0255 \\ 4.9528 \\ 4.0215 \\ -0.1658 \end{pmatrix}$$

as a final state. The above phenomenon indicates that

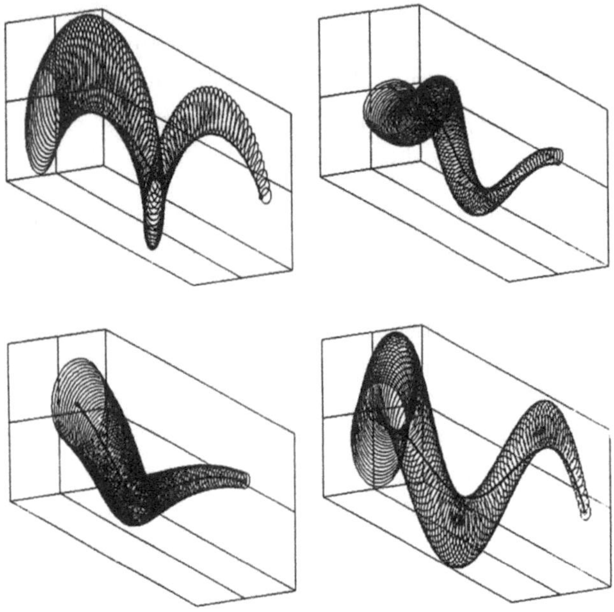

Figure 3.7.1.

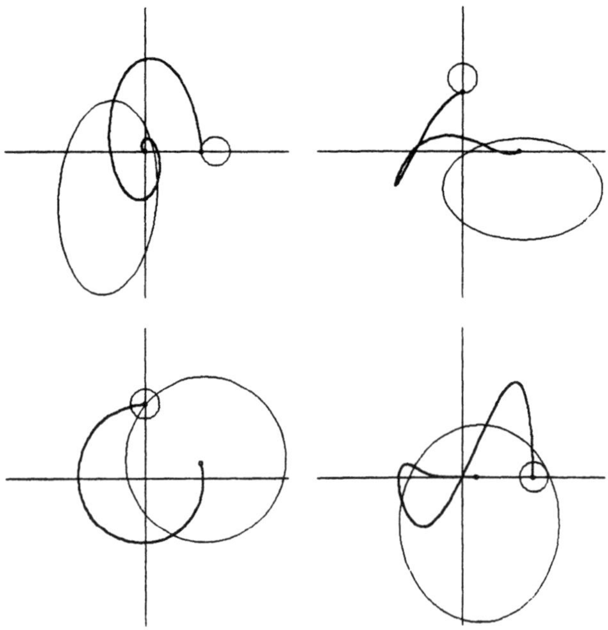

Figure 3.7.2.

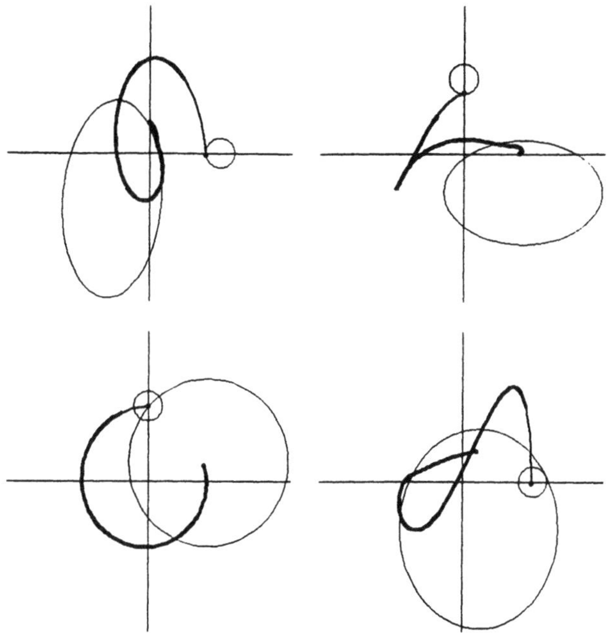

Figure 3.7.3.

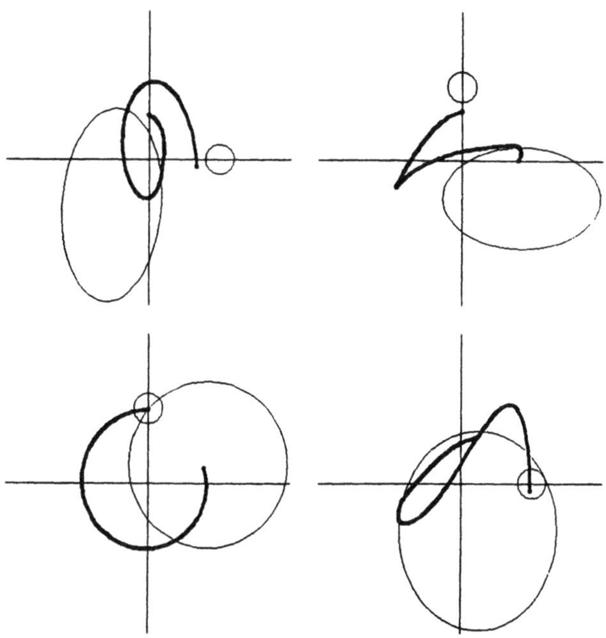

Figure 3.7.4.

- the initial state must be inside the solvability set $\mathcal{W}[0] = \mathcal{W}(0, 5, \mathcal{M})$, that is actually

$$x_0 \in \mathcal{W}(0, 5, \mathcal{M}) \setminus \mathcal{E}_-[0],$$

  as *it was possible* to steer the solution to (3.7.3),(3.7.1) into the target set $\mathcal{M}$,

- *in this particular numerical example* the control rule works beyond the tube $\mathcal{E}_-[t]$.

In the third example, we move the initial state x_0 further away, so that the control rule does not work any more (Figure 3.7.4):

$$x_0 = \begin{pmatrix} 4 \\ 1 \\ 0 \\ 3 \end{pmatrix},$$

and obtain as final state

$$x[5] = \begin{pmatrix} 0.0460 \\ 4.9150 \\ 3.3668 \\ -0.5540 \end{pmatrix}.$$

Figures 3.7.5 and 3.7.6 show the effect of changing the target set. We take the data of the first example except for the matrix M in the target set $\mathcal{M} = \mathcal{E}(m, M)$ by setting the radius to be 2:

$$M = \begin{pmatrix} 4 & 0 & 0 & 0 \\ 0 & 4 & 0 & 0 \\ 0 & 0 & 4 & 0 \\ 0 & 0 & 0 & 4 \end{pmatrix},$$

resulting in a final state

$$x[5] = \begin{pmatrix} 0.5875 \\ 4.8914 \\ 3.0158 \\ -0.0536 \end{pmatrix}.$$

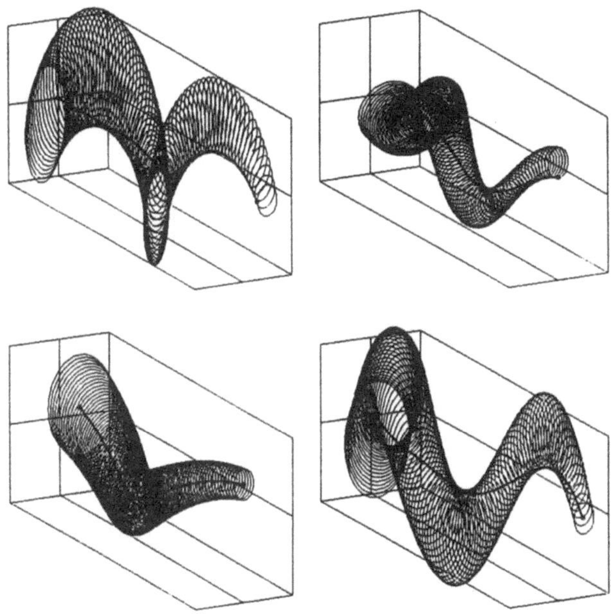

Figure 3.7.5.

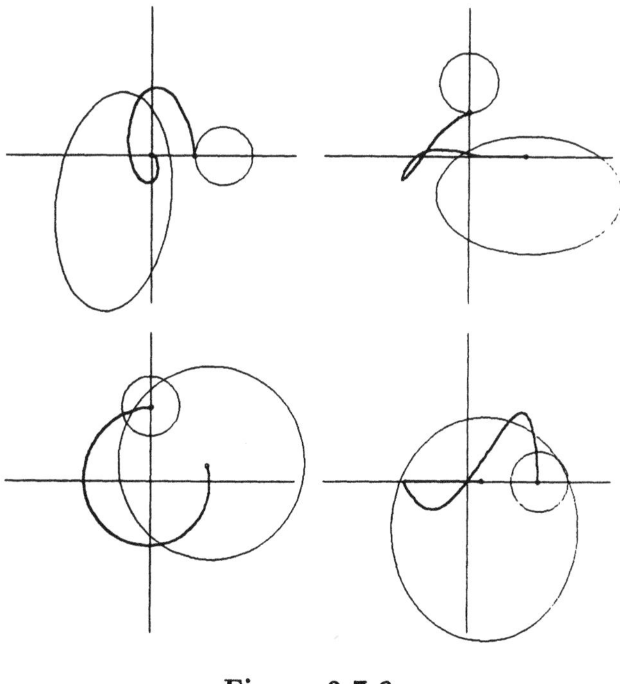

Figure 3.7.6.

The switching of the control, due to the specific form of (3.7.1), is clearly seen in Figure 3.7.6 and later in Figure 3.7.8. Taking again the data of the first example, we allow more freedom for the controls, changing the matrix $P(t)$ in the bounding set $\mathcal{P} = \mathcal{E}(p(t), P(t))$ again by setting the radius to be 2:

$$P(t) \equiv \begin{pmatrix} 4 & 0 & 0 & 0 \\ 0 & 4 & 0 & 0 \\ 0 & 0 & 4 & 0 \\ 0 & 0 & 0 & 4 \end{pmatrix}$$

with a final state

$$x[5] = \begin{pmatrix} 0.0235 \\ 4.9565 \\ 4.0536 \\ -0.1308 \end{pmatrix}.$$

Numerical simulations were made on a SUN Sparc Station.

Finally we shall consider *two coupled oscillators*, represented by a system with parameters

$$x_0 = \begin{pmatrix} -5 \\ 0 \\ -10 \\ 10 \end{pmatrix},$$

with target set $\mathcal{M} = \mathcal{E}(m, M)$ defined by

$$m = \begin{pmatrix} 10 \\ 0 \\ 0 \\ 10 \end{pmatrix}$$

and

$$M = \begin{pmatrix} 1 & 0 & 0 & 0 \\ 0 & 1 & 0 & 0 \\ 0 & 0 & 1 & 0 \\ 0 & 0 & 0 & 1 \end{pmatrix}$$

at final instant $t_1 = 3$.

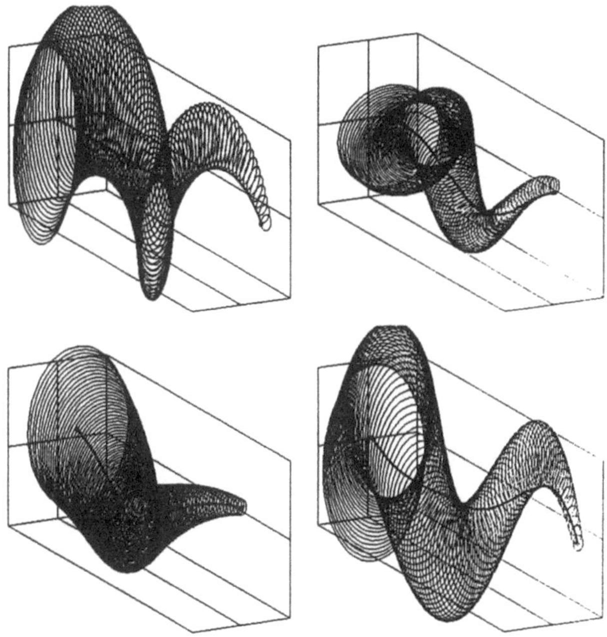

Figure 3.7.7.

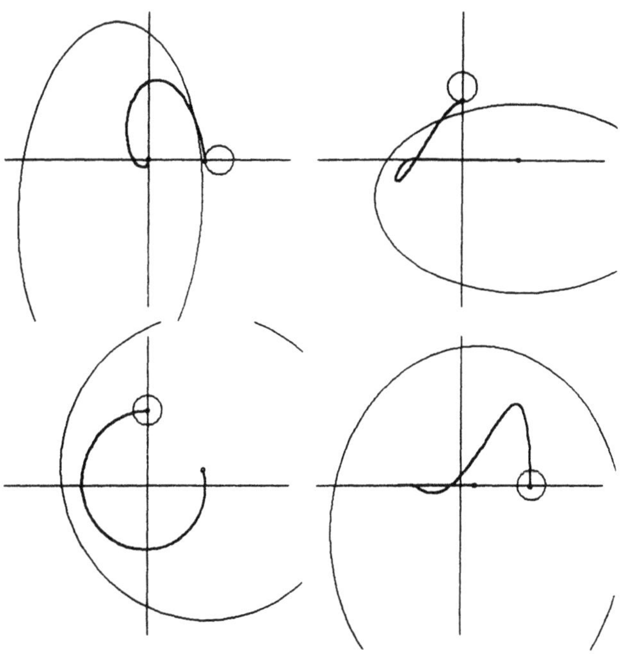

Figure 3.7.8.

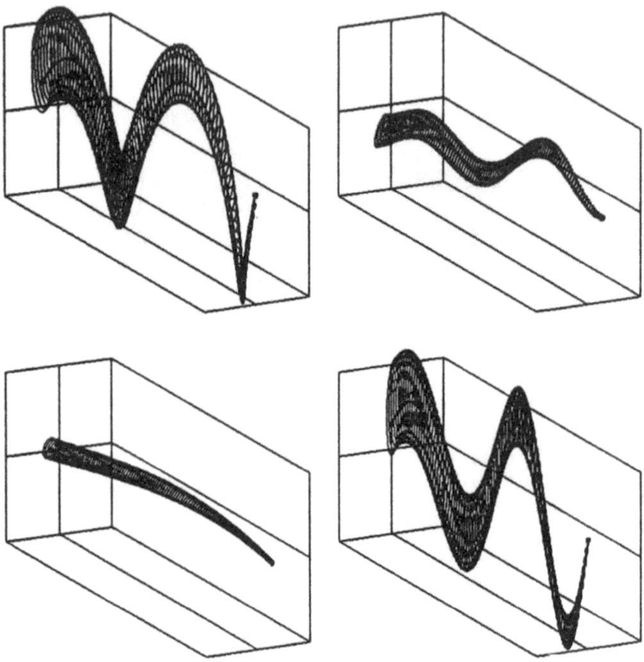

Figure 3.7.9.

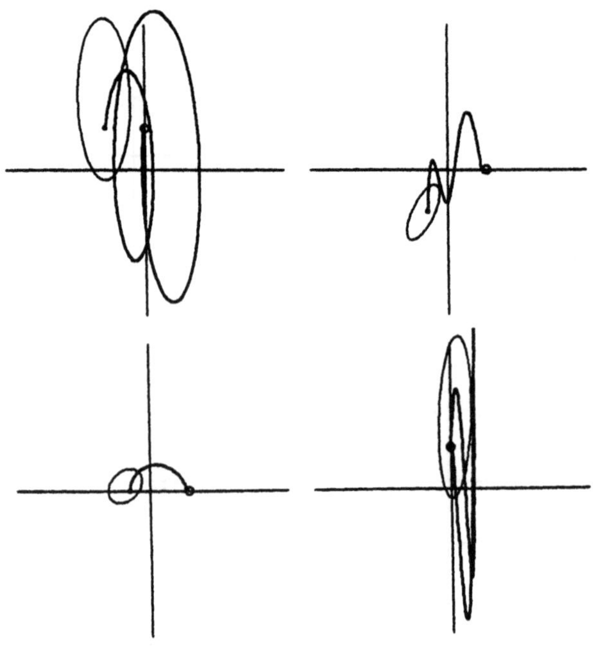

Figure 3.7.10.

The system matrix A is constant:

$$A(t) \equiv \begin{pmatrix} 0 & 1 & 0 & 0 \\ -1 & 0.25 & 0 & 0 \\ 0 & 0 & 0 & 1 \\ 16 & 0 & -16 & 0 \end{pmatrix},$$

and the constraint on the controls is defined by

$$p(t) \equiv \begin{pmatrix} 0 \\ 0 \\ 0 \\ 0 \end{pmatrix},$$

$$P(t) \equiv \begin{pmatrix} 9 & 0 & 0 & 0 \\ 0 & 0.1 & 0 & 0 \\ 0 & 0 & 9 & 0 \\ 0 & 0 & 0 & 0.1 \end{pmatrix}.$$

The target control problem is solved as before, in 100 steps, with synthesizing strategy calculated due to (3.7.1) through a difference scheme similar to the above. The four-dimensional ellipsoidal tubes and the synthesized control trajectory in phase space are shown in an appropriate scale in Figures 3.7.9 and 3.7.10 (here note the relatively small size of the target set).

3.8 Ellipsoidal Control Synthesis for Uncertain Systems

In this section we shall further apply the results of the previous paragraphs to the analytical ellipsoidal design of synthesizing control strategies, this time constructing them for *uncertain systems.*

Let us consider the Problem of Control Synthesis Under Uncertainty of Section 1.8 (Definition 1.8.1). There the idea was that the respective synthesizing control strategy $\mathcal{U}(t,x)$ should ensure that all the solutions $x[t] = x(t,\tau,x_\tau)$ to the differential inclusion

$$\dot{x}(t) \in \mathcal{U}(t,x(t)) + \mathcal{E}(q(t),Q(t)), \qquad \tau \leq t \leq t_1,$$

with initial state $x[\tau] = x_\tau \in \mathcal{W}^*[\tau]$, would satisfy the inclusion

$$x[t] \in \mathcal{W}^*[t], \ \ \tau \leq t \leq t_1,$$

and would therefore ensure the desired terminal condition $x[t_1] \in \mathcal{M}$.

Here $\mathcal{W}^*[t] = \mathcal{W}_*[t]$ is the *solvability set* of Definition 1.8.2 which, under Assumptions 1.7.1 or 1.7.2 presumed here, could be specified through the Alternated Integral (1.7.8), so that the set-valued function $\mathcal{W}^*[t]$ would satisfy the evolution equation (1.7.9), (1.7.10).

The exact solution scheme requires, as we have seen, to calculate the tube $\mathcal{W}^*[t]$ and then, for each instant of time t, to solve an extremal problem of type (1.8.10) whose solution finally allows us to specify the desired strategy $\mathcal{U}(t,x) = \mathcal{U}^0(t,x)$ according to (1.8.9). The strategy $\mathcal{U}^0(t,x)$ is again actually defined *as an algorithm* which, due to the presence of uncertain items, is more complicated, of course, than in the absence of these.

To obtain a simpler scheme, we shall substitute $\mathcal{W}^*[t]$ by one of its *internal ellipsoidal approximations* $\mathcal{E}_-[t] = \mathcal{E}(x^*, X_-(t))$. The conjecture is that once $\mathcal{W}^*[t]$ is substituted by $\mathcal{E}_-[t]$, we should just copy the scheme of Section 1.8. Namely, we should construct the new approximate strategy $\mathcal{U}^0_-(t,x)$ such that for every solution $x[t] = x(t,\tau,x_\tau)$ to the system

$$\text{(3.8.1)} \quad \dot{x}[t] = \mathcal{U}^0_-(t, x[t]) + f(t), \ \tau \leq t \leq t_1, \ x[\tau] = x_\tau, \ x_\tau \in \mathcal{E}_-[\tau],$$

the inclusion

$$\text{(3.8.2)} \qquad x[t] \in \mathcal{E}_-[t], \quad \tau \leq t \leq t_1$$

would be true, whatever is the function $f(t) \in \mathcal{E}(q(t), Q(t)$. This would ensure the terminal condition

$$x[t_1] \in \mathcal{E}(m, M) = \mathcal{M} = \mathcal{E}[t_1].$$

It will be proven again that once the approximation $\mathcal{E}_-[t]$ is selected appropriately, namely, due to relations (3.5.9), (3.5.30), (3.5.31), the desired strategy $\mathcal{U}^0_-(t,x)$ may be constructed as in Section 1.8, except that $\mathcal{W}^*[t]$ should be substituted by $\mathcal{E}_-[t]$. Namely,

$$\text{(3.8.3)} \quad \mathcal{U}^0_-(t,x) = \begin{cases} \mathcal{E}(p(t), P(t)) & \text{if } x \in \mathcal{E}_-[t] \\ p(t) - P(t)l^0(l^0, P(t)l^0)^{-1/2} & \text{if } x \notin \mathcal{E}_-[t], \end{cases}$$

where $l^0 = l^0(t,x) = \partial d(x, \mathcal{E}_-[t])/\partial x$ is the unit vector that solves the problem
(3.8.4)

$$d[t,x] = (l^0, x) - \rho(l^0 \mid \mathcal{E}_-[t]) = \max\{(l,x) - \rho(l|\mathcal{E}_-[t]) | \|l\| \leq 1\} \ ,$$

and as before

$$(3.8.5)\, d[t,x] = d(x, \mathcal{E}_-[t]) = h_+(x, \mathcal{E}_-[t]) = \min\{\|x - s\| | s \in \mathcal{E}_-[t]\} \ .$$

Remark 3.8.1 *We emphasize again that the given scheme follows the lines of Section 3.6, but* **the tube** $\mathcal{E}_-[t] = \mathcal{E}_-(t,x)$ **taken here is defined by relations (3.5.9), (3.5.30)** *rather than by (3.4.2), (3.4.4) as in Section 3.6. This reflects* **the uncertainty (3.1.6) in the inputs** f **of the system.**

The further reasoning is analogous to that of Section 3.6. Without repeating the similar elements in the scheme, we have to underline that the main new point here is the calculation of the derivative $dd[t,x]/dt$ due to the differential inclusion (3.8.1) with $\mathcal{E}_-[t] = \mathcal{E}_-(x^*, X_-(t))$ defined by (3.5.9), (3.5.30).

The desired solution strategy $\mathcal{U}^0_-(t,x)$ must satisfy a relation of type (1.8.9) which depends on vector $l^0 = l^0(t,x)$, the maximizer for problem (3.8.4), a direct analogue of problem (1.8.10) of Section 1.8. The respective relations may now be obtained in more detail than in the general case of Section 1.8, since $\mathcal{E}_-[t]$ is an ellipsoid.

The properties of l^0 are similar to those described in (3.6.7), (3.6.8) and in Lemma 3.6.1. Further on, we notice that again $\mathcal{P}(t) = \mathcal{E}(p(t), P(t))$ is an ellipsoid, so that problem (1.8.9) reduces to

$$(3.8.6) \qquad \arg\max\{(-l^0, u) | u \in \mathcal{E}(p(t), P(t))\} = \mathcal{U}^0_-(t,x),$$

and therefore relation (3.8.3) follows from Lemma 3.6.2.

We will now prove that the *ellipsoidal-based strategy* $\mathcal{U}^0_-(t,x)$ of (3.8.3) does solve the problem of control synthesis of Definition I.8.1, provided we start from a point $x_\tau = x(\tau) \in \mathcal{E}_-[\tau]$. Indeed, assume $x_\tau \in \mathcal{E}_-[\tau]$ and $x[t] = x(t, \tau, x_\tau)$, $\tau \leq t \leq t_1$ to be the respective trajectory. We will demonstrate that once $x[t]$ is a solution to (3.8.1), $\mathcal{U}(t,x) = \mathcal{U}^0_-(t,x)$, it will always satisfy (3.8.2).

Calculating

$$d[t] = d[t, x[t]] = \max\{(l, x[t]) - \rho(l \mid \mathcal{E}_-[t]) \| \|l\| \leq 1\},$$

we observe

$$\frac{d}{dt}d[t] = \frac{d}{dt}[(l^0, x[t]) - \rho(l^0|\mathcal{E}_-[t])] \ ,$$

and since l^0 is a unique maximizer, $(d[t] > 0)$,

$$\frac{d}{dt}d[t] = (l^0, \dot{x}[t]) - \frac{\partial}{\partial t}\rho(l^0|\mathcal{E}_-[t]) = \tag{3.8.7}$$

$$= (l^0, u[t] + f(t)) - \frac{d}{dt}[(l^0, x(t)) + (l^0, X_-(t)l^0)^{1/2}],$$

where $\mathcal{E}_-[t] = \mathcal{E}(x(t), X_-(t))$. For fixed functions $\pi(\cdot), H(\cdot)$ we have $\mathcal{E}_-[t] = \mathcal{E}(x(t), X_-(t))$, where $x(t), X_-(t)$ satisfy the system (3.5.9), (3.5.30).

Substituting this into (3.8.7) and differentiating the respective function of the maximum type due to equation

$$\frac{d}{dt}x[t] = u[t] + f(t),$$

where $u[t] \in \mathcal{U}(t, x[t])$ is a realization of the feedback control strategy $\mathcal{U}$ and $f(t)$ is an input disturbance, we have

$$\frac{d}{dt}d[t] = (l^0, u[t] + f(t)) - (l^0, p(t) + q(t)) - \frac{1}{2}(l^0, X_-(t)l^0)^{-1/2}\cdot$$

$$\cdot\{(l^0, (\pi(t)X_-(t) + \pi^{-1}(t)Q(t))l^0) -$$

$$(l^0, H^{-1}(t)([H(t)X_-(t)H(t)]^{1/2}[H(t)P(t)H(t)]^{1/2} +$$

$$+[H(t)P(t)S(t)]^{1/2}[H(t)X_-(t)H(t)]^{1/2})H^{-1}(t)l^0)\} \ .$$

Applying inequality $a^2 + b^2 \geq 2ab$ and the the Bunyakovsky-Schwartz inequality to the right-hand part of the previous formula, we come to

$$\frac{d}{dt}d[t] \leq \tag{3.8.8}$$

$$\leq (l^0, u[t] + f(t)) - (l^0, p(t) + q(t)) + (l^0, P(t)l^0)^{1/2} - (l^0, Q(t)l^0)^{1/2} \ ,$$

where

$$u[t] \in \mathcal{E}(p(t), P(t)), \;\; f(t) \in \mathcal{E}(q(t), Q(t),$$

In other terms we have

$$\frac{d}{dt}d[t] \leq (l^0, u[t] + f(t)) \; + \; \rho(-l^0 | \mathcal{E}(p(t), P(t))) \; - \; \rho(l^0 | \mathcal{E}(q(t), Q(t))) \; .$$

With $u[t] \in \mathcal{U}_-^0(t, x)$ and any feasible $f(t)$ this yields (almost everywhere)

$$\frac{d}{dt}d[t] \leq 0, \;\; x[t] \notin \mathcal{E}_-[t],$$

due to (3.8.6).

This also gives

$$\frac{d}{dt}V(t, x) \leq 0, \;\; V(t, x) = d^2[t].$$

Integrating $dd^2[t]/dt$ from τ to t_1, we come to the inequality ($d[t] = d^2_{\mathcal{W}^*}[t, x[t]]$, see notation of 1.8.19)

$$h_+^2(x[t_1], \mathcal{M}) \; = \; d^2_{\mathcal{W}^*}[t_1, x[t_1]] \; \leq \; d^2\mathcal{W}^*[\tau, x[\tau]] \; = \; h_+^2(x[\tau], X_-(\tau)) \;\; ,$$

so that $x[t_1] \in \mathcal{M}$ if $x[\tau] \in X_-(\tau)$. What follows is the assertion

Theorem 3.8.1 *Define an internal ellipsoidal approximation* $\mathcal{E}_-[t] = \mathcal{E}_-(x(t), X_-(t))$ *to the solvability set* $\mathcal{W}_*[t]$, *with given parametrization* $H(t), \pi(t)$ *in (3.5.30). Once* $x[\tau] \in \mathcal{E}_-[\tau]$ *and the synthesizing strategy is selected as* $\mathcal{U}_-^0(t, x)$ *of (3.8.6), (3.8.3), the following inclusion is true:*

$$x[t] \in \mathcal{E}_-[t], \quad \tau \leq t \leq t_1,$$

whatever is the solution $x[t]$ *to the differential inclusion*

(3.8.9) $$\frac{d}{dt}x \; \in \; \mathcal{U}_-^0(t, x) \; + \; \mathcal{E}(q(t), Q(t)),$$

and therefore

$$x[t_1] \in \mathcal{E}(m, M),$$

whatever is the disturbance $f(t) \in \mathcal{E}(q(t), Q(t))$ *in the synthesized system*

$$\frac{d}{dt}x \; = \; \mathcal{U}_-^0(t, x) \; + \; f(t) \;\; .$$

The ellipsoidal synthesis thus gives a solution strategy $\mathcal{U}_-^0(t,x)$ *for any internal approximation* $\mathcal{E}_-[t] = \mathcal{E}_-(x(t), X_-(t))$ *of the solvability tube* $\mathcal{W}^*[t]$. With $x \notin \mathcal{E}_-[t]$, the function $\mathcal{U}_-^0(t,x)$ is single-valued, while with $x \in \mathcal{E}_-[t]$ it is multivalued ($\mathcal{U}_-^0(t,x) = \mathcal{E}_-[t]$), being therefore upper-semicontinuous in x, measurable in t, and ensuring the existence of a solution to the differential inclusion (3.8.9). Theorem 3.8.1 indicates that each of the tubes $\mathcal{E}[t]$ is an ellipsoidal-valued bridge (see Remark 3.5.3).

Remark 3.8.2 *(i) Due to Theorem 3.5.2 (see 3.5.36), each element* $x \in int W^*[t]$ *belongs to a certain ellipsoid* $\mathcal{E}_-[t]$ *and may therefore be steered to the terminal set* $\mathcal{M}$ *by means of a certain ellipsoidal strategy* $\mathcal{U}_-(t,x)$. *Such a strategy may be specified in explicit form except for a scalar multiplier* $\lambda = \lambda(t,x)$, *which may be calculated in advance, as indicated in Remark 3.6.2. With this reservation, the suggested strategy* $\mathcal{U}_-^0$ *may be interpreted as* **an analytical design.**

(ii) We emphasize once more that the constructions given in Sections 3.5 and 3.8 are derived here under Assumption 1.7.1, which implies that there exists an internal curve $\tilde{x}(t)$ *such that* $\tilde{x}(t) + \epsilon(t)S_1(0) \subseteq \mathcal{W}_*[t]$ *for all* t *with continuous* $\epsilon(t) > 0$. *Then clearly* $int\mathcal{W}_*[t] \not\equiv \emptyset$.

We shall now proceed with further numerical examples (this time for uncertain systems) that demonstrate the constructive nature of the suggested solution schemes.

3.9 Control Synthesis for Uncertain Systems: Numerical Examples

In this section our particular intention first is to illustrate through simulation the effect of introducing an unknown but bounded disturbance $f(t)$ into the system. We shall do this by considering *a sequence of three problems* where only *the size of the bounding sets* for the disturbances $f(t)$ increases from case to case, starting from no disturbance at all (that is where the sets $\mathcal{Q}(t) = \mathcal{E}(q(t), Q(t))$, $t \in [t_0, t_1]$ are singletons) to more disturbance allowed, so that the problem still remains solvable.

The result is that in the first case we obtain a large internal ellipsoidal estimate $\mathcal{E}_-[t]$ of the solvability set $\mathcal{W}^*[t_0] = \mathcal{W}^*(t_0, t_1, \mathcal{M})$, while in the last it shrinks to be small. We also indicate the behaviour of isolated trajectories of system (3.8.1) in the presence of various specific feasible disturbances $f(t) \in \mathcal{E}(q(t), Q(t))$.

For the calculations we use a standard first-order discrete scheme for equations (3.8.9) and (3.8.30) by dividing the time interval – chosen to be $[0, 5]$ – into 100 subintervals of equal length (the details of such schemes may found in [63], [272], [273]. Instead of the set valued control strategy (3.8.3) we apply a *single valued* selection:

$$(3.9.1)\ u(t,x) = \left\{ \begin{array}{ll} p(t) & \text{if } x \in \mathcal{E}_-[t] \\ p(t) - P(t)l^0(l^0, P(t)l^0)^{-1/2} & \text{if } x \notin \mathcal{E}_-[t]. \end{array} \right\} ,$$

again in its discrete version. (The discrete version obviously does not require any additional justification for using the single-valued selection.) We calculate the parameters of the ellipsoid $\mathcal{E}_-[t] = \mathcal{E}(x(t), X_-(t))$ by chosing a specific parametrization which is

$$H(t) = P^{-1/2}(t)$$

and

$$\pi(t) = \frac{\mathrm{tr}^{1/2}(X_-(t))}{\mathrm{tr}^{1/2}(Q(t))}$$

in equation (3.5.30). We consider a 4-dimensional system of type (1.1.1), (3.1.2)–(3.1.4) with the initial position $\{0, x_0\}$ given by

$$x_0 = \begin{pmatrix} 2 \\ -10 \\ 1 \\ -6 \end{pmatrix},$$

at the initial moment $t_0 = 0$ and target set $\mathcal{M} = \mathcal{E}(m, M)$ defined by

$$m = \begin{pmatrix} 10 \\ 0 \\ 0 \\ 10 \end{pmatrix}$$

and

$$M \equiv \begin{pmatrix} 100 & 0 & 0 & 0 \\ 0 & 100 & 0 & 0 \\ 0 & 0 & 100 & 0 \\ 0 & 0 & 0 & 100 \end{pmatrix}$$

at the final moment $t_1 = 5$. We suppose the right hand side to be constant:

$$A(t) \equiv \begin{pmatrix} 0 & 1 & 0 & 0 \\ -1 & 0 & 0 & 0 \\ 0 & 0 & 0 & 1 \\ 0 & 0 & -4 & 0 \end{pmatrix} ,$$

describing the position and velocity of two independent oscillators. (Through the constraints on the control and disturbance, however, the system becomes coupled.) The restriction $u(t) \in \mathcal{E}(p(t), P(t))$ on the control and $v(t) \in \mathcal{E}(q(t), Q(t))$ on the disturbance is also defined by time independent constraints:

$$p(t) \equiv \begin{pmatrix} 0 \\ 0 \\ 0 \\ 0 \end{pmatrix} ,$$

$$P(t) \equiv \begin{pmatrix} 9 & 0 & 0 & 0 \\ 0 & 1 & 0 & 0 \\ 0 & 0 & 9 & 0 \\ 0 & 0 & 0 & 1 \end{pmatrix} .$$

The center of the disturbance is the same in all cases:

$$q(t) \equiv \begin{pmatrix} 0 \\ 0 \\ 0 \end{pmatrix} .$$

The difference between the three cases $i = 1, 2, 3$ appear in the matrices:

$$Q^{(1)}(t) \equiv \begin{pmatrix} 0 & 0 & 0 & 0 \\ 0 & 0 & 0 & 0 \\ 0 & 0 & 0 & 0 \\ 0 & 0 & 0 & 0 \end{pmatrix} ,$$

$$Q^{(2)}(t) \equiv \begin{pmatrix} 1 & 0 & 0 & 0 \\ 0 & 9 & 0 & 0 \\ 0 & 0 & 1 & 0 \\ 0 & 0 & 0 & 9 \end{pmatrix} ,$$

$$Q^{(3)}(t) \equiv \begin{pmatrix} 1 & 0 & 0 & 0 \\ 0 & 13.1 & 0 & 0 \\ 0 & 0 & 1 & 0 \\ 0 & 0 & 0 & 13.1 \end{pmatrix} .$$

Clearly, case $i = 1$ is the one treated in Section 3.7, but note that in the cases $i = 2, 3$ the data are chosen in such a way that neither the controls, nor the disturbances dominate the other, that is, both $\mathcal{P} \dot{-} \mathcal{Q} = \emptyset$ and $\mathcal{Q} \dot{-} \mathcal{P} = \emptyset$.

Obviously, in these cases the problem cannot be reduced to simpler situations without disturbances. More precisely, in these cases *Assumption 1.6.1* that allows such a reduction *is not fulfilled.* At the same time, the solvability set $W_*[t]$ contains an internal trajectory so that $\text{int}W_*[t] \neq \emptyset$ (see Remark 3.8.2(ii)). Its internal ellipsoidal approximations $\mathcal{E}_-[t]$ exist and may be calculated due to schemes of Section 3.5.

The calculations give the following internal ellipsoidal estimate $\mathcal{E}_-^{(i)}[0] = \mathcal{E}(x(0), X_-^{(i)}(0))$ of the solvability set $\mathcal{W}^{(i)}(0, \mathcal{M})$, $i = 1, 2, 3$:

$$x(0) = \begin{pmatrix} 2.4685 \\ -8.4742 \\ 1.5685 \\ -5.2087 \end{pmatrix},$$

and

$$X_-^{(1)}(0) = \begin{pmatrix} 323.9377 & 30.2735 & 0 & 0 \\ 30.2735 & 341.4382 & 0 & 0 \\ 0 & 0 & 147.0094 & 61.1077 \\ 0 & 0 & 61.1077 & 469.5488 \end{pmatrix},$$

$$X_-^{(2)}(0) = \begin{pmatrix} 46.3661 & 25.5502 & 0 & 0 \\ 25.5502 & 66.4791 & 0 & 0 \\ 0 & 0 & 45.3047 & 28.3397 \\ 0 & 0 & 28.3397 & 132.7509 \end{pmatrix},$$

$$X_-^{(3)}(0) = \begin{pmatrix} 12.2863 & 21.2197 & 0 & 0 \\ 21.2197 & 37.8930 & 0 & 0 \\ 0 & 0 & 33.6241 & 22.3911 \\ 0 & 0 & 22.3911 & 98.7732 \end{pmatrix}.$$

Now, as is easy to check, $x_0 \in \mathcal{E}(x(0), X_-^{(i)}(0))$ for $i = 1, 2, 3$ and therefore Theorem 3.8.1 is applicable, implying that the control strategy of (3.8.3) steers the solution of (3.8.1) into $\mathcal{M}$ under any admissible disturbance $f(t) \in \mathcal{E}(q(t), Q^{(i)}(t))$ in all three cases. Also, as it can be proved on the basis of their construction, we have the inclusions

$$\mathcal{E}(x(0), X_-^{(3)}(0)) \subset \mathcal{E}(x(0), X_-^{(2)}(0)) \subset \mathcal{E}(x(0), X_-^{(1)}(0))$$

holding, analogously to the corresponding inclusions between the original (nonellipsoidal) solvability sets $\mathcal{W}^{(i)}(0,5,\mathcal{M})$.

Since the ellipsoids appearing in this problem are *four dimensional*, and since the objective is to describe the solutions also through graphical representations, we present their *two dimensional projections*. The figures are therefore divided into four *windows*, showing projections of the original ellipsoids onto the planes spanned by the first and second ($\{x_1, x_2\}$), third and fourth ($\{x_3, x_4\}$), first and third($\{x_1, x_3\}$), and second and fourth ($\{x_2, x_4\}$) coordinate axes, in a clockwise order starting from bottom left. The drawn segments of coordinate axes corresponding to the state variables range from -30 to 30. The skew axis in Figures 3.9.1 to 3.9.3 is time, ranging from 0 to 5. Figures 3.9.1 to 3.9.3 show the graph of the ellipsoidal valued maps $\mathcal{E}_-^{(i)}[t]$, $t \in [0,5]$, $i = 1,2,3$, respectively, and of the solutions to equation

$$x(t_{k+1}) - x(t_k) = \sigma(A(t_k)x(t_k) + u(t_k, x(t_k)) + f(t_k)) \tag{3.9.2}$$

$$x[0] = x_0,\ 0 = t_0 \leq t \leq t_{100} = 5, \sigma = t_{k+1} - t_k > 0,\ \ k = 1, ..., 100\ \ ,$$

which is a discrete version of the equation

$$\dot{x}[t] = A(t)x[t] + u(t, x[t]) + f(t)\ \ .$$

(There may be problems with defining the existence of solutions to the last equation , however, since function $u(t,x)$ may turn to be discontinuous in x. We will therefore avoid this last equation and refer only to (3.9.2) and (3.7.1), see analogous situation in Section 3.7.)

Here $u(t,x)$ is defined by (3.9.1) and we consider three different choices of the disturbance $f(t)$, one being $f(t) \equiv 0$ and two other – so-called extremal bang-bang type – feasible disturbances. The construction of these disturbances is the following. The time interval $[0,5]$ is divided into subintervals of constant lengths. A value f is chosen randomly at the boundary of $\mathcal{E}(q(t), Q^{(i)}(t))$ and the disturbance is then defined by

$$f(t) = f$$

over all the first interval and

$$f(t) = -f$$

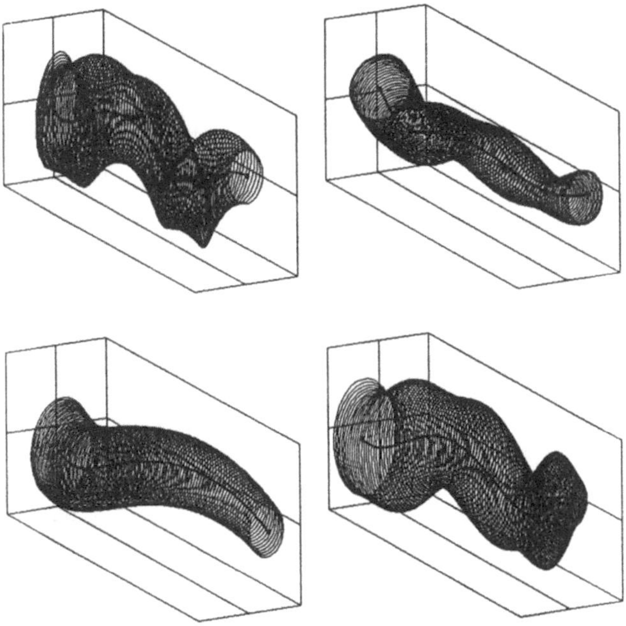

Figure 3.9.1.

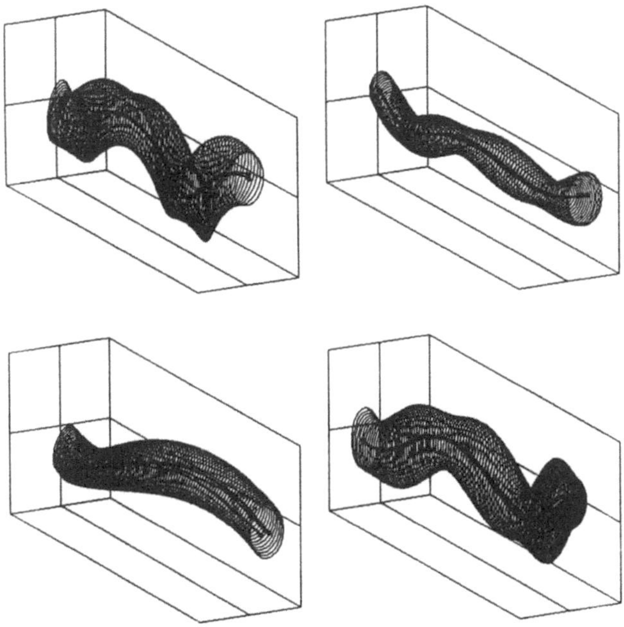

Figure 3.9.2.

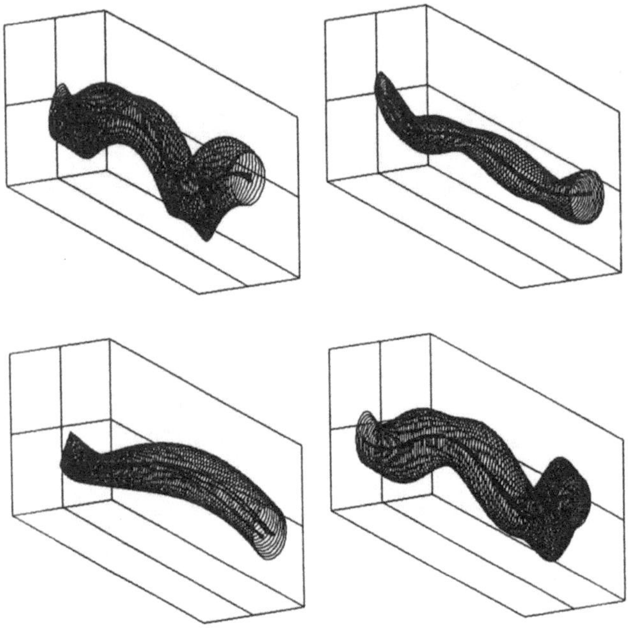

Figure 3.9.3.

over the second. Then a new value for f is selected and the above procedure is repeated for the next pair of intervals, etc. The controlled trajectory, that is the solution to (3.9.1), (3.9.2), is drawn in a thin line if it is inside the current ellipsoidal solvability set, and by a thick line if it is outside. So the statement of Theorem 3.8.1 is that the control ensures that a thin line cannot change into thick.

Figures 3.9.4 to 3.9.6 show the target set $\mathcal{M} = \mathcal{E}(m, M)$, (projections appearing as circles of radius 10), the solvability set $\mathcal{E}_-^{(i)}[0] = \mathcal{E}(w(0), W_-^{(i)}(0))$ at $t = 0$, and trajectories of the same solutions of (3.9.1), (3.9.2) in phase space.

The ellipsoids $\mathcal{E}_-[0]$ are only subsets of the respective solvability sets $\mathcal{W}^*(0, 5, \mathcal{M})$; therefore from Theorem 3.8.1 there does not follow a negative statement like, if the initial state is not contained in $\mathcal{E}_-[t_0]$, then it is not true that the trajectory can be steered into the target set $\mathcal{M}$ under any disturbance $f(t) \in \mathcal{Q}(t)$. However, if the ellipsoidal approximation $\mathcal{E}_-[0] \subset \mathcal{W}^*(0, 5, \mathcal{M})$ is appropriate, then it may occur that such a behaviour can be illustrated on the ellipsoidal approximations. To

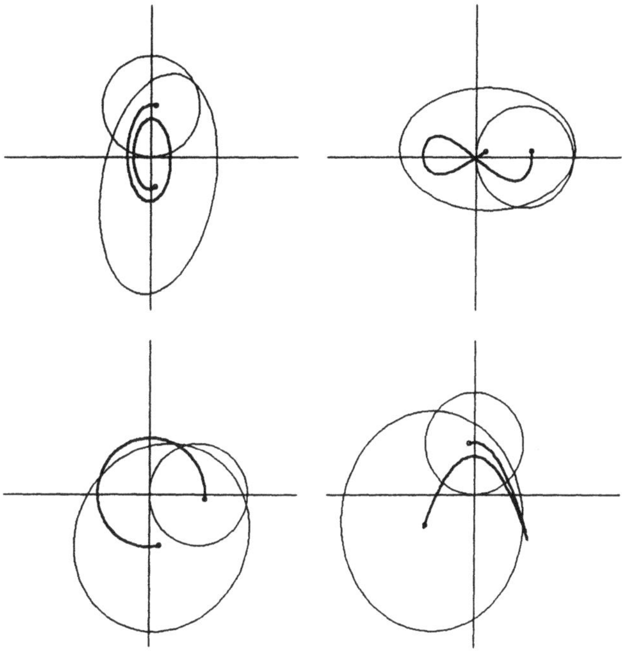

Figure 3.9.4.

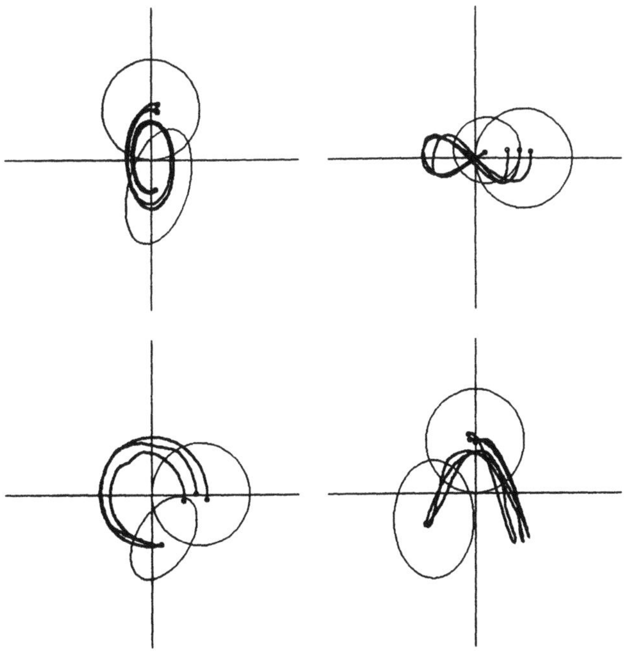

Figure 3.9.5.

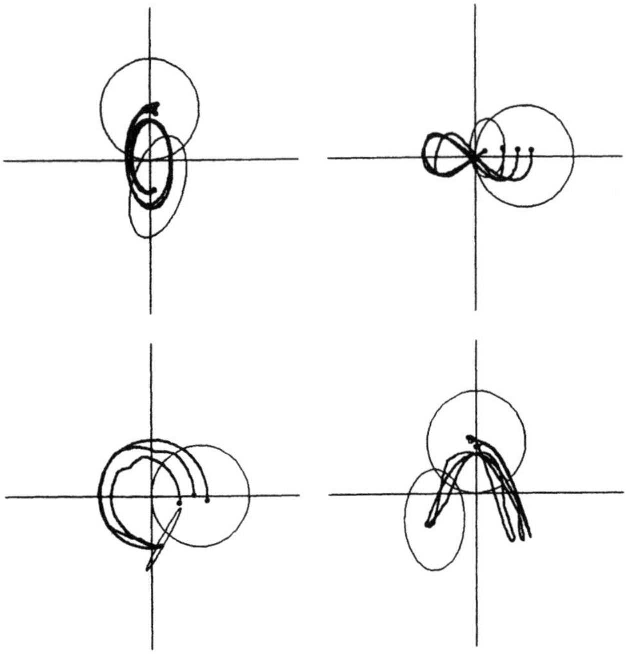

Figure 3.9.6.

show this, we return to the parameter values of the previous examples and change the initial state only, by moving it in such a way that

$$x_0 \in \mathcal{E}_-^{(1)}[0] \setminus \mathcal{E}_-^{(2)}[0] \tag{3.9.3}$$

holds, taking

$$x_0 = \begin{pmatrix} -12 \\ 0 \\ 3 \\ 0 \end{pmatrix} .$$

In Figures 3.9.7 and 3.9.8 it can be seen that relation (3.9.3) holds indeed. The trajectory in Figure 3.9.7 successfully hits the target set $\mathcal{M}$ at $t = 5$. (This is case $i = 1$, so there is no disturbance.)

Figure 3.9.8 shows two trajectories under two simulated feasible disturbances $f(t) \in \mathcal{E}(q(t), Q(t))$. In one case the control rule defined using the ellipsoidal tube $\mathcal{E}_-^{(2)}[t]$ steers the trajectory into the target $\mathcal{M}$, while under the other disturbance, it does not succeed. (One thick trajectory

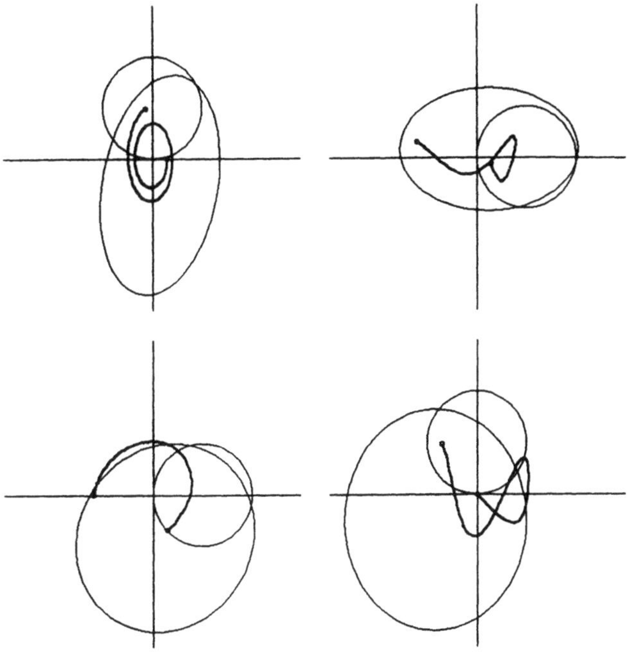

Figure 3.9.7.

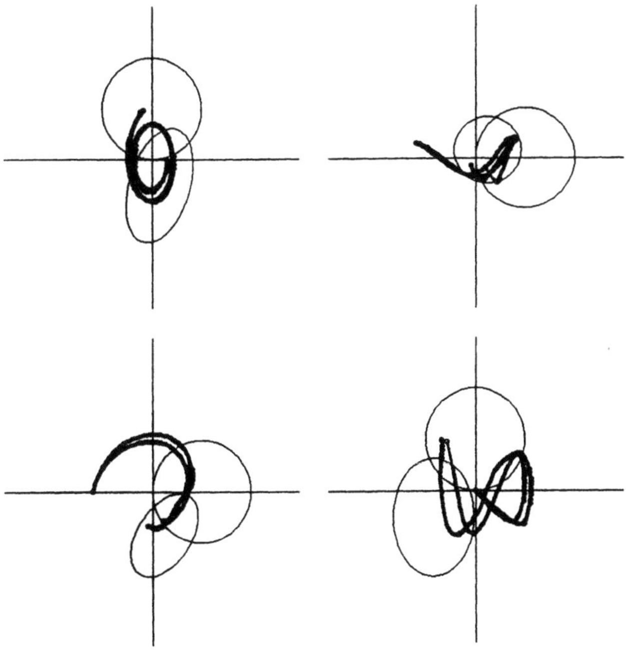

Figure 3.9.8.

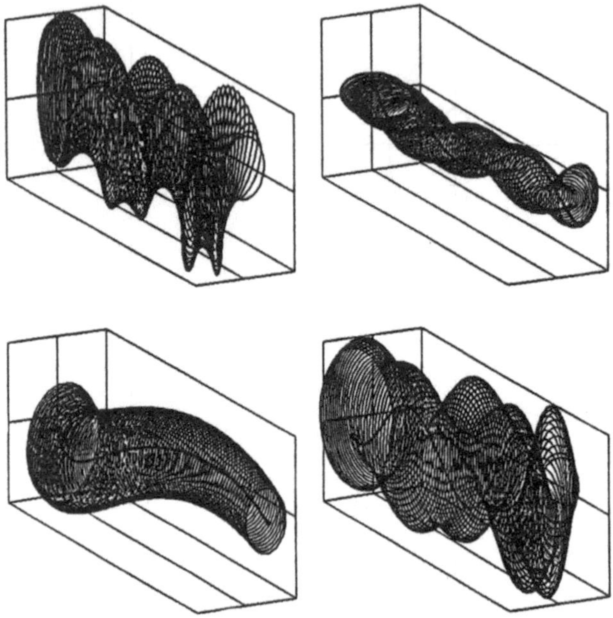

Figure 3.9.9.

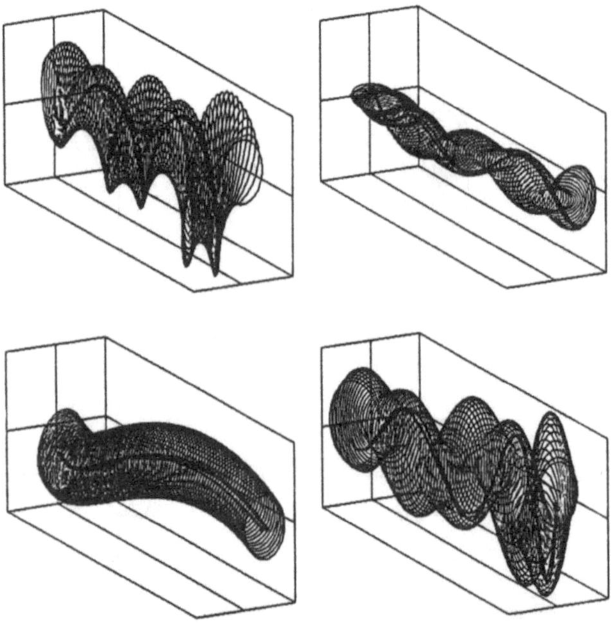

Figure 3.9.10.

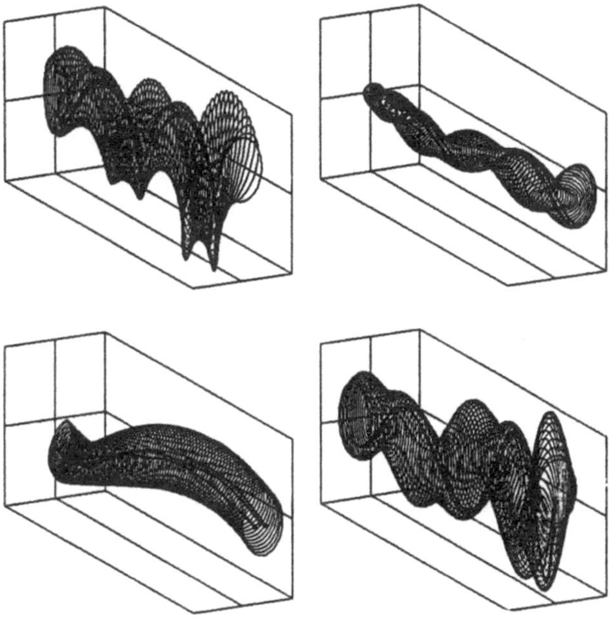

Figure 3.9.11.

changing into thin is clearly seen in the right hand side windows, and the projection of the endpoint of the other is outside in the lower left window. Compare these examples with those of Section 3.7.) There may of course be other control rules, like the one based on the exact (nonellipsoidal) solvability sets $\mathcal{W}^*[0] = \mathcal{W}^*(t, t_1, \mathcal{M})$, that could be successful, once $x(0) \in \mathcal{W}^*[0]$.

Finally we again consider a system that describes *two coupled oscillators* with matrix

$$A(t) \equiv \begin{pmatrix} 0 & 1 & 0 & 0 \\ -1 & 0 & 0 & 0 \\ 0 & 0 & 0 & 1 \\ -1 & 0 & -9 & 0 \end{pmatrix} ,$$

and with the other parameters (x^0, P, p, M, m, q) same as in the previous figures. Taking the disturbances to be restricted by $Q^{(1)}, Q^{(2)}, Q^{(3)}$ of the above and simulating the respective target control synthesis problem, we come to results shown in Figures 3.9.9 to 3.9.11 accordingly.

3.10 Target Control Synthesis Within Free Time Interval

Considering again the Problem of Control Synthesis Under Uncertainty of Definition 1.8.1, we shall modify this definition by deleting the requirement that the terminal instant t_1 is fixed. Thus, we shall require that the terminal inclusion $x(t) \in \mathcal{M}$ can be reached *at any instant* $t \in (t_0, t_1]$ (namely, not later than at t_1 rather than at fixed t_1, as before). We shall look for an ellipsoidal control synthesis solution to this problem within a scheme similar in nature to the one of Section 3.8. We have in view that the constraints on u, f and the target set $\mathcal{M}$ are all ellipsoidal-valued, as in Sections 3.1 and 3.8.

We shall now briefly describe this problem without going into specific details with the main aim to demonstrate a numerical example of a *nonconvex solvability set.*

Recall the solvability set of Section 1.8. For time interval $[\tau, t]$ it should be denoted, according to the respective notations, as $\mathcal{W}^*(\tau, t, \mathcal{M})$. Our new problem *with free terminal time* t will then be solvable for a given position $\{\tau, x\}$ if and only if $x \in W_f(\tau, \mathcal{M})$, where

$$W_f(\tau, \mathcal{M}) = \bigcup \{\mathcal{W}^*(\tau, t, \mathcal{M}) : t \in [\tau, t_1]\} \ .$$

Clearly, set $W_f(\tau, \mathcal{M})$ is not bound to be convex. The results of the previous sections allow us to formulate the following assertion, see [306]. (Here the earlier symbols Π^+, Σ of Sections 3.2 to 3.5 for the classes of functions $\pi(\cdot), H(\cdot)$ are complemented by $[t_0, t_1]$ that symbolizes the interval where these functions are defined.)

Theorem 3.10.1 *Fix continuous functions* $\pi(t) \in \Pi^+[t_0, t_1]$ *and* $H(t) \in \Sigma[t_0, t_1]$ *and define an internal approximation*

$$\mathcal{E}_-[\tau, t, \mathcal{M}] = \mathcal{E}(x(\tau), X_-(\tau|\pi(\cdot), H(\cdot)); t, \mathcal{M})$$

of $\mathcal{W}(\tau, t, \mathcal{M})$ *for* $\tau \in [t_0, t]$.[24]

[24] Here symbol $\mathcal{E}_-[\tau, t, \mathcal{M}] = \mathcal{E}(x(\tau), X_-(\tau|\pi(\cdot), H(\cdot)), t, \mathcal{M})$ stands for the internal ellipsoid described by equations (3.5.30) or (3.5.41), but with boundary condition (3.5.31) taken at instant t instead of t_1.

Once

(3.10.1) $$x_\tau = x(\tau) \in \mathcal{E}_-[\tau, t, \mathcal{M}]$$

for some $\tau \in [t_0, t]$ *and* $x[t'] = x(t', \tau, x_\tau)$ *is a solution to (3.8.1),(3.8.3), where* $\mathcal{E}_-[\tau]$ *is substituted by* $\mathcal{E}[\tau, t, \mathcal{M}]$, *the inclusion*

$$x[t'] \in \mathcal{E}(t', \tau, \mathcal{M})$$

shall be true for all $t' \in [\tau, t]$ *and in particular*

$$x[t] \in \mathcal{E}(m, M) \quad .$$

Hence, the strategy (3.8.3) taken for $\mathcal{E}_-[\tau] = \mathcal{E}[\tau, t, \mathcal{M}]$ *solves the terminal control problem by time* t, *whatever is the disturbance* $f(t) \in \mathcal{E}(q(t), Q(t)$.

Proof. Follows from the fact that

$$\bigcup \{\mathcal{E}_-[\tau, t, \mathcal{M}] | t \in [\tau, t_1]\} \subseteq W_f(\tau, \mathcal{M}) \quad ,$$

where $\mathcal{E}_-[\tau, t, \mathcal{M}] = \mathcal{E}(x(\tau), X_-(\tau|\pi(\cdot), H(\cdot)), t, \mathcal{M})$ and the pair $\pi(\cdot), H(\cdot)$ is fixed. Q.E.D.

Denote

$$\bigcup \{\mathcal{E}_-[\tau, t, \mathcal{M}] | t \in [\tau, t_1]\} = E_f(\tau, \mathcal{M} | \pi(\cdot), H(\cdot)) \quad .$$

Then, if

$$x_\tau \in E_f(\tau, \mathcal{M} | \pi(\cdot), H(\cdot)) \quad ,$$

there exists a minimal value $t = t^*$ among those t that ensure $x_\tau \in \mathcal{E}[\tau, t, \mathcal{M}]$. This is due to the continuity of the distance function

$$d(x_\tau, \mathcal{E}[\tau, t, \mathcal{M}]) = h_+(x_\tau, \mathcal{E}[\tau, t, \mathcal{M}]) = d[x_\tau, t]$$

in t (check this assertion), so that t^* is the minimal root of the equation

(3.10.2) $$d[x_\tau, t] = 0 \quad .$$

Denote $d_e(x_\tau, W^*(\tau, t, \mathcal{M})) = d_e[x_\tau, t]$ and t_e^* to be the minimal root of equation

(3.10.3) $$d_e[x_\tau, t] = 0 \quad ,$$

(the latter function d_e is also continuous in t). Time t^* shall then be the exact optimal time. But since $0 \leq d_e[x_\tau, t] \leq d[x_\tau, t]$, we further come to the following fact

Lemma 3.10.1 *The optimal time $t_e^* \leq t^*$, whatever is the internal tube $\mathcal{E}[\tau, t, \mathcal{M}]$ that generates the value t^*.*

Remark 3.10.1 *One should be aware that in general the functions $d[x_\tau, t], d_e[x_\tau, t]$ are not monotonous in t, so that the practical calculation of the roots of equations (3.10.1) and (3.10.2) may lead to unstable numerical procedures that require additional regularization.*

Exercise 3.10.1. Check the the following assertion.

Fix continuous functions $\pi(\cdot) \in \Pi^+[t_0, t]$ and $H(\cdot) \in \Sigma[t_0, t]$ for all $t \in [t_0, t_1]$ and define an external approximation $\mathcal{E}_+[\tau, t, \mathcal{M}] = \mathcal{E}_+(x(\tau), X_+(t|\pi(\cdot), H(\cdot)); t, \mathcal{M})$ of $\mathcal{W}(\tau, t, \mathcal{M})$ for $\tau \in [t_0, t]$.

Once

$$x_{t_0} = x[t_0] \notin \mathcal{E}_+[t_0, t, \mathcal{M}] \tag{3.10.4}$$

for all $t \in [t_0, t_1]$, then the problem of target control synthesis of this section (under uncertainty, with free target time), cannot be solved.

We shall now proceed with numerical examples. For the calculations we use the same discrete scheme as in Section 3.9 (dividing the time interval – chosen to be $[0, 5]$ – into 100 subintervals of equal lengths) and the control strategy of type (3.9.1) is found here through the same parametrization.

The parameters A, M, p, P, q are the same as in the examples of Figures 3.9.1 to 3.9.3, except that the initial position is given by

$$x_0 = \begin{pmatrix} 0 \\ -20 \\ 0 \\ 5 \end{pmatrix},$$

at the initial instant $t_0 = 0$ and for the target set $\mathcal{M} = \mathcal{E}(m, M)$ we have:

$$m = \begin{pmatrix} 20 \\ 0 \\ 0 \\ -20 \end{pmatrix}.$$

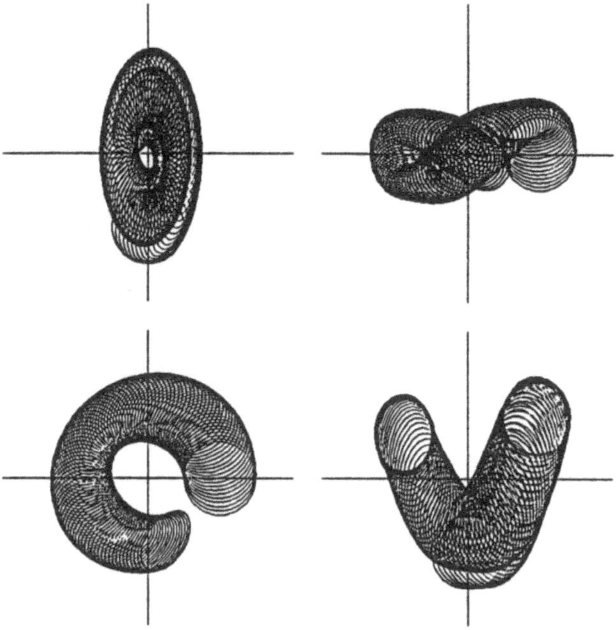

Figure 3.10.1.

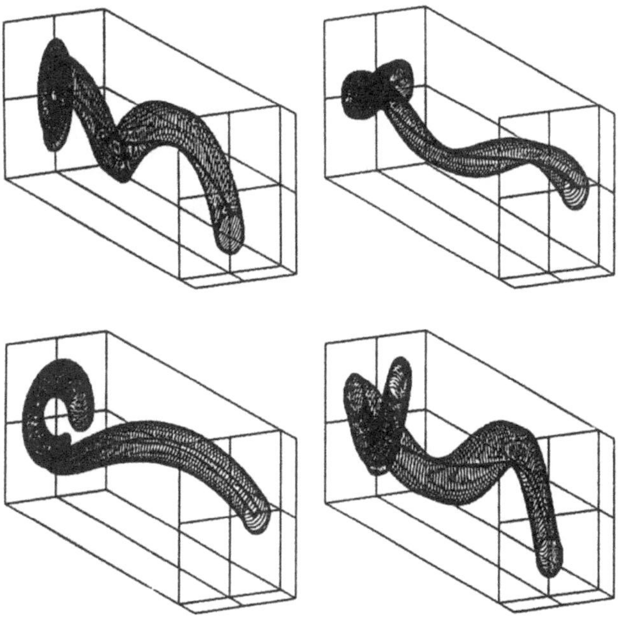

Figure 3.10.2.

For the constraint $\mathcal{E}(q, Q)$ on the adversary f here the matrix $Q = Q^{(2)}$ of Section 3.9. Note that the data are chosen in such a way that neither the controls, nor the disturbances dominate the other, that is, neither $\mathcal{E}(p, P) = \mathcal{P} \supset \mathcal{Q} = \mathcal{E}(q, Q)$ nor $\mathcal{Q} \supset \mathcal{P}$ holds. Obviously, in this case the problem cannot be reduced to simpler situations without disturbances.

The numerical calculation on the basis of Theorem 3.10.1. is carried out in the following way: after creating the internal estimate $E_f(t_0, \mathcal{M})$, we check whether $x(t_0) = x^0 \in \mathcal{E}_-(t_0, t, \mathcal{M})$, taking increasing values of $t \in [t_0, t_1]$. In such a way we find that this relation holds for

$$t = t^* = 4.6$$

i.e.,

$$x_0 \in \mathcal{E}_-[0, 4.6, \mathcal{M}] = \mathcal{E}[0, t^*, \mathcal{M}] \ .$$

So $t^* = 4.6$ is an upper estimate of t_e^* - the closest time instant by which the set $\mathcal{M}$ can be hit for any disturbance f.

According to Theorem 3.10.1 we keep the trajectory in the ellipsoidal valued map starting from the above ellipsoid $\mathcal{E}(0, t^*, \mathcal{M})$.

Figure 3.10.1 shows the internal estimate of the set $W_f(\tau, \mathcal{M})$ at $\tau = 0$ in the form of

$$\bigcup\{\mathcal{E}_-[0, t, \mathcal{M}] | t \in [0, t_1]\} \ .$$

In Figure 3.10.2 we see again the above set, the ellipsoidal valued map $\mathcal{E}_-[t, t^*, \mathcal{M}], t \in [0, t^*]$, as well as the controlled trajectories under two simulated disturbances f resulting in that the trajectories arrive to the target set $\mathcal{M}$ at time $t = t^* = 4.6$.

The layout of the two last figures is the same as before, with the drawn segments of coordinate axes corresponding to the state variables ranging from -40 to 40.

Part IV. ELLIPSOIDAL DYNAMICS: STATE ESTIMATION and VIABILITY PROBLEMS

Introduction

This last Part IV of the present book is concentrated around *state estimation and viability problems*, emphasizing constructive techniques for their solution worked out in the spirit of the earlier parts.

We emphasize that here the uncertain items – the initial states, system inputs and measurement noise are assumed to be unkown in advance, with no statistical information on them being available. The problem may then be further treated in two possible settings. The first one is when *the bounds on the unknowns are specified in advance.* This leads to the problem of guaranteed state estimation introduced in Section 1.12. A natural move in this setting is to use the *set-membership (bounding) approach.* A key element here is the notion of *information set* of states consistent with the system equations, the realization of the measurement and the constraints on the uncertain items. The information set always includes the unknown actual state of the system and thus gives *a set-valued guaranteed estimate* of this state. It may also be useful to find a *single vector-valued state estimator*, which may be selected, for example, as the center of the smallest ball that includes the information set (which is the so-called *the Chebyshev center* of this set). One of the main problems here is to give an appropriate description of the evolution of the information sets in time and of the dynamics of the vector-valued estimators. A detailed description of the bounding approach can be found in monographs [277], [181], [225], and reviews [226], [187], [186].

The calculation of information sets, even for the linear-convex problems of this book, is not a simple problem, though. Indeed, it requires us to describe more or less arbitrary types of convex compact sets, which actually are *infinite-dimensional* elements. One may try to approximate them by *finite-dimensional elements* however, particularly, by ellipsoids, as in the present book.

The approximation of information sets by *only one or few ellipsoids* is described in [278], [73]. This approximation may turn out to be useful in applied problems where computational simplicity stands above accuracy of solution. On the other hand, in sophisticated applications (to some types of pursuit-evasion games, for example), this rather rough approximation may be misleading. As mentioned above, among the objectives of this book is to produce an ellipsoidal approximation by *a parametrized variety of ellipsoids*, which, in the limit, gives an exact representation of the information sets.[25]

The parameters of the approximating external ellipsoids are described here as solutions to systems of ordinary differential equations. Two types of such equations are given in Sections 4.3 and 4.5. The latter is derived through the relations of Section 2.6, while the former follows from Dynamic Programming (DP) considerations. The DP techniques allow to link the bounding approach with another deterministic approach to state estimation.

This second approach to state estimation assumes that *no bounds* on the uncertain items are known. Given is a measure of uncertainty for the uncertain items and the vector-valued estimator is generated through a system which realizes the minimal norm of a certain input-output map or a saddle point of an appropriate dynamic game. The estimators are then calculated through the knowledge of *the information state* – the value function of a certain problem in dynamic optimization calculated as a forward solution of an appropriate H-J-B equation. This second scheme is often referred to as the so-called H_∞ approach.[26]

The important connection between the two approaches is that *the information sets are the level sets for the information states* – the solutions to the H-J-B equation of the H_∞ approach (Section 4.3, see also [32]). Since systems with magnitude constraints on the inputs generate H-J-B equations with no classical solutions, the latter equations could be analyzed within the notions of generalized solutions (of the viscosity or minmax types, for example, [82], [290]). In this book these generalized solutions are not calculated explicitly, but are rather approximated by classical solutions to systems of H-J-B equations constructed for adequate classes of *linear-quadratic extremal problems*. In terms of level

[25] The idea of such representations was indicated in [181], §§ 12.2, 15.1.

[26] The H_∞ approach to estimation and feedback control has been studied in many papers. Here we mention [94], [231] and especially those of J. Baras and M. James who introduced the notion of information state [30].

sets the last construction is again an ellipsoidal approximation. It is thus observed that the connection between the two approaches to the deterministic treatment of uncertainty in dynamics lies, basically, in the incorporation of the same DP equations to both settings.

The DP approach may as well be applied to the calculation of *attainability domains.* Particularly, if one deals with magnitude constraints on the inputs, then the ellipsoidal approximations to these domains may again be achieved through the construction of level sets for value functions of appropriate linear-quadratic extremal problems. However, Section 4.4 indicates that the respective ellipsoids could be transformed to be the same as those obtained through the purely geometrical considerations of Parts II and III, as described in Sections 2.7 and 3.2. Similar assertions are also proved for the calculation of *viability kernels* [15].

Among the problems of viability and state estimation are those, where the viability restriction or the state constraint induced by the measurement equation are not continuous in time. (This particularly happens, when the noise in the observations is modelled by discontinuous functions, that may turn to be only Lebesgue-measurable, for example.) A possible scheme for handling such situations lies in imbedding the original problem into one with *singular perturbations* (Section 4.6). The new problem is constructed such that it is free of the inadequacies of the original problem on one hand, and allows an approximation of the original one, on the other. A detailed description of this scheme for state estimation and viability problems of general type is given in references [191], [192]. Section 4.6 presents an ellipsoidal version of the technique.

4.1 Guaranteed State Estimation: A Dynamic Programming Perspective

We shall begin this section by discussing the two basic approaches to the *deterministic treatment of uncertainty* in the dynamics of controlled processes, as mentioned in the previous Introduction, treating them in the context of the problem of *state estimation* with a further aim of using ellipsoidal techniques.

The first of these, as we have seen in Section 1.12, is the *bounding approach* based on set-membership techniques. Here the uncertain items are taken to be *unknown but bounded with given bounds* on the performance range. The estimate is then sought for in the form of a set – the information domain, which was described by funnel equations (1.12.10) or (1.12.11).

The second one is the so-called H_∞ *approach* based in its linear version on the calculation of the minimal-norm input-output map for the investigated system and the error bound for the system performance expressed through this norm.

Although formally somewhat different, these two approaches appear to have close connections. These may be demonstrated particularly through the techniques of *Dynamic Programming* that are the topic of this section. Namely, it will be indicated that both approaches may be handled through *one and the same* equation of the H-J-B type.

For the case of problems with *ellipsoidal magnitude constraints* on the uncertain items that are treated in the next section and are among the the main points of emphasis in the present book, we shall indicate an *approximation technique* for solving the respective H-J-B equation. The technique is based on an approximation of the original problem with magnitude constraints by a parametrized variety of problems with quadratic integral constraints. Such a scheme shall then allow a turn to ellipsoidal approximations of attainability domains.

Let us start with a slightly more general problem than in Section 1.12. Consider again the system (1.12.1), (1.12.5) with $u(t) \equiv 0$, rewriting it as

$$\dot{x}(t) = A(t)x(t) + f(t), \quad x(t_0) = x^0 \quad , \tag{4.1.1}$$

$$y(t) = G(t)x(t) + v(t), \quad t_0 \le t \le \tau \quad . \tag{4.1.2}$$

We shall assume that the unknown items $\zeta(\cdot) = \{x^0, f(t), v(t), t_0 \le t \le \tau\}$ are now bounded by the inequality

$$\Psi(\tau, \zeta(\cdot)) = \int_{t_0}^{\tau} \psi(t, f(t), v(t))dt + \phi(x^0) \le \mu^2 \quad , \tag{4.1.3}$$

where $\Psi(\tau, \zeta(\cdot))$ reflects the accepted *uncertainty index* for the unknown items.

Particularly, the bounds may be of *the quadratic integral type*, namely, such that

$$\phi(x^0) = (x^0 - a, L(x^0 - a)) \quad , \tag{4.1.4}$$

(4.1.5) $\psi(t, f(t), v(t)) = (f(t) - f^*(t), M(t)(f(t) - f^*(t))) +$
$+(v(t) - v^*(t), N(t)(v(t) - v^*(t)))$,

where (p, q) $(p, q \in \mathbb{R}^k)$, stands for the scalar product in the respective space $\mathbb{R}^k$; $a \in \mathbb{R}^n$ is a given vector; $f^*(t), v^*(t)$ are given vector functions of respective dimensions, square-integrable in $t \in [t_0, \tau]$; $M(t), N(t)$ are positive definite, continuous, and $L > 0$.

Another common type of restriction is given by *magnitude bounds*, a particular case of which is described by *ellipsoidal-valued constraints* – the inequalities[27]

(4.1.6) $$I_0(x^0) = (x^0 - a, L(x^0 - a)) \leq \mu^2 \quad ,$$

(4.1.7) $$I_1(\tau, f(\cdot)) = \text{esssup}_t(f(t) - f^*(t), M(t)(f(t) - f^*(t)) \leq \mu^2$$

(4.1.8) $$I_2(\tau, v(\cdot)) = \text{esssup}_t(v(t) - v^*(t), N(t)(v(t) - v^*(t)) \leq \mu^2$$
$$t \in [t_0, \tau] \quad .$$

In this case the functional

(4.1.9) $$\Psi(\tau, \zeta(\cdot)) = \max\{I_0, I_1, I_2\} \quad .$$

As we shall observe in the sequel, the number μ in the restriction (4.1.1) may be or may not be given in advance, and the corresponding solution will, of course, depend on this specificity of the problem. Despite of the latter fact, the aim of the *state estimation* (filtering) problem could be described as follows:

(a) *determine an estimate* $x^0(\tau)$ *for the unknown state* $x(\tau)$ on the basis of the available information:

the system parameters, the measurement $y(t)$, $t \in [t_0, \tau]$, and the restrictions on the uncertain items $\zeta(\cdot)$ (if these are specified in advance);

(b) *calculate the error bounds* for the estimate $x^0(\tau)$ on the basis of the same information,

[27] In the coming Sections 4.1–4.3 the notations for the bounds on the unknowns are independent of those introduced earlier, emphasizing that the treatment of the state estimation (filtering) problem, as given here, is independent of the earlier material. In Section 4.4 we shall synchronise these notations with the earlier ones.

(c) *describe the evolution of the estimate* $x^0(\tau)$ *and the error bound in* τ, preferably through a dynamic recurrence-type relation, an ordinary differential equation, for example, if possible.

Let us discuss the problem in some more detail. Suppose that the constraints (4.1.1) with specified μ are given together with the available measurement $y = y(t), t \in [t_0, \tau]$. The bounding approach then requires that the solution be given through *the information domain* $\mathcal{X}(\tau)$ of Definition 1.12.1. With $\mathcal{X}(\tau)$ calculated, one may be certain that for the unknown actual value $x(\tau)$ we have: $x(\tau) \in \mathcal{X}(\tau)$, and may therefore find a certain point $\hat{x}(\tau) \in \mathcal{X}(\tau)$ that serves as the required estimate $x^0(\tau)$. As mentioned above, at the end of the previous section, this point $\hat{x}(\tau)$ may be particularly selected as the *"Chebyshev center"* for $\mathcal{X}(\tau)$, defined through the relation

(4.1.10)

$$\min_x \max_z (x - z, x - z) = \max_z (\hat{x}(\tau) - z, \hat{x}(\tau) - z), \; z \in \mathcal{X}(\tau) \; .$$

It is obviously the center of the smallest ball that includes the domain $\mathcal{X}(\tau)$. The inclusion

$$\hat{x}(\tau) \in \mathcal{X}(\tau)$$

will be secured as $\mathcal{X}(\tau)$ is convex. (This may not be the case for the general nonlinear problem, however, when the configuration of $\mathcal{X}(\tau)$ may be quite complicated.) The set $\mathcal{X}(\tau)$ gives an *unimprovable estimate* of the state-space variable $x(\tau)$, provided the bound on the uncertain items (the number μ) is given in advance.

On the other hand, in the second or H_∞ approach, the value μ for *the bound on the uncertain items is not assumed to be known*, while the value of the estimation error

$$e^2(\tau) = (x(\tau) - \hat{x}(\tau), x(\tau) - \hat{x}(\tau))$$

is then estimated, in its turn, merely through the smallest number σ^2 that ensures the inequality

$$e^2(\tau) \leq \sigma^2 \Psi(\tau, \zeta(\cdot)) \tag{4.1.11}$$

under restrictions (4.1.2) or (1.12.5).

Since we deal with the linear case, the smallest number σ^2 is clearly the square of *the minimal norm-type index of the input-output mapping* T, where

$$e(\tau) = \|T(\zeta(\cdot)) - \hat{x}(\tau)\|$$

with $y = y(t)$ given. It obviously depends on the type of norm (the type of functional $\Psi(\zeta(\cdot))$ selected to evaluate $\eta(\cdot)$). The latter worst-case estimate is less precise than in the first approach (since, as one may observe, it actually indicates a larger error bound). However, this may sometimes suffice for the specific problem under discussion.

We shall use the upcoming discussion in Section 4.3 to emphasize the connections between the two approaches and to indicate, through a Dynamic Programming (DP) technique, a general framework that incorporates both of these, producing either of them, depending on the a priori information, as well as on the required accuracy of the solutions.

Let us start by introducing a scheme for describing the information domains $\mathcal{X}(\tau)$, presuming $y(\cdot)$ to be given and restriction (4.1.3) to be of the quadratic integral type (4.1.3)–(4.1.5), to start with.

Denote

$$\eta(\cdot) = \{x^0, f(\cdot)\}, \quad x(t, t_0, x^0, f(\cdot)) = x(t, t_0, \eta(\cdot))$$

and

$$\Phi(\tau, \eta(\cdot)) = (x^0 - a, P_0(x^0 - a)) + \int_{t_0}^{\tau} ((f(t) - f^*(t), M(t)(f(t) - f^*(t)) +$$

$$+(y(t) - G(t)x(t, t_0, \eta(\cdot)) - v^*(t), N(t)(y(t) - G(t)x(t, t_0, \eta(\cdot)) - v^*(t)))dt \ .$$

Clearly,

(4.1.12) $$\Phi(\tau, \eta(\cdot)) = \{\Psi(\tau, \zeta(\cdot)) | v(t) \equiv y(t) -$$

$$-G(t)x(t, t_0, \eta(\cdot)) - v^*(t)\} \ .$$

Define

(4.1.13) $$V(\tau, x) = \inf_{\eta(\cdot)}\{\Phi(\tau, \eta(\cdot)) \,|\, x(\tau, t_0, \eta(\cdot)) = x\} \ .$$

With $L, N(t) > 0$ the operation "inf" in the line above may be substituted for "min", which will be attained at a unique element $\eta(\cdot) = \eta^0(\cdot)$.

Definition 4.1.1 *Given measurement $y(t), t \in [t_0, \tau]$ and functional $\Phi(\tau, \eta(\cdot))$ of (4.1.12), the respective function $V(\tau, x)$ will be referred to as* **the information state** *of system (4.1.1), (4.1.2),* **relative to measurement $y(\cdot)$ and criterion Φ.**

The given Definition holds not only for the quadratic index Ψ of (4.1.3)–(4.1.5) but for any other one as well (for Ψ of (4.1.9), for example.

An obvious assertion is given by

Lemma 4.1.1 *The information domain $\mathcal{X}(\tau)$ is the level set*

(4.1.14) $$\mathcal{X}(\tau) = \{x : V(\tau, x) \leq \mu^2\}$$

for the information state $V(\tau, x)$.

It should be emphasized here that both $V(\tau, x) \geq 0$ and $\mathcal{X}(\tau)$ depend on the given measurement $y(t)$ as well as on the type of functionals Ψ, Φ and that $\mathcal{X}(\tau) \neq \emptyset$, provided

(4.1.15) $$V^0(\tau) = \inf\{V(\tau, x) \mid x \in R^n\} \leq \mu^2 \ .$$

Since Lemma 4.1.1 indicates that the $\mathcal{X}(\tau)$ is *a level set* for $V(\tau, x)$, the knowledge of $V(\tau, x)$ will thus allow to calculate the sets $\mathcal{X}(\tau)$.

We emphasize once more the main conclusions:

(i) *the information domain $\mathcal{X}(\tau)$is the level set for the information state $V(\tau, x)$ that corresponds to the given number μ.*

(ii) *the information state depends both on $y(\cdot)$ and on the type of functional Φ.*

The crucial difficulty here is the calculation of the sets $\mathcal{X}(\tau)$, the function $V(\tau, x)$ and further on, of the estimate $x^*(\tau)$ for the unknown state $x(\tau)$. The calculations are relatively simple for an exceptional situation – *the linear-quadratic case.*

As already emphasized above, apart from their separate significance, the linear-quadratic solutions will be important in organizing ellipsoidal approximations for systems with magnitude constraints.

Let us therefore introduce a DP – type of equation, taking $V(\tau, x)$ to be *the value function* for the linear-quadratic problem (4.1.13) when $\Psi(\tau, \zeta(\cdot))$ is given by (4.1.3)–(4.1.5). The respective function $\Phi(\tau, \eta(\cdot))$ then obviously satisfies the Optimality Principle of Dynamic Programming (DP) [109].

Applying standard techniques [53], [109], we may observe that the DP equation for the value function is

$$\frac{\partial V}{\partial t} + \max_f\{(\frac{\partial V}{\partial x}, A(t)x + f) - (f - f^*(t), M(t)(f - f^*(t)))-$$
$$-(y(t) - G(t)x - v^*(t), N(t)(y(t) - G(t)x - v^*(t)))\} = 0$$

so that, after an the elimination of f, the respective forward H-J-B equation is as follows

$$\frac{\partial V}{\partial t} + (\frac{\partial V}{\partial x}, A(t)x + f^*(t)) + \frac{1}{4}(\frac{\partial V}{\partial x}, M^{-1}(t)\frac{\partial V}{\partial x})- \tag{4.1.16}$$
$$-(y(t) - G(t)x - v^*(t), N(t)(y(t) - G(t)x - v^*(t))) = 0$$

with boundary condition

$$V(t_0, x) = (x - a, L(x - a)) \ . \tag{4.1.17}$$

Its unique solution is a *quadratic form*

$$V(\tau, x) = (x - z(\tau), \mathcal{P}(\tau)(x - z(\tau))) + k^2(\tau) \tag{4.1.18}$$

where $\mathcal{P}(t), z(t), k^2(t)$ are the solutions to the following well-known equations [149], [57], [277], [181].[28]

$$\dot{\mathcal{P}} = -\mathcal{P}A(t) - A'(t)\mathcal{P} - \mathcal{P}M^{-1}(t)\mathcal{P} + G'(t)N(t)G(t) \ , \tag{4.1.19}$$
$$\mathcal{P}(t_0) = L \ ,$$

$$\dot{z} = A(t)z + \mathcal{P}^{-1}G'(t)N(t)(y(t) - G(t)z - v^*(t)) + f^*(t) \ , \tag{4.1.20}$$
$$z(t_0) = a \ ,$$

[28] Here we formally followed the standard scheme assuming $V(t, x)$ differentiable. The unicity of the solution $V(t, x)$ to this linear-quadratic problem with strictly convex cost and of the solution to the following equations (4.1.19)–(4.1.21) justify this procedure.

$$(4.1.21) \quad \dot{k}^2 = (y(t) - G(t)z - v^*(t), N(t)(y(t) - G(t)z - v^*(t))) \ ,$$

$$k^2(t_0) = 0.$$

Equations (4.1.19)–(4.1.21) are derived by direct substitution of $V(t,x)$ into equation (4.1.16).[29]

An obvious consequence of the given reasoning is the following assertion

Lemma 4.1.2 *Under restrictions (4.1.3)–(4.1.5) on the uncertain inputs $\zeta(\cdot) = \{\eta(\cdot), v(\cdot)\}$ the information domain $\mathcal{X}(\tau)$ for the system (4.1.1), (4.1.2) is the level set (4.1.14) for the information state $V(\tau, x)$, being an ellipsoid $E(z(\tau), (\mu^2 - k^2(\tau))\mathcal{P}^{-1}(\tau))$ given by the relation*

$$(4.1.22) \qquad \mathcal{X}(\tau) = E(z(\tau), (\mu^2 - k^2(\tau))\mathcal{P}^{-1}(\tau)) =$$

$$= \{x : (x - z(\tau), \mathcal{P}(\tau)(x - z(\tau))) \leq \mu^2 - k^2(\tau)\},$$

where $z(\tau), \mathcal{P}(\tau) > 0$, $k^2(\tau)$ are defined through equations (4.1.19)–(4.1.21).

Remark 4.1.1 *Note that the matrix-valued function $\mathcal{P}(t)$ does not depend on the measurement $y(\cdot)$, while the scalar function $k^2(t)$ depends on the measurement. The estimation error is given by an* **error set** *$\mathcal{R}(\tau) = \mathcal{X}(\tau) - z(\tau)$ which therefore depends only on $k^2(\tau)$.*

Formula (4.1.21) immediately indicates the worst-case realization $y^*(t)$ of the measurement $y(t)$ which yields the largest set $\mathcal{X}(t)$ (with respect to inclusion). Namely, if it possible to obtain the specific measurement $y^*(t)$ through the triplet

$$\eta(\cdot) = \eta^*(\cdot) = \{a, f^*(\cdot)\}, v(\cdot) = v^*(\cdot),$$

(among other possible triplets), then

$$y^*(t) = G(t)x(t, t_0, \eta^*(\cdot)) + v^*(t)$$

[29] One may easily observe that the first two equations (4.1.19) and (4.1.20) are the same as in stochastic "Kalman" filtering theory. However, the third one, (4.1.21), is not present in stochastic theory. It is specific for the set-membership approach and reflects the dynamics of the size of the information set.

is the *worst-case realization* of the measurement and the respective value

$$V^*(\tau) = V^0(\tau)|_{y(\cdot)=y^*(\cdot)} = 0 \ .$$

In order to check the last assertions, let us introduce an equation for the function $h(t) = x(t) - z(t)$, where $x(t)$ is the realization of the actual trajectory generated due to equation

$$\dot{x} = A(t)x + f, \ \ x(t_0) = x^0 \ . \tag{4.1.23}$$

Subtracting (4.1.23) from (4.1.20), we come to

$$\dot{h} = \mathcal{A}(t)h(t) - K(t)(v(t) - v^*(t)) + \tag{4.1.24}$$

$$+f(t) - f^*(t), \ \ h(t_0) = x^0 - a \ ,$$

where

$$\mathcal{A}(t) = A(t) - \mathcal{P}^{-1}(t)G'(t)N(t)G(t), K(t) = \mathcal{P}^{-1}(t)G'(t)N(t) \ .$$

If the actual realization $x(t)$ is generated by $x^0 = a, f(t) \equiv f^*(t)$, so that $x(t) = x(t, t_0, \eta^*(\cdot))$, and the realization of the measurement noise is $v(t) \equiv v^*(t)$, then (4.1.24), (4.1.21) yield $h(t) \equiv 0, k^2(t) \equiv 0$. We therefore come to

Lemma 4.1.3 *The worst-case realization $y(t) = y^*(t)$ of the measurement is a function that (among other possible triplets $\zeta(\cdot)$) may be generated by the triplet $\{x^0 = a, f(t) \equiv f^*(t), v(t) \equiv v^*(t)\}$ which yields $k^2(\tau) = 0$.*

The worst-case error set is the ellipsoid

$$\mathcal{R}(\tau) = \mathcal{X}(\tau) - z(\tau) = E(0, \mu^2\mathcal{P}^{-1}(\tau)) = \{e : (e, \mathcal{P}(\tau)e) \leq \mu^2\} \ .$$

The other extreme situation is when the measurement is *the best possible.*

Lemma 4.1.4 *There exists a function (measurement noise) $v = \bar{v}(t)$, such that the triplet $\zeta(\cdot) = \{a, f^*(t), \bar{v}(t)\}$ generates, due to system (4.1.1), (4.1.2), a measurement $\bar{y}(\cdot)$ that ensures $k^2(\tau) = \mu^2$, so that in this case the informational set $\mathcal{X}(\tau)$ is a singleton and*

$$\mathcal{X}(\tau) = \{x(\tau, t_0, \eta^*(\cdot))\} \ .$$

Returning to equation (4.1.24) and rewriting (4.1.21) in view of measurement equation

$$y(t) = G(t)x(t) + \bar{v}(t) \quad ,$$

we come to

(4.1.25)

$$\dot{k}^2(t) = (G(t)h(t) + \bar{v}(t) - v^*(t), N(t)(G(t)h(t) + \bar{v}(t) - v^*(t))) \quad ,$$

and (with $f(t) \equiv f^*(t)$)

$$\dot{h}(t) = \mathcal{A}(t)h(t) - K(t)(v^*(t) - \bar{v}(t)) \tag{4.1.26}$$

we shall require that $k(t), h(t)$ satisfy the following boundary-value problem

$$k^2(t_0) = 0, k^2(\tau) = \mu^2; \quad h(t_0) = 0, h(\tau) = 0 \quad . \tag{4.1.27}$$

The solution $\bar{v}(t)$ to this problem obviously satisfies the requirements of the last Lemma, ensuring particularly, at instant τ, the equalities

$$x(\tau) = x(\tau, t_0, \eta^*(\cdot)) = z(\tau), \ \{x(\tau)\} = \mathcal{X}(\tau) \quad .$$

We leave to the reader to verify that such a solution $\bar{v}(t)$ does exist.

Finally, let us assume that there is no measurement equation, so that we simply have the standard system

$$\dot{x} \ = \ A(t)x \ + \ f(t), \ \ x(t_0) = x^0 \quad , \tag{4.1.28}$$

with quadratic constraint (4.1.3)–(4.1.5), $N(t) \equiv 0$ Then the set $\mathcal{X}(\tau)$ is merely *the attainability domain* for system (4.1.28) under the constraint (4.1.3)–(4.1.5), $N(t) \equiv 0$. We may therefore follow the calculations of the above, setting $N(t) \equiv 0$.

The procedure then automatically gives the following result

Lemma 4.1.5 *Under restrictions (4.1.3)–(4.1.5), $N(t) \equiv 0$ on the inputs $\eta(\cdot) = \{x^0, f(\cdot)\}$ the attainability $\mathcal{X}(\tau)$ for system (4.1.28) is the level set (4.1.14) for the function*

$$V(\tau, x) = (x - z(\tau), \mathcal{P}^{-1}(\tau)(x - z(\tau))) \quad ,$$

where $\mathcal{P}(t), z(t)$ are the solutions to the equations

$$\dot{z} = A(t)z + f^*(t) \quad , \tag{4.1.29}$$

$$z(t_0) = a \quad ,$$

(4.1.30)
$$\dot{\mathcal{P}} + \mathcal{P}A(t) + A'(t)\mathcal{P} + \mathcal{P}M^{-1}(t)\mathcal{P} = 0 \quad ,$$
$$\mathcal{P}(t_0) = L \quad ,$$

being an ellipsoid $E(z(\tau), \mathcal{P}^{-1}(\tau))$ given by relation

(4.1.31)
$$\mathcal{X}(\tau) = E(z(\tau), \mathcal{P}^{-1}(\tau)) =$$
$$= \{x : (x - z(\tau), \mathcal{P}(\tau)(x - z(\tau))) \leq \mu^2\} \quad .$$

We are now prepared to extend the results of this section to problems with magnitude constraints.

4.2 From Dynamic Programming to Ellipsoidal State Estimates

Let us now specify the *information state* $V(\tau, x)$ of (4.1.13) for the case of *magnitude constraints*, presuming Φ is defined through relations (4.1.6)–(4.1.9). One may observe that $\Phi(\tau, \eta)$ again satisfies the Optimality Principle (and is thus a *positional functional* in terms of [170]). One may therefore again calculate $V(\tau, x)$ through the H-J-B equation or, if necessary, through its generalized versions that deal with nondifferentiable functionals (see [83], [290], [109]). We shall not pursue the last direction, but shall rather apply yet another scheme which will be of direct further use in this book.

Denote

$$\Lambda(\tau, \eta(\cdot), \alpha, \beta(\cdot), \gamma(\cdot)) = \alpha(x^0 - a, L(x^0 - a)) +$$
$$+ \int_{t_0}^{\tau} (\beta(t)(f(t) - f^*(t), M(t)(f(t) - f^*(t))) +$$
$$+\gamma(t)(y(t) - x(t, t_0, \eta(\cdot)) - v^*(t), N(t)(y(t) - x(t, t_0, \eta(\cdot)) - v^*(t)))dt \quad .$$

Lemma 4.2.1 *Assuming $M(t), N(t), t \in [t_0, \tau]$ continuous, the restrictions (4.1.6)–(4.1.8) are equivalent to the system of inequalities*

(4.2.1)
$$\Lambda(\tau, \eta(\cdot), \alpha, \beta(\cdot), \gamma(\cdot)) \leq \mu^2$$

whatever are the parameters

(4.2.2) $$\alpha \geq 0, \beta(t) \geq 0, \gamma(t) \geq 0$$

where

(4.2.3) $$\alpha + \int_{t_0}^{\tau} (\beta(t) + \gamma(t))dt = 1 \quad .$$

The functions $\beta(\cdot), \gamma(\cdot)$ are taken to be measurable, with (4.2.2) being true for almost all t.

We further denote the triplet $\{\alpha, \beta(\cdot), \gamma(\cdot)\} = \omega(\cdot)$ and the variety of triplets ω that satisfy (4.2.1)–(4.2.3), as $\Omega = \{\omega(\cdot)\}$.

Proof. With (4.2.1)–(4.2.3) given, take any triplet $\omega(\cdot)$, then multiply (4.1.6) by α, (4.1.7) by $\beta(t)$ and (4.1.8) by $\gamma(t)$, then integrate the last two relations over $t \in [t_0, \tau]$. Adding the results, we obtain (4.2.1) due to (4.2.3).

Conversely, assume (4.2.1) to be true for any $\omega(\cdot) \in \Omega$. Taking $\alpha = 1, \beta(t) \equiv 0, \gamma(t) \equiv 0$, one comes to (4.1.4). Further on , assume for example, that (4.1.7) is false and therefore that $I_1(t, f(\cdot)) \geq \epsilon > 0$ on a set e of measure $mes(e) > 0$.Then taking $\alpha = 0$; $\beta(t) \equiv (mes(e)^{-1}$, $t \in e; \beta(t) \equiv 0, t \notin e$; $\gamma(t) \equiv 0$, one comes to a contradiction with (4.2.4) and thus (4.2.7) turns out to be true. Similarly, the third condition (4.2.8) also follows from (4.2.1).[30] Q.E.D.

Using a similar reasoning the reader may now verify the following assertion

Lemma 4.2.2 *The function $\Phi(\tau, \eta(\cdot))$ of (4.1.12), (4.1.9) may be expressed as*

(4.2.4) $$\Phi(\tau, \eta(\cdot)) = \sup\{\Lambda(\tau, \eta(\cdot), \omega(\cdot)) \mid \omega(\cdot) \in \Omega\} \quad .$$

The proof of an analogous fact may be also found in [181].

For further calculations we emphasize the following obvious property

[30] With slight modifications the present Lemma 4.2.1 may as well be proved if $\beta(t), \gamma(t)$ are taken to be continuous.

Lemma 4.2.3 *The functional $\Lambda(\tau,\eta(\cdot),\omega(\cdot))$ is convex in $\eta(\cdot) = \{x^0, f(\cdot)\}$ on the set of elements $\eta(\cdot)$ restricted by the equality $\{x : x(\tau,\eta(\cdot)) = x\}$.*

Due to Lemma 4.2.2 we have

$$V(\tau,x) = \inf_{\eta}\{\Phi(\tau,\eta(\cdot)) \mid x(\tau,\eta(\cdot)) = x\} = \tag{4.2.5}$$

$$= \inf_{\eta}\sup_{\omega(\cdot)}\{\Lambda(\tau,\eta(\cdot),\omega(\cdot))\} \ .$$

under restriction $\omega(\cdot) \in \Omega$.

$$x(\tau,\eta(\cdot)) = x \tag{4.2.6}$$

The functional $\Lambda(\tau,\eta(\cdot),\omega(\cdot))$ is linear (therefore, concave) in ω and convex in η , according to Lemma 4.1.3. In view of *minmax-type theorems* (see [101], [86]), the order of operations *inf, sup* may be interchanged. We therefore come to the relation

$$V(\tau,x) = \sup_{\omega}\min_{\eta(\cdot)}\Lambda(\tau,\eta(\cdot),\omega(\cdot)) \tag{4.2.7}$$

under restrictions $\omega(\cdot) \in \Omega$ and (4.2.6).

The internal problem of finding

$$V(\tau,x,\omega(\cdot)) = \min\{\Lambda(\tau,\eta(\cdot),\omega(\cdot)) \mid \eta(\cdot), x(\tau,\eta(\cdot)) = x\} \tag{4.2.8}$$

may be solved through equation (4.1.16) (see Remark 4.2.2) with $V(\tau,x)$ substituted for $V(\tau,x,\omega(\cdot))$ and $M(t), N(t)$ for $\beta(t)M(t), \gamma(t)N(t)$, rmeespectively, with boundary condition being

$$V(t_0,x,\omega(\cdot)) \ = \ \alpha(x-a, L(x-a)) \ . \tag{4.2.9}$$

This leads to

Lemma 4.2.4 *The information state (4.2.7) is given by*

$$V(\tau,x) = \sup\{V(\tau,x,\omega(\cdot)) \mid \omega(\cdot) \in \Omega\} \tag{4.2.10}$$

where $V(\tau,x,\omega)$ is the solution to equation (4.1.16), under boundary condition (4.2.9), with $M(t), N(t)$ substituted for $\beta(t)M(t), \gamma(t)N(t)$.

Solving problem (4.2.10), we observe

(4.2.11)
$$V(\tau, x, \omega(\cdot)) =$$
$$= (x - z(\tau, \gamma(\cdot)), \mathcal{P}(\tau, \omega(\cdot))(x - z(\tau, \gamma(\cdot)))) + k^2(\tau, \gamma(\cdot)) \ ,$$

where $\mathcal{P} = \mathcal{P}(t, \omega(\cdot)), z = z(t, \gamma(\cdot)), k = k(t, \gamma(\cdot))$ satisfy the equations

(4.2.12) $$\dot{\mathcal{P}} = -\mathcal{P}A - A'\mathcal{P} - \beta^{-1}(t)\mathcal{P}M^{-1}(t)\mathcal{P} + \gamma(t)G'(t)N(t)G(t) \ ,$$

(4.2.13)
$$\dot{z} = A(t)z + \gamma(t)\mathcal{P}^{-1}(t)G'(t)N(t)(y(t) - G(t)z - v^*(t)) + f^*(t) \ ,$$

(4.2.14)
$$\dot{k}^2(t) =$$
$$= \gamma(t)(y(t) - G(t)z - v^*(t), N(t)(y(t) - G(t)z - v^*(t)))$$

(4.2.15) $$P(t_0) = \alpha L, z(t_0) = a, k(t_0) = 0 \ .$$

Finally this develops into the assertion

Theorem 4.2.1 *For the system (4.1.1), (4.1.2) the information state $V(\tau, x)$ relative to measurement $y(\cdot)$ and nonquadratic (magnitude) criterion (4.1.9), is the upper envelope*

(4.2.16) $$V(\tau, x) = \sup\{V(\tau, x, \omega(\cdot)) \mid \omega(\cdot) \in \Omega\}$$

of a parametrized family of quadratic forms $V(\tau, x, \omega(\cdot))$ of type (4.2.11) over the functional parameter $\omega(\cdot) = \{\alpha, \beta(\cdot), \gamma(\cdot)\}$, where $\omega(\cdot) \in \Omega$.

As we have observed in the previous sections, the information domain $\mathcal{X}(\tau) = E(z(\tau), \mu^2\mathcal{P}^{-1}(\tau))$ is defined by $V(t, x)$ through inequality (4.1.14) with μ given. Moreover, for each of the ellipsoidal level sets

(4.2.17)
$$\mathcal{X}(\tau, \omega(\cdot)) =$$
$$= E(z(\tau, \omega(\cdot)), (\mu^2 - k^2(\tau))\mathcal{P}^{-1}(\tau, \omega(\cdot)) = \{x : V(\tau, x, \omega(\cdot)) \leq \mu^2\}$$

where $V(\tau, x, \omega(\cdot))$ is a *nondegenerate quadratic form*(!), we obviously have

$$\mathcal{X}(\tau) \subseteq \mathcal{X}(\tau, \omega(\cdot)) = E(z(\tau, \omega(\cdot)), (\mu^2 - k^2(\tau))\mathcal{P}^{-1}(\tau, \omega(\cdot))), \forall \omega(\cdot) \in \Omega \ ,$$

so that (4.2.16) yields the following fact

Theorem 4.2.2 *For the system (4.1.1), (4.1.2), with criterion (4.1.9), the information set $\mathcal{X}(\tau)$ is the intersection of ellipsoids*

$$\mathcal{X}(\tau,\omega(\cdot)) = E(z(\tau,\omega(\cdot)),(\mu^2 - k^2(\tau))\mathcal{P}^{-1}(\tau,\omega(\cdot)))$$

namely

$$(4.2.18)\mathcal{X}(\tau) = \cap\{E(z(\tau,\omega(\cdot)),(\mu^2 - k^2(\tau))\mathcal{P}^{-1}(\tau,\omega(\cdot))) \mid \omega(\cdot) \in \Omega\}$$

where

$$z(t) = z(t,\gamma(\cdot)), \mathcal{P}(t) = \mathcal{P}(t,\omega(\cdot)), k^2(t) = k^2(t,\gamma(\cdot))$$

are defined through equations (4.2.12)–(4.2.15).

The worst case measurement $y(t) = y^*(t)$ is a function that may be generated (among other possible triplets) by triplet $\zeta^*(\cdot) = \{\eta^*(\cdot), v^*(\cdot)\}$, where $x^0 = a$, $f(t) = f^*(t)$, $v(t) = v^*(t)$. This yields $k^2(\tau) = 0$ and

$$V^*(\tau) = V^0(\tau)|_{y(\cdot)=y^*(\cdot)}$$

where

$$V^0(\tau) = inf\{V(\tau,x)|x \in \mathbb{R}^n\} = 0 \ .$$

The last part of the Theorem that deals with the worst-case measurement $y^*(\cdot)$ may be checked by substituting $\zeta^*(\cdot)$ into (4.2.13), (4.2.14) and following the reasoning of the previous section.

Remark 4.2.1 *Observe that function $k^2(t,\gamma(\cdot))$ depends upon the measurement $y(\cdot)$, while $\mathcal{P}(t,\omega(\cdot))$ does not depend upon $y(\cdot)$.*

Remark 4.2.2 *The fact that functions $\beta(t),\gamma(t)$ are taken measurable does not forbid us to use equation (4.1.16) and the further schemes of Section 4.1 for the function $V(t,x,\omega(\cdot))$. This particularly is due to the unicity of the solution to the extremal problem (4.2.8). Besides that, $\alpha(t),\beta(t)$ may be assumed continuous (see footnote after Lemma 4.2.1).*

In the absence of state constraints induced by the measurement ($N(t) \equiv 0$), one should simply delete the restriction (4.1.8) and set $\gamma(\cdot) = 0$ in the previous Theorem. This also gives $k(t) \equiv 0$.

Corollary 4.2.1 *In the absence of the state constraint (4.1.8) relations (4.2.17) and (4.2.18) generated by equations (4.2.12)–(4.2.15) remain true, provided $\gamma(t) \equiv 0$. The set $\mathcal{X}(t)$ is then the attainability domain of Section 1.2 for system (4.1.1) under ellipsoidal magnitude constraints (4.1.6) and (4.1.7).*

Further, in Section 4.4, we shall rearrange the results obtained here in terms of earlier notations and compare them with those obtained in Parts II–III. But prior to that we shall discuss the calculation of error bounds for the estimation problems.

4.3 The State Estimates, Error Bounds, and Error Sets

Let us now pass to the discussion of *the estimates and the error bounds.* Consider the informational domain $\mathcal{X}(\tau)$ to be specified. Under the assumptions of Sections 4.1 and 4.2, set $\mathcal{X}(\tau)$ will be closed and bounded. Let us calculate the *Chebyshev center* of $\mathcal{X}(\cdot)$. Following formula (4.1.10), we have to minimaximize the function

$$\min_z \max_x \,(x - z, x - z) = \max_x \,(x - \hat{x}, x - \hat{x})$$

under the restriction

$$V(\tau, x) \leq \mu^2 \ .$$

Applying the conventional generalized Lagrangian technique [69], [260], [265] to the internal maximization problem, we have

$$\min_z \max_x \{(x - z, x - z) \ - \ \lambda^2{}_\mu V(\tau, x)\} \ . \tag{4.3.1}$$

Since $\hat{x}(\tau)$ is the center of the smallest ball that includes $\mathcal{X}(\tau)$, a convex and compact set, the inclusion $\hat{x}(\tau) \in \mathcal{X}(\tau)$ is always true.

Here the number λ_μ^2 is the Lagrange multiplier which generally depends on μ as also does $\hat{x}(\tau) = \hat{x}_\mu(\tau)$.[31] With $V(\tau, x)$ being a quadratic form of

[31] The multiplier λ_μ for the internal maximization problem should also depend on x. However, in this section it is presumed that it is the one that corresponds to the vector $\hat{x}$.

type (4.1.6), the solution to (4.3.1) is the center of the ellipsoid (4.1.22), namely, $\hat{x}(\tau) = z(\tau)$, whatever is the value of μ.

Summarizing the results, we have

Lemma 4.3.1 *The minmax estimate $\hat{x}(t)$ (the Chebyshev center) for the informational domain $\mathcal{X}(\tau)$ of Section 1.12, satisfies the property*

$$\hat{x}(\tau) \in \mathcal{X}(\tau)$$

and in general depends on μ: $\hat{x}(\tau) = \hat{x}_\mu(\tau)$.

In the linear-quadratic case (4.1.3)-(4.1.5) the vector

$$\hat{x}(\tau) = z(\tau)$$

is the center $z(\tau)$ of the ellipsoid $E(\tau, \mathcal{P}^{-1}(\tau))$ described by the (4.1.22) and does not depend on the number μ.

In order to compare the set-membership (bounding) and the H_∞ approaches, let us find the estimate $\bar{x}(\tau)$ for the H_∞ approach to state estimation. Then we have to solve the following problem:

Find the smallest number σ^2 that ensures

$$\min_z \max_{\zeta(\cdot)} \{(x - z, x - z) - \sigma^2 \Psi(\tau, \zeta(\cdot))\} \leq 0$$

under the conditions

$$x(\tau, t_0, \eta(\cdot)) = \ x; \ G(t)x(t, t_0, \eta(\cdot)) + v(t) \equiv y(t); \ t_0 \leq t \leq \tau \ .$$

This, however, is equivalent to the problem of finding the smallest number $\sigma^2 = \sigma_0^2$ that ensures

$$\min_z \max_x \{(x - z, x - z) - \sigma^2 \{\inf_{\eta(\cdot)} \Phi(\tau, \eta(\cdot)) | x(\tau, t_0, \eta(\cdot)) = x\} \leq 0 \tag{4.3.2}$$

or, equivalently,

$$\min_z \max_x \{(x - z, x - z) - \sigma^2 V(\tau, x)\} \leq 0 \ . \tag{4.3.3}$$

It is not difficult to observe the following

Lemma 4.3.2 *In the quadratic case (4.1.3)–(4.1.5) the Lagrange multiplier λ_μ that corresponds to the maximum over x in (4.3.1) (with $z = \hat{x}$), satisfies the equality*

$$\lambda_\mu^2 = \sigma_0^2, \ \forall \mu$$

and the solution $\bar{x}(\tau)$ to (4.3.2), (4.3.3)) satisfies

$$\bar{x}(\tau) = \hat{x}(\tau) = z(\tau) = z_\mu(\tau), \ \forall \mu.$$

This conclusion follows from standard calculations in quadratic programming and linear-quadratic control theory.

As an exercise in optimization we ask the reader to check the next proposition:

Proposition 4.3.1 In the case (4.1.3), (4.1.9) of magnitude constraints the Lagrange multiplier λ_μ of (4.3.1) that corresponds to the maximum over x with $z = \hat{x}$ and the number $\sigma = \sigma_0$ of (4.3.2), (4.3.3) are related as follows

$$\lambda_\mu^2 \to \sigma_0^2, \ (\mu \to \infty)$$

with estimate

$$\hat{x}_\mu(\tau) \to \bar{x}(\tau), \ (\mu \to \infty)$$

Remark 4.3.1 *Among the recommended estimates for deterministic state estimation problems of the above one may encounter the following one [129]:*

$$z^*(\tau) = \text{argmin}\{V(\tau, x) \mid x \in R^n\} \ .$$

Such a selection of the estimate is certainly justified for the linear-quadratic problem since then, as we have seen,

$$z^*(\tau) = \hat{x}(\tau) = \bar{x}(\tau) = z(\tau), \tag{4.3.4}$$

and all the estimate types coincide (!).

However, as soon as we apply a nonquadratic functional $\Phi(\tau, \eta(\cdot))$, like the one given by (4.1.3), (4.1.9,), one may observe that all of the estimates (4.3.4) may turn to be different, despite the linearity of the system. (Provide an example for the last assertion.)

One of the basic elements of the solutions to the state estimation problem are the *error bounds* for the estimates. For the set-membership (bounding) approach, when the bounds on the uncertain items $\zeta(\cdot)$ are specified in advance, these are naturally given in the form of *error sets.* Here the error set may be taken as

$$\mathcal{R}(\tau) = \mathcal{X}(\tau) - \hat{x}(\tau) \ .$$

As indicated above, the set $\mathcal{R}$ will be the *largest possible* (with respect to inclusion) if the realizations of the uncertain items $\zeta(\cdot)$ will generate *the worst-case measurement* $y^*(t)$. As we have seen, for the problems treated here these are $\zeta(\cdot) = \{x_0^* = a; v(t) \equiv v^*(t); f(t) \equiv f^*(t)\}$.

On the other hand, the set $\mathcal{R}$ is the *smallest possible* if it a singleton $\mathcal{R} = \{0\}$, in which case it is generated by the *best-case measurement* $\bar{y}(\tau)$.

For the quadratic integral constraint the best-case measurement is described in Lemma 4.1.4. The principles for identifying such measurements for magnitude constraints are indicated in references [186], [187].

As for the H_∞ approach, the estimation error $e^2(\tau)$ will depend upon the number σ^2 in the inequality (4.3.2). The smallest possible value σ_0^2 of this of number depends in general on the given measurement $y(t)$ that determines the restriction (4.3.2).

Among all possible measurements, the largest possible value of σ_0^2 will be attained again at the worst- case measurement $y^*(t)$ specified in Lemmas 4.1.3 and 4.2.2.

Exercise 4.3.1. In the H_∞ setting, for the case of quadratic integral index $\Psi(\tau, \zeta(\cdot))$, check whether the best-case function $\bar{y}(\cdot)$ of Lemma 4.1.4 does yield the value $\sigma_0^2 = 0$.

Remark 4.3.2 *Given measurement $y(\cdot)$, suppose we have calculated number σ_0^2 for the H_∞ approach. If, moreover, we are also given the number μ in (4.1.3), then, in the quadratic integral case (4.1.3)–(4.1.5), the number $\mu\sigma_0 \geq 0$ will be the radius of the smallest sphere that surrounds the error set $\mathcal{R}(\tau)$:*

$$\mathcal{R}(\tau) = \mathcal{X}(\tau) - \hat{x}(\tau) \subseteq \{x : (x, x) \leq \mu^2\sigma_0^2\} \ .$$

The properties of the Chebyshev centers for the set-membership and the H_∞ solutions in the nonlinear case yield yet more diversity in the estimates. This however leads us beyond the scope of the present book.

We shall now compare the ellipsoidal relation derived in Sections 4.1 and 4.2 with those introduced earlier, in Parts II and III.

4.4 Attainability Revisited: Viability Through Ellipsoids

In this section we again deal with the main object of this book, namely, with systems restricted by magnitude constraints. We shall first rearrange the relations of Section 4.2 using the notations of the earlier parts.

Let us start with system

$$\dot{x} = A(t)x + u \quad , \tag{4.4.1}$$

under constraints

$$u \in \mathcal{E}(p(t), P(t)), x(t_0) \in \mathcal{E}(x^*, X_0) \quad .$$

We also take $A(t) \equiv 0$, which gives no loss of generality, as indicated in Section 1.1. Then, due to Corollary 4.2.1, the attainability domain $\mathcal{X}[\tau] = \mathcal{X}(\tau, t_0, \mathcal{X}^0)$ at time τ, from set $\mathcal{X}^0 = \mathcal{E}(a, X_0)$ (see also Definition 2.1) may be derived from Theorem 4.2.2 if one substitutes $A(t) \equiv 0, L^{-1} = X_0, P(t) \equiv M^{-1}(t), \gamma(t) \equiv 0$.

Reformulating the Theorem for this particular case, we have, with $\mu = 1$,

Theorem 4.4.1 *The attainability domain $\mathcal{X}[\tau] = \mathcal{X}(\tau, t_0, \mathcal{X}^0)$ is the intersection of ellipsoids $\mathcal{X}[\tau] = \mathcal{X}(\tau, \chi(\cdot)) = \mathcal{E}(z(\tau), \mathcal{P}^{-1}(\tau, \chi(\cdot)))$, namely,*

$$\mathcal{X}[\tau] = \cap\{\mathcal{E}(z(\tau), \mathcal{P}^{-1}(\tau, \chi(\cdot)))|\chi(\cdot)\}, \tag{4.4.2}$$

where $\chi(\cdot) = \{\alpha, \beta(\cdot)\}$,

$$\alpha > 0, \beta(t) > 0, \;\; \alpha + \int_{t_0}^{\tau} \beta(t)dt = 1 \quad ,$$

and $z(t), \mathcal{P}(t) = \mathcal{P}(t, \chi(\cdot))$ *are defined through equations*

$$\dot{z} = p(t), \;\; z(t_0) = x^* \;\; , \tag{4.4.3}$$

$$\dot{\mathcal{P}} = -\beta^{-1}(t)\mathcal{P}P^{-1}(t)\mathcal{P}, \;\; \mathcal{P}_{t_0} = \alpha X_0^{-1} \;\; . \tag{4.4.4}$$

In order to compare the last result given in terms of equations (4.4.3),(4.4.4) to the one given in Theorem 3.2.5 in terms of equation (3.2.5), we make the substitution $X(t) = \mathcal{P}^{-1}(t)$, using also the relation $\dot{\mathcal{P}}^{-1} = -\mathcal{P}^{-1}\dot{\mathcal{P}}\mathcal{P}^{-1}$. This gives

$$\dot{X}(t) = \beta^{-1}(t)P(t), \;\; X(t_0) = (1 - \int_{t_0}^{\tau} \beta(t)dt)^{-1} X_0 \;\; , \tag{4.4.5}$$

or, after integration from t_0 to τ,

$$X(\tau) = (1 - \int_{t_0}^{\tau} \beta(t)dt)^{-1} X_0 + \int_{t_0}^{\tau} \beta^{-1}(t)P(t)dt \;\; . \tag{4.4.6}$$

Let us now take equation (3.2.5) of Section 3.2 and also integrate it from t_0 to τ with same initial set X_0. This gives

$$X^+(\tau) = \int_{t_0}^{\tau} \kappa(\tau, t)\pi^{-1}(t)P(t)dt + \kappa(\tau, t_0)X_0 \;\; , \tag{4.4.7}$$

where

$$\kappa(\tau, t) = exp\Big(\int_t^{\tau} \pi(s)ds\Big) \;\; .$$

Comparing (4.4.6) and (4.4.7), one may observe, by direct calculation, that by setting

$$\pi(t) = \beta(t)(1 - \int_t^{\tau} \beta(s)ds)^{-1}, \;\; t_0 \leq t \leq \tau \;\; , \tag{4.4.8}$$

relation (4.4.7) is transformed into (4.4.6).

Indeed, taking

$$exp\Big(-\Big(\int_t^{\tau} \pi(s)ds\Big)\Big) = 1 - \int_t^{\tau} \beta(s)ds \;\; ,$$

and differenting both parts in t, we have

$$\pi(t)\; exp\Big(-\Big(\int_t^{\tau} \pi(s)ds\Big)\Big) = \beta(t) \;\; , \tag{4.4.9}$$

which gives (4.4.8), on one hand, and transforms (4.4.7) into (4.4.6) on the other.

One may observe from (4.4.9) that for any function $\pi(t) > 0$ defined on the interval $t_0 \leq t \leq \tau$ there always exists a function $\beta(t)$ of the type

$$\beta(t) > 0,\ t_0 \leq t \leq \tau,\ (1 - \int_{t_0}^{\tau} \beta(t)dt) > 0\ ,$$

Since for any $\beta(t)$ of the last type there exists a function $\pi(t)$, due to (4.4.8), we are in the position to formulate

Theorem 4.4.2 *The attainability domain $\mathcal{X}[\tau] = \mathcal{X}(\tau, t_0, \mathcal{X}^0)$ of (4.4.2) allows an equivalent representation*

$$(4.4.10)\quad \mathcal{X}[\tau] = \cap\{\mathcal{E}(z(\tau), X(\tau))|\beta(\cdot)\} = \cap\{\mathcal{E}(x^*(\tau), X^+(\tau))|\pi(\cdot)\}$$

where $z(\tau) = x^*(\tau)$ and the parametrized varieties of matrices $\{X(\tau)\}, \{X^+(\tau)\}$ are described by equivalent relations (4.4.6), (4.4.7). The respective parameters $\beta(\cdot), \pi(\cdot)$ are related due to (4.4.8), (4.4.9).

The respective differential equations for $X[t], X^+[t]$ $t_0 \leq t \leq \tau$ are given by (4.2.5), (4.2.6), and (4.4.5). These equations may be transformed one into another through the substitutions given above, so that at instant τ they would yield the same solutions.

Let us now pass to the discussion of *Dynamic Programming techniques for the viability problem.*

Consider system (4.1.1) under constraints (4.1.2), (4.1.4) on u, x_0 and viability constraint

$$(4.4.11)\qquad x(t) \in \mathcal{E}(q(t), Q(t)),\ t_0 \leq t \leq \tau\ ,$$

with function $f(t) \equiv 0$. Constraint (4.4.11) follows from (4.1.2), (4.1.8) if $y(t) \equiv 0, G(t) \equiv I, N^{-1}(t) \equiv Q(t), v^*(t) = -q(t)$ (however, (4.4.11) may now be taken to be true *everywhere*).

We shall look for the *viability set* $W[\tau]$ at given instant τ which is the set of all points $x = x(\tau)$ for each of which there exists a control $u =$

$u(t)$ that ensures the respective trajectory $x(t, t_0, x|u(\cdot))$ to satisfy the viability constraint:

$$x[t] = x(t, t_0, x|u(\cdot)) \in \mathcal{E}(q(t), Q(t)), \quad \tau \leq t \leq t_1 \ ,$$

Set $W[\tau]$ coincides with the *solvability set* $\mathcal{W}[\tau] = \mathcal{W}(\tau, t_1, \mathcal{M})$ of Definition 1.9.4 if we take $G(t) \equiv I, \mathcal{Y}(t) \equiv \mathcal{E}((t), Q(t)), \mathcal{M} = \mathcal{E}(q(t_1), Q(t_1))$.

We shall determine $W[\tau]$ as *the level set*

$$W[\tau] = \{x : V_v(\tau, x) \leq 1\}$$

for *the viability function* $V_v(\tau, x)$, which we define as the solution to the following problem:

$$V_v(\tau, x) = \min_{u(\cdot)}\{\Phi(\tau, u(\cdot))|x[t] = x(t, \tau, x|u(\cdot))\} \tag{4.4.12}$$

where

$$\Phi(\tau, u(\cdot)) = \max\{J_0, J_1, J_2\} \ ,$$

and

$$J_0(x[t_1]) = (x[t_1] - q(t_1), Q(t_1)(x[t_1] - q(t_1))) \ , \tag{4.4.13}$$

$$J_1(\tau, u(\cdot)) = \text{esssup}_t(u(t) - p(t), P(t)(u(t) - p(t))) \ , \tag{4.4.14}$$

$$J_2(\tau, x[t]) = \max_t(x[t] - q(t), Q(t)(x[t] - q(t)) \ , \tag{4.4.15}$$

with $t \in [\tau, t_1]$ and $x[t] = x(t, \tau, x|u(\cdot))$ being the trajectory of system (4.4.1) that starts at position $\{\tau, x\}$ and is steered by control $u(t)$.

The solution to this problem may be described by a certain forward dynamic programming (H-J-B) equation [109]. In order to avoid generalized solutions of this equation, we shall follow the scheme of Section 4.4.2 by solving a linear-quadratic control problem. It is to minimize

$$\Lambda(\tau, x, u(\cdot), \omega(\cdot)) = \int_\tau^{t_1} (\gamma(t)(x[t] - q(t), Q(t)x[t] - q(t)) +$$

$$+\beta(t)(u(t) - p(t), P(t)(u(t) - p(t))))dt + \alpha(x[t_1] - q(t_1), Q(t_1)x[t_1] - q(t_1))$$

over $u(\cdot)$, with $x[t] = x(t, \tau, x|u(\cdot)$. Here $\omega(\cdot) = \{\alpha, \beta(\cdot), \gamma(\cdot))\}$ and

$$\alpha > 0, \beta(t) > 0, \gamma > 0, \ \ \alpha + \int_\tau^{t_1} (\beta(t) + \gamma(t))dt = 1 \ . \tag{4.4.16}$$

The variety of such elements $\omega(\cdot)$ is further denoted as Ω.

Then, in analogy with the previous Section 4.2, we have

$$(4.4.17) \qquad V_v(\tau, x) = \min_{u(\cdot)}\{\Phi(\tau, u(\cdot))|x[t] = x(t, \tau, x|u(\cdot))\} =$$

$$= \min_{u(\cdot)} \sup_{\omega(\cdot)} \Lambda(\tau, x, u(\cdot), \omega(\cdot)) \ .$$

The operations of min and sup may again be interchanged. Doing this, we denote

$$(4.4.18) \qquad V_v(\tau, x) = \sup_{\omega(\cdot)} V_v(\tau, x, \omega(\cdot)) \ ,$$

where

$$V_v(\tau, x, \omega(\cdot)) = \min_{u(\cdot)} \Lambda(\tau, x, u(\cdot), \omega(\cdot)) \ .$$

We again look for this function as a quadratic form

$$(4.4.19) \qquad V_v(\tau, x, \omega(\cdot)) =$$

$$= (x - z(\tau, \gamma(\cdot), \mathcal{P}(\tau, \omega(\cdot))(x - z(\tau, \gamma(\cdot))) \ ,$$

where $\mathcal{P}[t] = \mathcal{P}(t, \omega(\cdot)), z[t] = z(t, \gamma(\cdot)), k = k(t, \gamma(\cdot))$ satisfy the equations

$$(4.4.20) \; \dot{\mathcal{P}} = -\mathcal{P}A(t) - A'(t)\mathcal{P} + \beta^{-1}(t)\mathcal{P}P(t)\mathcal{P} - \gamma(t)Q^{-1}(t) \ ,$$

$$(4.4.21) \qquad \dot{z} = A(t)z - \gamma(t)\mathcal{P}^{-1}Q^{-1}(t)(z + q(t)) + p(t) \ ,$$

$$(4.4.22) \qquad k^2(t) = -\gamma(t)(z + q(t), Q^{-1}(t)(z + q(t))) \ ,$$

$$(4.4.23) \qquad \mathcal{P}(t_1) = \alpha Q^{-1}(t_1), \; z(t_1) = q(t_1), \; k(t_1) = 0 \ .$$

It may be more convenient to deal with matrix $X_v(t) = \mathcal{P}^{-1}[t]$ which satisfies equation

$$(4.4.24) \; \dot{X}_v = A(t)X_v + X_vA'(t) + \gamma(t)X_vQ^{-1}(t)X_v - \beta^{-1}(t)P(t) \ ,$$

$$(4.4.25) \qquad X_v(t_1) = \alpha^{-1}Q(t_1) \ .$$

Following the reasoning of Section 4.2, we formulate the following assertion

Lemma 4.4.1 *The viability function $V_v(\tau, x)$ is the upper envelope*

(4.4.26) $$V_v(\tau, x) = \sup\{V_v(\tau, x, \omega(\cdot)) | \omega(\cdot) \in \Omega\}$$

of a parametrized variety of quadratic forms $V_v(\tau, x, \omega(\cdot))$ of type (4.4.19) over the functional parameter $\omega(\cdot) = \{\alpha, \beta(\cdot)), \gamma(\cdot))\}$, where $\omega(\cdot) \in \Omega$.

Since the level sets for $V_v(\tau, x, \omega(\cdot))$ are ellipsoids, namely,

$$W[\tau, \omega(\cdot)] = \mathcal{E}(z[\tau], (1 - k^2[\tau])^{-1} X_v[\tau])$$

and since $W[\tau]$ is a level set for $V_v(\tau, x)$, we are able, due to Lemma 4.4.1, to come to

Theorem 4.4.3 *The viability set $W[\tau]$ is the intersection of ellipsoids, namely,*

(4.4.27) $$W[\tau] = \{\cap \mathcal{E}(z[\tau], (1 - k^2[\tau])^{-1} X_v[\tau]) | \omega(\cdot) \in \Omega\},$$

where z, k, X_v are defined through equations (4.4.21)–(4.4.25).

The set-valued function $W[t]$, $\tau \leq t \leq t_1$ is known as *the viability tube* which may therefore also be approximated by ellipsoids along the schemes of this section.

4.5 The Dynamics of Information Domains: State Estimation as a Tracking Problem

As we have remarked before , the information domains of Section 4.1.1 or 4.1.12 are nothing else than *attainability domains under state constraints* when the last are given, for example, by inequalities (4.1.8), (4.1.2). These domains $\mathcal{X}(\tau)$ were therefore already described through Dynamic Programming approaches in Sections (4.1.1)–(4.1.3). However, some other types of ellipsoidal estimates and their dynamics may be derived for $\mathcal{X}(\tau)$ directly, through the funnel equations of Sections 1.9 and 1.12 and the elementary formulae of Part II.

In this section we consider the attainability domain $\mathcal{X}[\tau]$ for system

$$\dot{x} = u(t) + f(t) \quad , \tag{4.5.1}$$

under the constraints and notations of Section 4.4 for $u(t), x^0$ and state constraint

$$x(t) \in \mathcal{E}(y(t), K(t)) \quad , \tag{4.5.2}$$

where the matrix-valued function $K(t) > 0, K(t) \in \mathcal{L}(\mathbb{R}^n, \mathbb{R}^n)$ and the function $y(t) \in \mathbb{R}^n$ (the observed output in the state estimation problem) sassumed to be continuous.[32] To treat this case we shall first follow the funnel equation of type (1.9.21). For the attainability domain $\mathcal{X}(t)$ under state constraint (4.5.2) this gives

$$\mathcal{X}(t+\sigma) = (\mathcal{X}(t) + \sigma(\mathcal{E}(p(t), P(t)) +$$

$$+ f(t))) \cap \mathcal{E}(y(t+\sigma), \mathcal{K}(t+\sigma)) + o(\sigma)\mathcal{S}, \quad , \sigma > 0 \quad .$$

Presuming $\mathcal{X}(t) = \mathcal{E}(x(t), X(t))$, we shall seek for the external ellipsoidal estimate $\mathcal{E}(x(t+\sigma), X(t+\sigma))$ of $\mathcal{X}(t+\sigma)$. To do this we shall use relations (2.3.1) and (2.7.14).

Namely, using (2.3.1), we first take the estimate

$$\mathcal{E}(x(t), X(t)) + \sigma(\mathcal{E}(p(t), P(t)) + f(t)) \in \mathcal{E}(\tilde{x}(t), \tilde{X}(t)) \quad ,$$

where $\tilde{x}(t) = x(t) + \sigma p(t) + \sigma f(t)$, and

$$\tilde{X}(t) = (1+q)X(t) + (1+q^{-1})\sigma^2 P(t), \quad q > 0 \quad . \tag{4.5.3}$$

Further, using (2.7.4), we have

$$\mathcal{E}(\tilde{x}, \tilde{X}) \cap \mathcal{E}(y(t), K(t)) \in \mathcal{E}(x(t+\sigma), X(t+\sigma)) \quad ,$$

where

$$x(t+\sigma) = (I-M)(x(t) + \sigma p(t) + \sigma f(t)) + My(t+\sigma) \quad , \tag{4.5.4}$$

and

$$X(t+\sigma) = (1+\pi)(I-M)\tilde{X}(t)(I-M)' \tag{4.5.5}$$

---

[32] The case of measurable functions $y(t)$ which allows more complicated discontinuities in $y(t)$ and is of special interest in applications is considered lower in Section 4.6.

$$+(1+\pi^{-1})MK(t+\sigma)M', \ \pi > 0 \ .$$

Making the substitutions

$$q = \sigma\bar{q}, \ \pi = \sigma\bar{\pi}, \ M = \sigma\bar{M} \ ,$$

collecting (4.4.3) -(4.4.5) together and leaving the terms of order ≤ 1 in σ, we come to

$$x(t+\sigma) - x(t) = \sigma p(t) + \sigma f(t) + \sigma\bar{M}(y(t+\sigma) - x(t))$$

and also

$$X(t+\sigma) - X(t) =$$

$$= \sigma((\bar{\pi}+\bar{q})X(t) - \bar{M}X - X\bar{M}' + \bar{q}^{-1}P + \bar{\pi}^{-1}\bar{M}K(t+\sigma)\bar{M}') \ .$$

Dividing both parts of the previous equations by $\sigma > 0$ and passing to the limit $\sigma \to +0$, we further come, in view of the continuity of $y(t), K(t)$, to differential equations (deleting the bars in the notations)

(4.5.6) $$\dot{x} = p(t) + f(t) + M(t)(y(t) - x) \ ,$$

(4.5.7) $$\dot{X} = (\pi(t) + q(t))X + q(t)^{-1}P - M(t)X - XM'(t) + \pi^{-1}M(t)K(t)M'(t) \ ,$$

where

(4.5.8) $$x(t_0) = x^0, \ \ X(t_0) = X_0 \ ,$$

and $\pi(t) > 0, q(t) > 0, M(t)$ are continuous functions.

What further follows from Theorem 1.3.3 (formula 1.3.31) and Lemma 2.7.3 (formula 2.7.11) is the assertion

Theorem 4.5.1 *The attainability domain $\mathcal{X}(\tau)$ for system (4.5.1) under restrictions (4.1.2), (4.1.4) and state constraints (4.5.2) (with $y(t), K(t)$ continuous) satisfies the inclusion $\mathcal{X}(\tau) \in \mathcal{E}(x(\tau), X(\tau))$, where $x(t), X(t)$ satisfy the differential equations (4.5.6)–(4.5.8) within the interval $t_0 \leq t \leq \tau$.*

Moreover, the following relation is true

(4.5.9) $$\mathcal{X}(\tau) = \cap\{\mathcal{E}(x(t), X(t)) | \pi(\cdot), q(\cdot), M(\cdot)\}$$

where $\pi(t) > 0, q(t) > 0, M(t)$ are continuous functions.

Exercise 4.5.1. Suppose the state constraint (4.5.2) is substituted by relation

$$G(t)x(t) \in \mathcal{E}(y(t), K(t)) \tag{4.5.10}$$

where $y(t) \in \mathbb{R}^m, K(t) \in \mathcal{L}(\mathbb{R}^m, \mathbb{R}^m), G(t) \in \mathcal{L}(\mathbb{R}^n, \mathbb{R}^m)$ and $G(t)$ is continuous.

Prove that in this case the previous relations together with Theorem 4.5.1 still hold with obvious changes. Namely, (4.1.6), (4.1.7) should be substituted by

$$\dot{x} = p(t) + f(t) + M(t)(y(t) - G(t)x) \ , \tag{4.5.11}$$

$$\dot{X} = (\pi(t) + q(t))X + q(t)^{-1}P(t) - \tag{4.5.12}$$

$$-M(t)G(t)X - XG'M'(t) + \pi^{-1}M(t)K(t)M'(t) \ ,$$

with same boundary conditions (4.5.8).

Remark 4.5.1 *To obtain the equations for $\mathcal{X}(\tau)$ through Dynamic Programming, we just have to take relations (4.2.12)-(4.2.15) and set $v^*(t) \equiv 0, N(t) \equiv K^{-1}(t)$.*

We thus have two sets of relations for $\mathcal{X}(\tau)$, namely, the one given in Section 4.1.2 and the one given in the present section.

Then each of the approaches leads to a variety of ellipsoidal sets that include $\mathcal{X}(\tau)$, on the one hand, and allow exact representations of types (4.2.18) (variety 1) and (4.5.9) (variety 2), on the other.

It is not difficult to observe that variety 2 of ellipsoids given in (4.5.9) depends on more parameters than in (4.2.18) and is therefore richer than variety 1. This has the following implication: if among the varieties 1 or 2 we are to select an ellipsoid optimal in some conventional sense (see, for example, Section 4.2), then we may expect that variety 2 (the richer one) will produce a tighter optimal ellipsoid than variety 1. Elementary examples of such kind are given above, in Section 2.6.

Remark 4.5.2 *System (4.5.6)-(4.5.8) was derived under the assumption that function $y(t)$ is continuous. However, we may as well assume that $y(t)$ is allowed to be piecewise continuous (from the right). Then the respective value in equations (4.5.6), (4.5.11) should be $y(t) = y(t+0)$.*

One could observe, that the funnel equation used earlier in the proof of Theorem 4.5.1, is the one given in (1.12.10). A similar derivation is possible, however, if we use funnel equation (1.12.11), which for constraint (4.5.2) is as follows:[33]

$$\lim_{\sigma \to +0} \sigma^{-1} \cdot h_+(\mathcal{X}(t+\sigma), \mathcal{X}(t) \cap \mathcal{E}(y(t), K(t)) + \sigma\mathcal{E}(p(t), P(t))) = 0 \ .$$

Then for the maximal solution of this equation, with $\mathcal{X}[t_0] = \mathcal{X}_0$, we have

$$\mathcal{X}[t+\sigma] = \mathcal{X}[t] \cap \mathcal{E}(y(t), K(t)) + \sigma\mathcal{E}(p(t), P(t)) + \mathcal{S}o(\sigma) \ ,$$

and subsequently follow the operations: first take the estimate

$$\mathcal{X}[t] \cap \mathcal{E}(y(t), K(t)) \subseteq \mathcal{E}(\hat{x}(t), \hat{X}(t)) \ ,$$

then the estimate

$$\mathcal{E}(\hat{x}(t), \hat{X}(t)) + \sigma\mathcal{E}(p(t), P(t)) \subseteq \mathcal{E}(x(t+\sigma), X(t+\sigma)) \ .$$

With obvious modifications the futher reasoning is similar to proof of Theorem 4.5.1 (see (4.5.3)–(4.5.5) and following relations). The conclusion is that finally we again come to equations (4.5.6)–(4.5.8), except that this time we did not use the continuity of $y(t)$, having implicitly used its piecewise-continuity from the right. (We have implicitly required that at each point t we have

$$\sigma^{-1} \int_t^{t+\sigma} y(s)ds \to y(t), \ \ \sigma \to +o \ ,$$

This is ensured by the latter piecewise continuity.)

Lemma 4.5.1 *Theorem 4.5.1 remains true if $y(t)$ is piecewise continuous from the right.*

Exercise 4.5.1a. Solve Exercise 4.5.1 under conditions of Lemma 4.5.1.

---

[33] Though agreed in Section 3 that we do not indicate the sign + in the limit transition $\sigma \to +0$ when $h = h_+$, in this specific relation we emphasize the sign

We shall now give a control interpretation of the state estimation problem. Consider again the state estimation (attainability) problem for system (4.5.1), (4.5.10), (4.1.2), (4.1.4), $f(t) \equiv 0$. Though the set $\mathcal{X}[\tau]$ may be approximated both externally and internally by ellipsoidal-valued functions, we shall again further deal only with the former case. (An indication on a general scheme for internal ellipsoidal approximations of intersections of ellipsoids is given at the end of Section 2.6.)

As in Section 3.3, let us introduce an ellipsoidal funnel equation, but for the present problem now. In doing this we shall seek for the external ellipsoidal approximation of the respective tube $\mathcal{X}[t]$. Consider the evolution equation

$$\lim_{\sigma \to +0} \sigma^{-1} \cdot h_{-}(\mathcal{E}(t+\sigma), \mathcal{E}(t) \cap \mathcal{E}(y(t), K(t)) + \sigma \mathcal{E}(p(t), P(t))) = 0 \ ,$$

(4.5.13)

$$t_0 \leq t \leq t_1, \quad \mathcal{E}[t_0] = \mathcal{E}(x^0, X_0) \ .$$

A set-valued function $\mathcal{E}_{+}[t]$ will be defined as a solution to (4.5.13) if it satisfies (4.5.13) for almost all t and is *ellipsoidal-valued.* Obviously the solution $\mathcal{E}_{+}[t]$ is *nonunique* and satisfies the inclusion

$$\mathcal{E}_{+}[t] \supset \mathcal{X}[t], \quad t_0 \leq t \leq t_1, \qquad \mathcal{E}_{+}[t_0] = \mathcal{X}[t_0] \ .$$

Moreover, as a consequence of Lemmas 2.2.1 and 2.6.3, one may come to

Theorem 4.5.2 *For any $t_0 \leq t \leq t_1$ the following equality is true*

$$\mathcal{X}[t] = \bigcap \{\mathcal{E}_{+}[t] \mid \mathcal{E}_{+}[\cdot] \text{ is a solution to (4.5.13) } \} \ .$$

The ellipsoidal solutions $\mathcal{E}_{+}[t] = \mathcal{E}(x_{-}(t), X_{-}(t))$ to (4.5.13) allow explicit representations through appropriate systems of ODEs for the centers $x_{-}(t)$ and the matrices $X_{-}(t) > 0$ of these ellipsoids. One may check that among these are, particularly, the solutions $x(t, M), X(t, M, \pi, q)$ to system (4.5.6)–(4.5.8). A more complicated problem is to find the tightest estimates, or, in other terms, the minimal (with respect to inclusion) ellipsoidal solutions to (4.5.13).

Exercise 4.5.2. Check whether it is possible to select parameters $M(t), \pi(t), q(t)$ in (4.5.6)–(4.5.7), so as to produce an inclusion-minimal external ellipsoidal estimate $\mathcal{E}(x(\tau), X(\tau)) \supseteq \mathcal{X}[\tau]$ at time τ.

The indications on how to select such estimates for the static case are given in Sections 2.3 and 2.6.

As we have earlier observed in similar situations, parameters M, π, q *may be interpreted as controls* and the problem of specifying the best ellipsoids *as control problems.* This also leads to the following considerations.

Denote the external ellipsoid of system (4.5.11), and (4.5.12) as $\mathcal{E}_\omega[t] = \mathcal{E}(x(t, M), X(t, M, \pi, q))$, where $\omega = \{M(\cdot), \pi(\cdot), q(\cdot)\}$. The center $x(t, M)$ of the tube $\mathcal{E}_\omega[t]$, $t_0 \leq t \leq t_1$, satisfies equation (4.5.11) with $x(t_0) = x^0$.

Let us denote the actual trajectory to be estimated, as $x^*(\cdot)$. By construction, the inclusions

$$\mathcal{E}_\omega[t] \supseteq \mathcal{X}^*[t] \supseteq x^*(t), \; t_0 \leq t \leq t_1 \quad ,$$

are true. Therefore the approximate estimation procedure is that *the estimator $x(t, M)$ tracks the unknown trajectory $x^*(t)$*, and the ellipsoid $\mathcal{E}_\omega[t]$ around it plays the role of a guaranteed confidence region. The set $\mathcal{E}(0, X(t, M, \pi, q))$ then estimates the *error set* $\{\mathcal{X}^*[t] - x(t, M)\}$ of the estimation process.

The trajectory of the estimator $x(t, M)$ depends on the measurement output $y(s), t_0 \leq s \leq t$, and therefore realizes *a feedback procedure.* (The parameter M may be also chosen through feedback from $y(\cdot)$.)

The tracking procedure described here is similar in nature to a *differential game of observation.*[34]

Example 4.5.1

Given is a 4-dimensional system

$$\text{(4.5.14)} \qquad \dot{x} = A(t)x + u(t), \quad G(t)x = y(t) + v(t) \quad ,$$

[34] A feedback duality theory for differential games of observation and control is described in [179].

$$u(t) \in \mathcal{E}(p(t), P(t)), \quad v(t) \in \mathcal{E}(0, K(t)) \ ,$$

over the time interval $[0, 5]$. We first describe the the ***attainability domain under state constraints*** (or, interpreting $y(t)$ as the observation, the *information domain*).

The initial state is bounded by the ellipsoid $\mathcal{X}_0 = \mathcal{E}(x^0, X_0)$ at the starting time $t_0 = 0$ with

$$x^0 = \begin{pmatrix} 1 \\ 0 \\ 1 \\ 0 \end{pmatrix} \quad \text{and} \quad X_0 = \begin{pmatrix} 1 & 0 & 0 & 0 \\ 0 & 1 & 0 & 0 \\ 0 & 0 & 1 & 0 \\ 0 & 0 & 0 & 1 \end{pmatrix} .$$

The matrix $A(t)$ is constant:

$$A(t) \equiv \begin{pmatrix} 0 & 1 & 0 & 0 \\ -8 & 0 & 0 & 0 \\ 0 & 0 & 0 & 1 \\ 0 & 0 & -4 & 0 \end{pmatrix} .$$

It describes the position and velocity of two independent oscillators. The unknown inputs $u(t) \in \mathcal{E}(p(t), P(t))$ are bounded by constant constraints where

$$p(t) \equiv \begin{pmatrix} 0 \\ 0 \\ 0 \\ 0 \end{pmatrix}, \quad \text{and} \quad P(t) \equiv \begin{pmatrix} 1 & 0 & 0 & 0 \\ 0 & 0.01 & 0 & 0 \\ 0 & 0 & 1 & 0 \\ 0 & 0 & 0 & 0.01 \end{pmatrix} ,$$

(this form of the bounding sets makes the system coupled).[35]

The state constraint is defined by the data

$$G(t) \equiv \begin{pmatrix} 0 & 1 & 0 & 0 \\ 0 & 0 & 0 & 0 \\ 0 & 0 & 0 & 1 \end{pmatrix}, \quad k(t) \equiv \begin{pmatrix} 0 \\ 0 \end{pmatrix}, \quad K(t) \equiv \begin{pmatrix} 16 & 0 \\ 0 & 25 \end{pmatrix} .$$

In Figure 4.5.1 we show the graph of external ellipsoidal estimates of the 4-dimensional *state space variable* $x(t)$ – with and without state

[35] Following Section 1.1, we could transform this system to an equivalent form, where $A = 0$ and $P = P(t)$ is time-dependent.

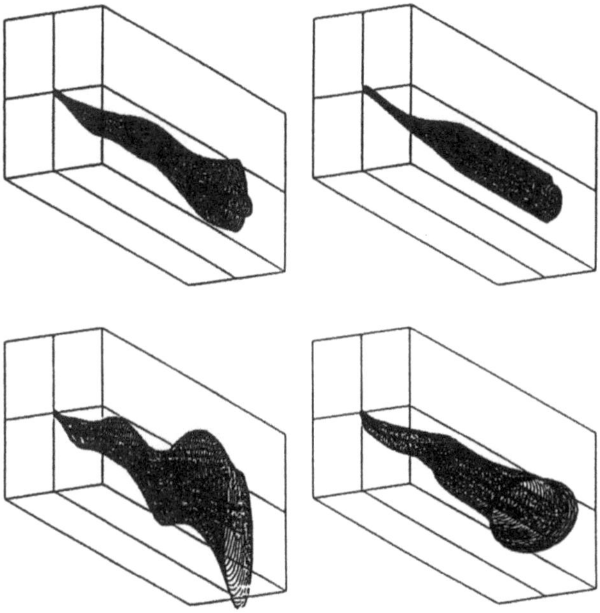

Figure 4.5.1.

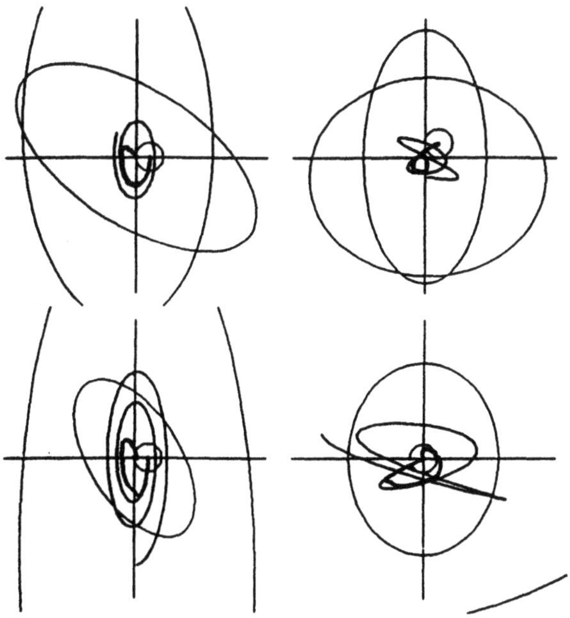

Figure 4.5.2.

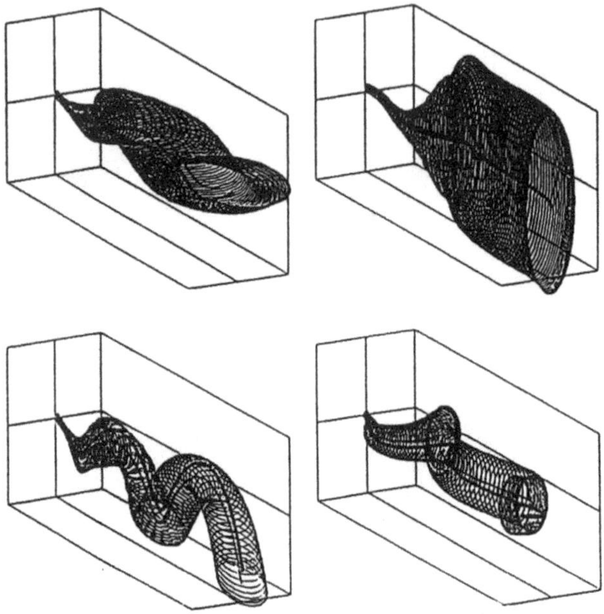

Figure 4.5.3.

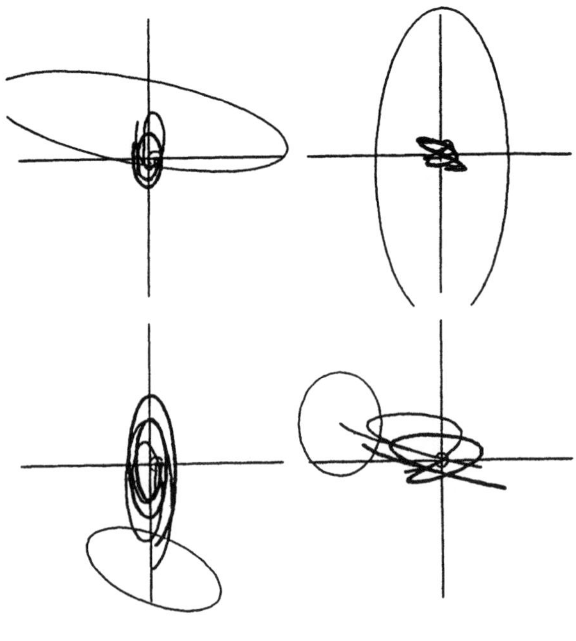

Figure 4.5.4.

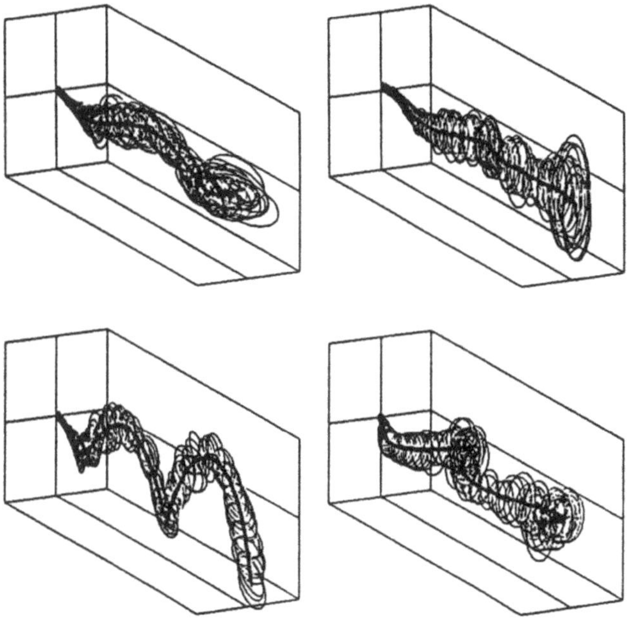

Figure 4.5.5.

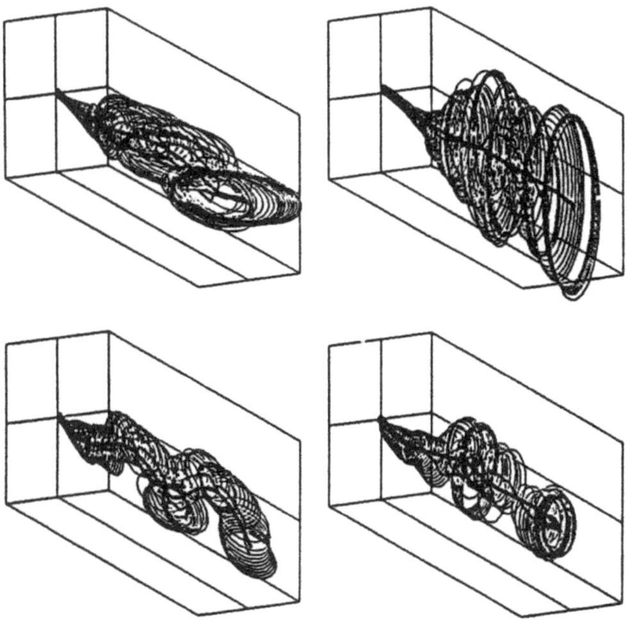

Figure 4.5.6.

constraints – presenting them in four *windows*, being confined to projections onto the planes spanned by the first and second, third and fourth, first and third, and second and fourth coordinate axes, in a clockwise order starting from bottom left. The drawn segments of coordinate axes corresponding to the output variables range from -30 to 30. The skew axis in Figure 4.5.1 is time, ranging from 0 to 5.

Calculations are based on the discretized version of the system (4.5.14) and the schemes of this section. The parameters m, π, q in each step are selected as trace-minimal along the results of Section 2.6. Figure 4.5.2 shows the trajectory of the centers, initial sets and the ellipsoidal estimates of the state space variables x, projected on to the planes spanned by two coordinate axes (chosen with the same arrangement of the four windows as in Figure 4.5.1), with drawn segments ranging from -10 to 10.

We turn now to the *guaranteed state estimation problem* interpreted as a *tracking problem*, as described above. We keep the above parameter values of the time interval, $A(t)$, $\mathcal{E}(x_0, X_0)$, $\mathcal{E}(p(t), P(t))$, and $G(t)$ and the same calculation schemes.

We model the trajectory $x^*(t)$, – the one to be tracked – by using the following construction for the triplet $\zeta^*(\cdot) = \{x_0^*, u^*(\cdot), v^*(\cdot)\}$. The initial value x_0^* is a (randomly selected) element at the boundary of the initial set $\mathcal{X}_0 = \mathcal{E}(x_0, X_0)$. The input $u^*(\cdot)$ is of the so called *extremal bang-bang type.* The time interval is divided into subintervals of constant lengths. A value u is chosen randomly at the boundary of the respective bounding set, that is in case of the input $u^*(t)$, of set $\mathcal{P}(t) = \mathcal{E}(p(t), P(t))$ and its value is then defined as $u^*(t) = u$ over all the first interval and as $u^*(t) = -u$ over the second. Then a new random value for u is selected and the above procedure is repeated for the next pair of intervals, etc. For modelling the measurement noise $v^*(\cdot)$ (generating together with x_0^* and $u^*(\cdot)$ the actual measurement $y^*(\cdot)$), we use a similar procedure.

As is well known, the size of the error set of the estimation depends on the nature of $v^*(\cdot)$. According to [181], if we choose it in such a way that it takes a constant value at the boundary of $\mathcal{E}(0, K(t))$ over all the time interval under study, then it corresponds to the worst case. This means in large confidence regions, while using, e.g., the extremal bang-bang construction, good noises are created, reducing the confidence regions' size.

Figure 4.5.3 shows the process developing over time – the drawn segments of coordinate axes corresponding to the output variables range from -20 to 20. In Figure 4.5.4 the initial sets of uncertainty (appearing as circles) are displayed in phase space, as well as the confidence region at the final instant. Coordinate axes range here from -10 to 10. The trajectory drawn with the thick line is the actual output $x^*(t)$. The thin line represents the trajectory of the centers $x(t, M)$ of the projections of the tracking ellipsoids. Figures 4.5.5 and 4.5.6 show how much the estimation can improve if the noise changes from worst to better – although we obtain here only external ellipsoidal estimates of the true error sets. Opposed to Figures 4.5.3 and 4.5.4, where the noise was constant, we chose its range to be within $[-0.5, 0.05]$. The range of the coordinate axes is again $[-20, 20]$.

4.6 Discontinuous Measurements and the Singular Perturbation Technique

The idea of applying singular perturbation techniques to the state estimation problem of the present book is motivated by the necessity to treat measurements $y(t)$ that are of a bad nature, possibly discontinuous. Indeed, in this section we shall allow Lebesgue-measurable realizations $y(t)$ of the measurement output.

Consider system (1.12.19), (1.12.3)–(1.12.5), where all the sets involved are ellipsoids:

(4.6.1) $$\dot{x} \in A(t)x + \mathcal{E}(p(t), P(t)) \ ,$$

(4.6.2) $$x(t_0) \in \mathcal{E}(x^0, X_0),$$

(4.6.3) $$G(t)x(t) \in y(t) + \mathcal{E}(0, K(t)), \ \ t_0 \leq t \leq \tau \ .$$

Here

$$p : [t_0, t_1] \to \mathbb{R}^n, \ \ y : [t_0, t_1] \to \mathbb{R}^m \ ,$$

$$P(t) \in \mathcal{L}(\mathbb{R}^n, \mathbb{R}^n), \ \ K(t) \in \mathcal{L}(\mathbb{R}^m, \mathbb{R}^m), \ \ x^0 \in \mathbb{R}^n,$$

the matrices $X_0, P(t), K(t)$ are symmetric and positively definite.

Our goal will be to find *the exact external ellipsoidal estimate* for the attainable set $X[\tau]$ for the system (4.6.1)–(4.6.3).

After collecting the preliminary results of Sections 1.12 and 2.2, and using the notations similar to those of Section 1.12, we are in a position to formulate the following result:

Theorem 4.6.1 *Given instant $\tau \in [t_0, t_1]$, the following exact formula is true*

$$X[\tau] = X(\tau, t_0, X_0) = \tag{4.6.4}$$

$$= \Pi_x(\cap\{\mathcal{E}(z(\tau, L), Z(\tau, L, \pi, \chi) | L(\cdot), \pi(\cdot), \chi(\cdot)\}$$

where

$$L(\cdot) \in \mathbf{L}; \quad \pi(t) > 0, \quad \chi(t) > 0, \quad t \in [t_0, \tau] \quad .$$

Here $z(t, L) = \{x(t), s(t)\}$ is a solution to the system

$$\dot{x} = A(t)x + p(t) \quad , \tag{4.6.5}$$

$$L(t)\dot{s} = -G(t)x + y(t) \quad , \tag{4.6.6}$$

$$x(t_0) = x^0, \quad s(t_0) = 0 \quad ,$$

and $Z_i(t), i = 1, 2, 3,$ of

$$Z(t, L, \pi, \chi) = \begin{pmatrix} Z_1(t) & Z_2'(t) \\ Z_2(t) & Z_3(t) \end{pmatrix}$$

being the solutions to the matrix differential equations

$$\dot{Z}_1 = A(t)Z_1 + Z_1A'(t) + \chi^{-1}(t)Z_1 + +\chi(t)(1 + \pi^{-1}(t))P(t) \quad ,$$

$$L(t)\dot{Z}_2 = -G(t)Z_1 + L(t)Z_2A'(t) + L(t)\chi^{-1}(t)Z_2 \quad ,$$

$$L(t)\dot{Z}_3 = -G(t)Z_2' - L(t)Z_2G'L'^{-1}(t) +$$

$$+\chi^{-1}(t)L(t)Z_3 + \chi(t)(1 + \pi(t))K(t)L'^{-1}(t) \quad ,$$

$$Z_1(t_0) = X_0, \quad Z_2(t_0) = 0, \quad Z_3(t_0) = I \quad ,$$

where the identity matrix $I \in \mathcal{L}(\mathbb{R}^m, \mathbb{R}^m)$.

Proof. Introducing the perturbed system

$$\dot{x} \in A(t)x + \mathcal{E}(p(t), P(t)) \quad , \tag{4.6.7}$$

$$L(t)\dot{s} \in -G(t)x + \mathcal{E}(y(t), K(t)) \tag{4.6.8}$$

$$\{x_0, s_0\} \in \mathcal{E}(\{x_0, 0\}, \tilde{X}_0) \ ,$$

$$\tilde{X}_0 = \begin{pmatrix} X_0 & 0 \\ 0 & I \end{pmatrix} \ .$$

Applying consequently theorem 1.12.5 and Corollary 3.2.2 to systems (4.6.5), (4.6.6) and (4.6.1)–(4.6.3), we come to the equality (4.6.4).

Q.E.D.

The proposed scheme *does not require the measurement $y(t)$ to be continuous.*

We shall further illustrate the procedures of this section through a numerical example.

Example 4.6.1[36]

To illustrate the *Singular Perturbation Technique*, we chose a system of two dimensions, and a scalar measurement equation, taking right-hand side constant:

(4.6.9) $$A(t) \equiv \begin{pmatrix} 0 & 1 \\ -8 & 0 \end{pmatrix} \ .$$

The unknown inputs $u(t)$ are bounded by time independent constraints $v(t) \in \mathcal{E}(p(t), P(t))$ with

(4.6.10) $$p(t) \equiv \begin{pmatrix} 0 \\ 0 \end{pmatrix}, \qquad \text{and} \qquad P(t) \equiv \begin{pmatrix} 1 & 0 \\ 0 & 0.01 \end{pmatrix} \ ,$$

and the initial state $x_(t_0)$ by the ellipsoid $\mathcal{E}(x_0, X_0)$, there

(4.6.11) $$x_0 = \begin{pmatrix} 1 \\ 0 \end{pmatrix} \qquad \text{and} \qquad X_0 = \begin{pmatrix} 1 & 0 \\ 0 & 10 \end{pmatrix} \ .$$

Further we take the measurement equation to be 1-dimensional:

$$G(t) \equiv (0 \quad 1), \qquad y(t) \equiv 1, \qquad K(t) \equiv (1) \ .$$

Additionally we suppose the initial condition:

$$s(0) \in \mathcal{S}_0, \quad \mathcal{S}_0 = [-10^{-5}, 10^{-5}] \ .$$

[36]The calculation of this Example belongs to K. Sugimoto.

Therefore, we have

$$\mathcal{Z}_0 = \mathcal{E}(x_0, X_0) \times \mathcal{S}_0 \subset \mathbb{R}^2 \times \mathbb{R} \ .$$

The time interval was divided into 100 subintervals of equal lengths and the calculations were based on a discretized version of system (4.6.1)–(4.6.3) with data (4.6.9)–(4.6.11).

We further calculate the ellipsoidal estimate

$$\mathcal{X}[\tau] = \Pi_x(\mathcal{E}(z(\tau, L_+), Z(\tau, L_+, \pi, \chi)) \cap \mathcal{E}(z(\tau, L_-), Z(\tau, L_-, \pi, \chi))) \ ,$$

for the following *two choices* for the function L:

$$L_+(t) = \begin{cases} 1 & \text{if } t \in [0, 3.5] \\ 0.3 & \text{if } t \in (3.5, 5] \end{cases} \ , \qquad L_-(t) = \begin{cases} 1 & \text{if } t \in [0, 3.5] \\ -0.3 & \text{if } t \in (3.5, 5] \end{cases} \ ,$$

with the range of coordinate axes being -30 to 30.

Parameters π, χ are chosen as

$$\pi(t) = \frac{tr^{1/2}(P(t))}{tr^{1/2}(K(t))}, \quad \chi(t) = \frac{tr^{1/2}(Z(t))}{tr^{1/2}(K(t))} \ .$$

It is useful to note that in general

$$\Pi_x(\mathcal{E}_1 \cap \mathcal{E}_2) \subseteq \Pi_x(\mathcal{E}_1) \cap \Pi_x(\mathcal{E}_2) \ . \tag{4.6.12}$$

An illustration of that is given in Figure 4.6.1, where the thin lines denote the projections of two 3-dimensional ellipsoids on the plane spanned by the first two coordinates (upper left window), the first and third coordinate (upper right) and second and third (lower right). The thicker line denotes the projection of their intersection on the same planes. Here (4.6.12) is *a proper inclusion*.

Returning to our numerical example, we illustrate it in Figure 4.6.2, where the upper left window shows the projections onto the plane spanned by the two state variables. Here they coincide as expected. In the upper right we see the projection of the two estimating tubes (corresponding to L_+, L_-) onto the plane of the measurement variable and the first state variable, while in the lower window, the tubes are projected onto the plane of the measurement variable and the second state

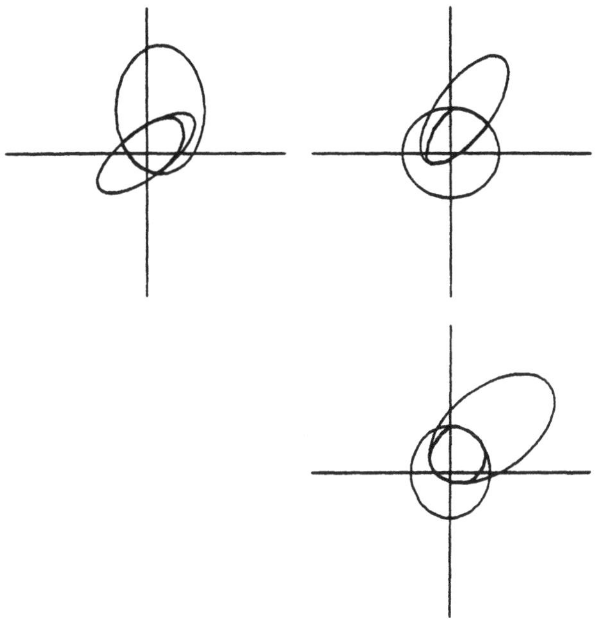

Figure 4.6.1.

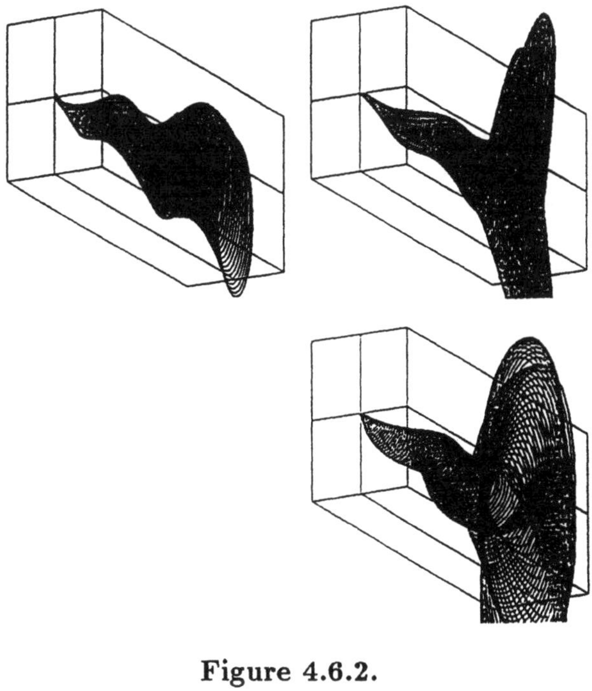

Figure 4.6.2.

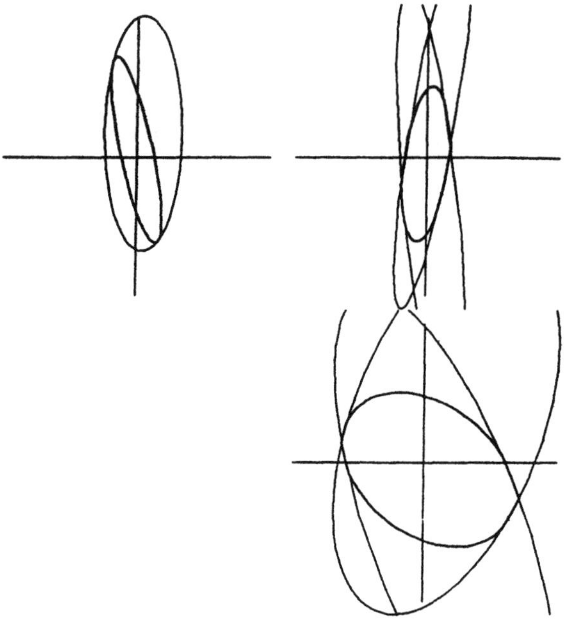

Figure 4.6.3.

variable. In Figure 4.6.3 we see the estimates (in the same arrangement of the windows and in the same scale) at instant $t = 4.25$, drawn by thin lines, and the *projection of their intersection*, drawn by a thicker line. It is to be noted here, that in the space of the first two variables, the projections of the two estimates coincide again, but the projection of their intersection *is a proper subset.* We leave to the reader to try these techniques with various types of discontinuous realizations $y(t)$.

Bibliography

[1] AGRACHEV A.A., GAMKRELIDZE R.V., Quadratic Mappings and Smooth Vector-Functions: Euler Characteristics of Level Sets, in ser.: *Modern Problems of Mathematics. Newest Achievements*, Vol. 35, R.V. Gamkrelidze, ed., VINITI, Moscow, 1989, pp. 179-239 (in Russian).

[2] ALEFELD G., HERZBERGER J., *Introduction to Interval Computation*, Academic Press, New York, 1983.

[3] AKIAN M., QUADRAT J-P, VIOT M., Bellman Processes, 11th International Conference on Systems Analysis and Optimization, Sophia-Antipolis, France, 1994.

[4] ALEKSANDROV A.D., On the Theory of Mixed Volumes of Convex Bodies, pp. I-IV, *Mat.Sbornik*, Vol. 2., NN5,6, 1937, Vol. 3, NN1,2, 1938 (in Russian).

[5] ANANIEV B.I., Minimax Mean-Square Estimates in Statistically Uncertain Systems, *Differencialniye Uravneniya*, N8, 1984, pp. 1291-1297 (in Russian), translated as "Differential Equations".

[6] ANANIEV B.I., A Guaranteed Filtering Scheme for Hereditary Systems with no Information on the Initial State, *Proceedings ECC-95*, Rome, Vol. 2, pp. 966-971.

[7] ANANIEV B.I., PISCHULINA I.Ya., Minimax Quadratic Filtering in Systems with Time-Lag, in: *Differential Control Systems*, A.V. Kryazhimski, ed., Ural Sci. Cent., 1979, pp. 3-12.

[8] ANTOSIEWICZ H.A., Linear Control Systems, *Archive for Rat. Mechanics and Analysis*, Vol. 12, N4, 1963.

[9] ARNOLD V.I., *The Theory of Catastrophes*, 3rd edition, Nauka, Moscow, 1990 (in Russian).

[10] ARTSTEIN Z., A Calculus for Set-Valued Maps and Set-Valued Evolution Equations, *Set-Valued Analysis*, Vol. 3, 1995, pp. 213-261.

[11] ATTOUCH H., AUBIN J-P., CLARKE F., EKELAND I., eds. *Analyse Nonlinéaire*, Gauthiers-Villars, C.R.M. Université de Montréal, 1989.

[12] AUBIN J-P., Motivated Mathematics, *SIAM News*, Vol. 18, NN1,2,3, 1985.

[13] AUBIN J-P., Differential Games: A Viability Approach, *SIAM Journal on Control and Optimization*, Vol. 28,1990, pp. 1294-1320.

[14] AUBIN J-P., Fuzzy Differential Inclusions, *Problems of Control and Information Theory*, Vol. 19, 1990, pp. 55-67.

[15] AUBIN J.-P., *Viability Theory*, Birkhäuser, Boston, 1991.

[16] AUBIN J.-P., *Initiation à l'Analyse Appliquée*, Masson, Paris, 1994.

[17] AUBIN J.-P., Mutational and Morphological Analysis: Tools for Shape Regulation and Optimization, Preprint, *CEREMADE*, Université Paris-Dauphine, September 1995.

[18] AUBIN, J.-P., CELLINA A., *Differential Inclusions*, Springer-Verlag, Berlin, 1984.

[19] AUBIN J.P., DA PRATO G., Stochastic Viability and Invariance, *Annali Scuola Normale di Pisa*, 27, 1990, pp. 595-694.

[20] AUBIN J.P., EKELAND I., *Applied Nonlinear Analysis*, Wiley-Interscience, 1984.

[21] AUBIN J.-P., FRANKOWSKA H., *Set-Valued Analysis*, Birkhäuser, Boston, 1990.

[22] AUBIN J.-P., FRANKOWSKA H., Inclusions aux Dérivées Partielles Gouvernant des Contrôles de Rétroaction, *Comptes-Rendus de l'Académie des Sciences*, Paris, 311, 1990, pp. 851-856.

[23] AUBIN J.-P., FRANKOWSKA H., Viability Kernels of Control Systems, in: *Nonlinear Analysis*, Ch. Byrnes, A. Kurzhanski, eds., ser. PSCT9, Birkhäuser, Boston, 1990.

[24] AUBIN J-P., SIGMUND K., Permanence and Viability, *Journal of Comput. and Appl. Mathematics*, Vol. 22, 1988, pp. 203-209.

[25] AUMANN R.J., Integrals of Set-valued Functions, *Journal of Math. Anal. and Appl.*, Vol. 12, 1965.

[26] BAIER R., LEMPIO F., Computing Aumann's Integral, in: *Modeling Techniques for Uncertain Systems*, Kurzhanski, A.B., Veliov, V.M., eds., ser. PSCT18, 1994, pp. 71-92.

[27] BALAKRISHNAN A.V., *Applied Functional Analysis*, Springer-Verlag, 1976.

[28] BANKS H.T., JACOBS M.Q., A Differential Calculus for Multifunctions, *Journal of Math. Anal. and Appl.*, Vol. 29, 1970, pp. 246-272.

[29] BARAS J.S., BENSOUSSAN A., JAMES M.R., Dynamic Observers as Asymptotic Limits of Recursive Filters: Special Cases, *SIAM Journal on Appl. Math.*, Vol. 48, N5, 1988, pp. 1147-1158.

[30] BARAS J., JAMES M., Nonlinear H_∞, Control, *Proceedings of the 33rd IEEE, CDC*, Lake Buena Vista, FL, Vol. 2, 1994, pp. 1435-1438.

[31] BARAS J.S., JAMES M.R., ELLIOT R.J., Output-Feedback, Risk-Sensitive Control and Differential Games for Continuous-Time Nonlinear Systems, *Proceedings of the 32nd IEEE, CDC*, San Antonio, Texas, Vol. 4., 1994, pp. 3357-3360.

[32] BARAS J.S., KURZHANSKI A.B., Nonlinear Filtering: The Set-Membership (Bounding) and the H_∞ Approaches, *Proceedings of the IFAC NOLCOS Conference*, Tahoe, CA, Plenum Press, 1995.

[33] BARBASHIN E.A., On the Theory of Generalized Dynamic Systems, Scientific Notes of the Moscow University, *Mathematics*, Vol. 2, N135, 1949 (in Russian).

[34] BARBASHIN E.A., *Liapunov Functions*, Nauka, Moscow, 1970 (in Russian).

[35] BARBASHIN E.A., ALIMOV Yu.I., On the Theory of Dynamical Systems with Multi-Valued and Discontinuous Characteristics, *Dokl. AN SSSR*, 140, 1961, pp. 9-11 (in Russian).

[36] BARBU V., DA PRATO G., Hamilton-Jacobi Equations in Hilbert Spaces: Variational and Semigroup Approach, *Annali di Matematica Pura e Applicata*, CXLII, 1985, pp. 303-349.

[37] BASAR T., BERNHARD P., *H^∞ Optimal Control and Related Minimax Design Problems*, ser. SCFA, Birkhäuser, Boston, 1991.

[38] BASAR T., KUMAR P.R., On Worst-Case Design Strategies, *Comput. and Math. with Appl.*, Vol. 13(1-3), 1978, pp. 239-245.

[39] BASAR T., DIDINSKY G., PAN Z., A New Class of Identifiers for Robust Parameter Identification and Control in Uncertain Systems, *Proceedings of Workshop on "Robust Control via Variable Structure and Liapunov Techniques"*, Benvenetto, Italy, September 1994.

[40] BARMISH B.R., LEITMANN G., On Ultimate Boundedness Control of Uncertain Systems in Absence of Matching Conditions, *IEEE Trans. Aut. Control*, AC-27, 1982, p. 1253.

[41] BARTOLINI G., ZOLEZZI T., Asymptotic Linearization of Uncertain Systems by Variable Structure Control, *Syst. Cont. Letters*, Vol. 10, 1988, pp. 111-117.

[42] BASAR T., MINTZ M., Minimax Terminal State Estimation for Linear Plants with Unknown Forcing Functions, *International Journal of Control*, Vol. 16(1), 1972, pp. 49-70.

[43] BASILE G., MARRO G., *Controlled and Conditioned Invariants in Linear Systems Theory*, Prentice Hall, Englewood Cliffs, NJ, 1991.

[44] BEHREND F., Über die kleinste umbeschriebene und die größte einbeschriebene Ellipse eines konvexen Bereichs, *Math. Annalen*, Vol. 115, 1938, pp. 379-411.

[45] BELL D.J., Singular Problems in Optimal Control: A Survey, *Int. J. Control*, 21, 1975, pp. 319-331.

[46] BELLMAN R., *Dynamic Programming,* Princeton University Press, 1957.

[47] BENSOUSSAN A., DA PRATO G., DELFOUR M.C., MITTER S.K., *Representation and Control of Infinite-Dimensional Systems*, ser. SCFA, Birkhäuser, Boston, Vol. 1, 1992, Vol. 2., 1993.

[48] BERGER M., *Géométrie*-3, CEDIC, Fernand Nathan, Paris, 1978.

[49] BERKOVITZ L.D., Characterization of the Values of Differential Games, *Appl. Math. and Optim.*, Vol. 17, 1988, pp. 177-183.

[50] BERKOVITZ L.D., Differential Games of Survival, *Journal of Math. Anal. Appl.* Vol. 129,2, 1988, pp. 493-504.

[51] BERNHARD P., A Discrete-Time Certainty Equivalence Principle, *Systems and Control Letters*, 1994.

[52] BERNHARD P., Expected Values, Feared Values and Partial Information Control, Preprint, *INRIA*, Sophia-Antipolis, January 1995.

[53] BERTSEKAS D.P., *Dynamic Programming Deterministic and Stochastic Models*, Prentice Hall, Englewood Cliffs, NJ, 1987.

[54] BERTSEKAS D.P., RHODES I.B., On the Minimax Reachability of Target Sets and Target Tubes, *Automatica*, N7, 1971, pp. 233-247.

[55] BERTSEKAS D.P., RHODES I.B., Recursive State Estimation for a Set-Membership Description of Uncertainty, *IEEE Trans. Aut. Control*, AC-16, 1971, pp. 117-128.

[56] BERTSEKAS D.P., RHODES I.B., Sufficiently Informative Functions and the Minmax Feedback Control of Uncertain Dynamic Systems, *IEEE Trans. Aut. Control*, AC-18(2), 1973, pp. 117-123.

[57] BITTANTI S., ed., *The Riccati Equation in Control, Systems and Signals*, Pitagora Editrice, 1989.

[58] BIRGE J.R., WETS R.J.-B., Sublinear Upper Bounds for Stochastic Programs with Recourse, *Math. Prog.*, Vol. 43, 1989, pp. 131-149.

[59] BLAGODATSKIH V.I., FILIPPOV A.F., Differential Inclusions and Optimal Control. *Trudy Mat. Inst. AN SSSR*, 169, 1985, pp. 194-252, translated as Proceedings of the Steklov Math. Inst., 4, North Holland, Amsterdam, 1986.

[60] BONNENSEN T., FENCHEL W., *Theorie der Konvexen Körper*, Springer, Berlin, 1934.

[61] BROCKETT R.W., *Finite-Dimensional Linear Systems*, John Wiley, New York, 1970.

[62] BROCKETT R.W., Pulse Driven Dynamical Systems, in: *Systems, Models and Feedback: Theory and Applications*, A. Isidori, T.J. Tarn, eds., 1992, ser. PCST 12, Birkhäuser, Boston.

[63] BRYSON A.E., HO Y.-C., *Applied Optimal Control*, Ginn, Waltham, MA, 1969.

[64] BURAGO Yu.D., ZALGALLER V.A., *Geometrical Inequalities*, Nauka, Leningrad, 1980 (in Russian).

[65] BUSEMANN H., A Theorem on Convex Bodies of the Brunn-Minkowski Type, *Proceedings of the National Acadamy of Sciences USA*, Vol. 35, 1949, pp. 27-31.

[66] BYRNES Ch.I., ISIDORI A., Feedback Design from the Nonzero Dynamics Point of View *Computation and Control*, ser. PSCT1, Birkhäuser, Boston.

[67] BYRNES Ch.I., KURZHANSKI A.B., eds., *Nonlinear Synthesis*, ser. PCST9, Birkhäuser, Boston, 1989.

[68] CASTAING C., VALADIER M., *Convex Analysis and Measurable Multifunctions*, Lecture Notes in Mathematics, Vol. 580, Springer-Verlag, 1977.

[69] CEA J., *Optimisation: Théorie et Algorithmes*, Dunod, Paris, 1971.

[70] CELLINA A., ORNELAS A., Convexity and the Closure of the Solution Set to Differential Inclusions. Preprint, *SISSA*, Trieste, 1988.

[71] CELLINA A., ORNELAS A., Representation of the Solution Set to Lipschitzian Differential Inclusions. Preprint, *SISSA*, Trieste, 1988.

[72] CHERNOUSKO F.L., Optimal Guaranteed Estimates of Indeterminacies with the Aid of Ellipsoids, I., II., III., *Izv. Acad. Nauk SSSR, Tekhn. Kibernetika*, 3,4,5, 1980, (in Russian), translated as Engineering Cybernetics.

[73] CHERNOUSKO, F.L., *State Estimation for Dynamic Systems,* CRC Press, 1994.

[74] CHEUNG M-F., YURKOVICH S., PASSINO K.M., An Optimal Volume Algorithm for Parameter Set Estimation, *IEEE, Trans. Aut. Cont.*, Vol. 39, N6, 1994, pp. 1268-1272.

[75] CHISCI L., GARULLI A., ZAPPA G., Recursive State Bounding by Parallelotopes, Preprint, Univ. Firenze, 1995.

[76] CHIKRII A.A., *Conflict-Controlled Processes*, Naukova Dumka, Kiev, 1992 (in Russian).

[77] CHIKRII A.A., ZHUKOVSKII V.I., *Linear-Quadratic Differential Games*, Naukova Dumka, Kiev, 1994 (in Russian).

[78] CLARKE F.H., *Optimization and Nonsmooth Analysis*, Wiley-Interscience, New York, 1983.

[79] COLOMBO G., Approximate and Relaxed Solutions of Differential Inclusions. Preprint, *SISSA*, Trieste, 1988.

[80] COMBA J.L.D., STOLFI J., Affine Arithmetic and its Applications to Computer Graphics, *Anais do SIBGRAPI VI,* 1993, pp. 9-18.

[81] CORLESS M., LEITMANN G., Adaptive Control for Uncertain Dynamical Systems, *Dyn. Syst. and Microphys., Contr. Theory and Mech.*, Acad. Press, 1984, pp. 91-158.

[82] CRANDALL M.G., LIONS P.L., Viscosity Solutions of Hamilton-Jacobi Equations, *Trans. Amer. Math. Soc.*, 277, 1983, pp. 1-42.

[83] CRANDALL M.G., ISHII H., LIONS P.L., User's Guide to Viscosity Solutions of Second Order Partial Differential Equations, *Bull. of the Amer. Math. Soc.*, Vol. 27, N1, 1992, pp. 1-67.

[84] DANZER L., LAUGWITZ D., LENZ H., Über das Löwnersche Ellipsoid und sein Analogon unter den Einem Eikörper Einbeschriebenen Ellipsoiden, *Arch. Math.* 8, 1957, pp. 214-219.

[85] DEIMLING K., Multivalued Differential Equations on Closed Sets, *Differential and Integral Equations*, 1, 1988, pp. 23-30.

[86] DEMIANOV V.F., *Minimax: Directional Differentiation*, Leningrad University Press, 1974.

[87] DEMIANOV V.F., LEMARECHAL C., ZOWE J., Approximation to a Set-Valued Mapping, I: A Proposal, *Appl. Math. Optim.*, 14, 1986, pp. 203-214.

[88] DEMIANOV V.F., RUBINOV A.M., *Foundations of Nonsmooth Analysis and Quasidifferential Calculus*, Nauka, Moscow, 1990 (in Russian).

[89] DEMIANOV V.F., VASILIEV L.V., *Nondifferentiable Optimization*, Nauka, Moscow, 1981 (in Russian)

[90] DI MASI G., GOMBANI A., KURZHANSKI A., eds., *Modeling, Estimation and Control of Systems with Uncertainty*, ser. PCST10, Birkhäuser, Boston, 1990.

[91] DONTCHEV A.L., *Perturbations, Approximations and Sensitivity Analysis of Optimal Control Systems*, Lecture Notes in Control and Inform. Sciences, 52, Springer-Verlag, 1986.

[92] DONTCHEV A.L., LEMPIO F., Difference Methods for Differential Inclusions: A Survey, *SIAM Rev.*, Vol. 34, 1992.

[93] DONTCHEV A.L., VELIOV V.M., Singular Perturbation in Mayer's Problems for Linear Systems, *SIAM Journal on Control and Optimization*, 21, 1983, pp. 566-581.

[94] DOYLE J.C., FRANCIS B.A., TANNENBAUM A.R., *Feedback Control Theory*, McMillan, New York, 1992.

[95] DUBROVIN B.A., NOVIKOV S.P., FOMENKO A.T., *Modern Geometry: Methods and Applications*, URSS Pub., Moscow, 1994.

[96] DUNFORD, N., SCHWARTZ, J.T., *Linear Operators, Part I*, Interscience Publishers Inc., New York, 1957.

[97] DWYER R.A., EDDY W.F., Maximal Empty Ellipsoids, in: *Proceedings of the Fifth Annual ACM-SIAM Symposium on Discrete Algorithms*, Arlington, VA, USA, New York, ACM, 1994, pp. 98-102.

[98] EGGLESTON H., *Convexity*, Cambridge University Press, 1963.

[99] EFIMOV N.V., *Advanced Geometry*, Nauka, Moscow, 1971 (in Russian).

[100] EKELAND I., TEMAM R., *Analyse Convexe et Problèmes Variationnelles*, Dunod, Paris, 1974.

[101] FAN KY, Minmax Theorems, *Proceedings of the National Acadamy of Sciences USA*, Vol. 39, N1, 1953, pp. 42-47.

[102] FILIPPOV A.F., On Some Problems of Optimal Control Theory. *Vestnik Mosk. Univ.*, Math., N2, 1958, pp. 25-32 (in Russian), English version in *SIAM Journal on Control and Optimization*, N1, 1962, pp. 76-84.

[103] FILIPPOV A.F., Classical Solutions of Differential Equations with a Multivalued Right-Hand Side, *SIAM Journal on Control and Optimization*, N5, 1967, pp. 609-621.

[104] FILIPPOV A.F., *Differential Equations with Discontinuous Right-hand Side*, Kluwer Acad. Pub., 1988.

[105] FILIPPOVA T.F., A Note on the Evolution Property of the Assembly of Viable Solutions to a Differential Inclusion, *Computers Math. Applic.*, 25, 1993, pp. 115-121.

[106] FILIPPOVA T.F., On the Modified Maximum Principle in the Estimation Problems for Uncertain Systems, Working Paper WP-92-32, IIASA, Laxenburg, Austria, 1992.

[107] FILIPPOVA T.F., KURZHANSKI A.B., SUGIMOTO K., VALYI I., Ellipsoidal Calculus, Singular Perturbations and the State Estimation Problem for Uncertain Systems, in: *Bounding Approaches to System Identification*, M. Milanese, J. Norton, H. Piet-Lahanier, E. Walter, eds., Plenum Press, 1995, also in Working Paper WP-92-51, IIASA, Laxenburg, Austria, 1992.

[108] FLEMING W.H., RISHEL R.W., *Deterministic and Stochastic Optimal Control*, Springer-Verlag, 1975.

[109] FLEMING W.H., SONER H.M., *Controlled Markov Processes and Viscosity Solutions*, Springer-Verlag, 1993.

[110] FLEMING W.H., SOUGANIDIS P.E., On the Existence of Value Function for a Two-Player Zero-Sum Differential Game, *Indiana Univ. Math. J.*, Vol. 38, 1989, pp. 293-314.

[111] FOGEL E., System Identification via Membership Set Constraints with Energy Constrained Noise, *IEEE, Trans. Aut. Contr.* Vol. 24, N5, 1979, pp. 752-757.

[112] FORMALSKI A.M., On the Corner Points of the Boundaries of Attainability Regions, *Prikl. Mat. Meh.*, Vol. 47, 1983, translated as Appl. Math. and Mech.

[113] FRANK M., WOLFE P., An Algorithm for Quadratic Programming, *Naval Res. Log. Quart.*, Vol. 3., 1956, pp. 95-110.

[114] FRANKOWSKA H., L'Équation d'Hamilton-Jacobi Contingente, *Comptes-Rendus de l'Académie des Sciences*, Paris, 304, ser. 1, 1987, pp. 295-298.

[115] FRANKOWSKA H., Contingent Cones to Reachable Sets of Control Systems, *SIAM Journal on Control and Optimization*, Vol. 27, 1989, pp. 170-198.

[116] FRANKOWSKA H., Lower-Semicontinuous Solutions of Hamilton-Jacobi-Bellman Equations, *SIAM Journal on Control and Optimization*, Vol. 31, N1, 1993, pp. 257-272.

[117] FRANKOWSKA H., PLASCACZ M., RZEZUCHOWSKI T., Measurable Viability Theorems and Hamilton-Jacobi-Bellman Equations, *J. Diff. Equat.*, Vol. 116, N2, 1995, pp. 265-305.

[118] FRANKOWSKA H., QUINCAMPOIX M., Viability Kernels of Differential Inclusions with Constraints, *Math. Syst., Est. and Control*, Vol. 1, N3, 1991, pp. 371-388.

[119] FRIEDMAN A., *Differential Games*, Interscience, New York, 1971.

[120] GANTMACHER F.R., *Matrix Theory, I-II,* Chelsea Publishing Co., New York, 1960.

[121] GRUBER P., Approximation of Convex Bodies, in: *Convexity and Applications*, P. Gruber, J. Wills, eds., Birkhäuser, 1993.

[122] GIARRÈ L., MILANESE M., TARAGNA M., H_∞, – Identification and Model Quality Evaluation, Preprint, Politecnico Torino, 1995.

[123] GRAVES C.H., VERES S.M., Using MATLAB Toolbox "GBT" in Identification and Control, in: *IEE Colloquium on Identification of Uncertain Systems*, Digest No. 1994/105.

[124] GRÜNBAUM B., *Convex Polytopes*, Interscience, London, 1967.

[125] GUSEINOV KH.G., USHAKOV V.N., Strongly and Weakly Invariant Sets with Respect to a Differential Inclusion, *Soviet Mathematical Doklady*, Vol. 38, 3, 1989 (in Russian).

[126] GUSEINOV KH.G., SUBBOTIN A.I., USHAKOV V.N., Derivatives for Multivalued Mappings with Applications to Game-theoretical Problems of Control, *Problems of Control and Inform. Theory*, Vol. 14, 3, 1985, pp. 155-167.

[127] HADWIGER H., *Vorlesungen Über Inhalt, Oberfläche und Isoperimetrie*, Springer-Verlag, Berlin, 1957.

[128] HALE J.K., LIN X.B., Symbolic Dynamics and Nonlinear Flows, *Annali Mat. Pura e Applicata*, Vol. 144, 1986, pp. 229-260.

[129] HIJAB O.B., *Minimum-Energy Estimation*, Ph.D. Thesis, UC-Berkeley, 1980.

[130] HINRICHSEN D., MARTINSEN B., eds., *Control of Uncertain Systems*, ser. PSCT6, Birkhäuser, Boston, 1990.

[131] HOFBAUER J., SIGMUND K., *The Theory of Evolution and Dynamical Systems* Cambridge University Press, 1988.

[132] HONIN V.A., Guaranteed Estimation of the State of Linear Systems by Means of Ellipsoids, in: *Evolutionary Systems in Estimation Problems*, A.B. Kurzhanski, T.F. Filippova, eds., Sverdlovsk, 1980 (in Russian).

[133] IOFFE A.D., TIKHOMIROV V.M., *The Theory of Extremal Problems*, Nauka, Moscow, 1979.

[134] ISAACS R., *Differential Games*, John Wiley, New York, 1965.

[135] ISIDORI A., *Nonlinear Control Systems*, Springer-Verlag, 1989.

[136] ISIDORI A., Attenuation of Disturbances in Nonlinear Control Systems, in: *Systems, Models and Feedback: Theory and Applications*, Isidori A., Tarn T.J., eds., ser. PCST 12, Birkhäuser, Boston, 1992.

[137] JOHN F., Extremum Problems with Inequalities as Subsidiary Conditions, in: K. Friedrichs, O. Neugebauer, J.J. Stoker, eds. *Studies and Essays, Courant Anniversary Volume*, John Wiley & Sons Inc., New York, 1948.

[138] JOULIN L., WALTER E., Guaranteed Estimation of Stability Domain via Set-Inversion, *IEEE Trans. Aut. Cont.*, Vol. 39, N4, 1994, pp. 886-889.

[139] JUNG H.E.W, Über die kleinste Kugel, die eine räumliche Figur einschließt, *J. Reine Angew. Math.*, Vol. 123, 1901, pp. 241-257.

[140] KAC I.Ya., KURZHANSKI A.B., Minimax Estimation in Multistage Systems, *Sov. Math. Dokl.*, Vol. 16, N2, 1975, pp. 374-379.

[141] KAHAN W., Circumscribing an Ellipsoid About the Intersection of Two Ellipsoids, *Can. Math. Bull.*, Vol. 11, 1968.

[142] KAILATH T., *Linear Systems* Prentice Hall, Englewood Cliffs, NJ, 1980.

[143] KAILATH T., The Innovations Approach to Detection and Estimation Theory, *Proceedings of the IEEE*, Vol. 58, 1970, pp. 680-695.

[144] KALL P., Stochastic Programming with Recourse: Upper Bounds and Moment Problems: A Review, in: *Advances in Math. Optimization (Dedicated to Prof. F. Nožička)*, J. Guddat, B. Bank, H. Hollatz, P. Kall, D. Klatte, B. Kummer, K. Lommatzch, K. Tammer, M. Vlach, eds. Akademie-Verlag, Berlin, 1988.

[145] KALL P., STOYAN D., Solving Stochastic Programming Problems with Recourse Including Error Bounds, *Math. Operationsforsch. Statist., Se. Opt.*, Vol. 13, 1982, pp. 431-447.

[146] KALL P., WALLACE S.W., *Stochastic Programming*, John Wiley, Chichester, 1994.

[147] KALMAN R.E., On the General Theory of Control Systems, *Proceedings of the 1st IFAC Congress, Vol. 1*, Butterworths, London, 1960.

[148] KALMAN R.E., *Nine Lectures on Identification,* Lecture Notes on Economics and Math. Systems, Springer-Verlag, 1993.

[149] KALMAN R.E., BUCY R., New Results in Linear Filtering and Prediction Theory, *Trans. AMSE*, 83, D, 1961.

[150] KANTOROVICH L.V., AKILOV G.P., *Functional Analysis*, Pergamon Press, 1982.

[151] KARL W.C., VERGHESE G.C, WILLSKY A.S., Reconstructing Ellipsoids from Projections, *Graph. Mod. and Image Proc.*, Vol. 56, N2, 1994, 124-139.

[152] KHACHIYAN L.D., A Polynomial Algorithm in Linear Programming, *Sov. Math. Dokl.*, Vol. 20, 1979, pp. 191-194.

[153] KHACHIYAN L.D., An Inequality for the Volume of Inscribed Ellipsoids, *Discr. and Comput. Geom.*, Vol. 5, N3, 1990, pp. 219-222.

[154] KHACHIYAN L.D., TODD M., *On the Complexity of Approximating the Maximal Inscribed Ellipsoid for a Polytope*, Cornell Univ. Tech. Rep. N893, Ithaca, NY, 1990.

[155] KISELEV O.N., POLYAK B.T., Ellipsoidal Estimation with Respect to a Generalized Criterion, *Avtomatika i Telemekhanika*, Vol. 52, N9, 1991, pp. 133-45, translated as Automation and Remote Control, September 1991, Vol. 52, N9, pp. 1281-92.

[156] KLEE V.L., Extremal Structure of Convex Sets, Part I, *Arch. Math.*, Vol. 8, 1957, pp. 234-240, Part II, *Math. Zeit.*, Vol. 69, 1958, pp. 90-104.

[157] KLIR G.J., FOLGER T.A., *Fuzzy Sets, Uncertainty and Information,* Prentice Hall, Englewood Cliffs, NJ, 1988.

[158] KNOBLOCH H., ISIDORI,A., FLOCKERZI D., *Topics in Control Theory,* Birkhäuser, DMV-Seminar, Band 22, 1993.

[159] KOENIG H., PALLASCHKE D., On Khachian's Algorithm and Minimal Ellipsoids, *Num. Math.*, Vol. 36, 1981, pp. 211-223.

[160] KOKOTOVIC P., BENSOUSSAN A., BLANKENSHIP G., eds., *Singular Perturbations and Asymptotic Analysis in Control Systems,* Lecture Notes in Contr. and Inform. Sci., 90, Springer-Verlag, 1986.

[161] KOLMOGOROV A.N., FOMIN S.V., *Elements of Theory of Functions and Functional Analysis,* Nauka, Moscow, 1968.

[162] KOMAROV V.A., The Equation for Attainability Sets of Differential Inclusions in the Problem with State Constraints, *Trudy Matem. Instit. Akad. Nauk SSSR,* Vol. 185, 1988, pp. 116-125.

[163] KOMAROV V.A., The Estimates of Attainability Sets of Differential Inclusions, *Mat. Zametki,* 37, 1985, pp. 916-925.

[164] KONSTANTINOV G.N., SIDORENKO G.V., Outer Estimates of Reachable Sets of Controlled Systems, *Izv. Akad. Nauk SSSR, Teh. Kibernet,* N3, 1986, translated as Engineering Cybernetics.

[165] KOSCHEEV A.S., KURZHANSKI A.B., On Adaptive Estimation of Multistage Systems Under Conditions of Uncertainty, *Izvestia AN SSSR, Teh. Kibernetika,* N4, 1983 (in Russian), pp. 72-93, translated as Engineering Cybernetics, N2, 1984, pp. 57-77.

[166] KRASOVSKII N.N., On the Theory of Controllability and Observability of Linear Dynamic Systems, *Priklad. Mat. i Meh.*, Vol. 23,N 4, 1964 (in Russian), translated as J. of Appl. Math. and Mech.

[167] KRASOVSKI N.N., *The Theory of Control of Motion,* Nauka, Moscow, 1968 (in Russian).

[168] KRASOVSKII N.N., *Game-Theoretic Problems on the Encounter of Motions* Nauka, Moscow, 1970 (in Russian), English translation: *Rendezvous Game Problems,* Nat. Tech. Inf. Serv., Springfield, VA, 1971.

[169] KRASOVSKII N.N., *The Control of a Dynamic System*, Nauka, Moscow, 1986 (in Russian).

[170] KRASOVSKI N.N., KRASOVSKI A.N., *Feedback Control Under Lack of Information*, Birkhäuser, Boston, 1995.

[171] KRASOVSKI N.N., SUBBOTIN A.N., *Positional Differential Games*, Springer-Verlag, 1988.

[172] KRASTANOV M., KIROV N., Dynamic Interactive System for Analysis of Linear Differential Inclusions, in: *Modeling Techniques for Uncertain Systems*, Kurzhanski, A.B., Veliov V.M., eds., ser. PSCT18, 1994, pp. 123-130.

[173] KRENER A., Necessary and Sufficient Conditions for Worst-Case H_∞ Control and Estimation, *Math. Syst. Estim. and Control*, N4, 1994.

[174] KUMAR P.R., VARAIYA P., *Stochastic Systems: Estimation, Identification and Adaptive Control*, Prentice Hall, Englewood Cliffs, NJ, 1986.

[175] KUNTSEVICH V.M., LYCHAK M., *Guaranteed Estimates, Adaptation and Robustness in Control Systems,* Letcure Notes in Contr. Inf. Sci., Vol. 169, Springer-Verlag, 1992.

[176] KURATOWSKI R., *Topology*, Vols. 1,2, Academic Press, New York, 1966.

[177] KURZHANSKI A.B., On the Duality of Problems of Optimal Control and Observation, *Prikl. Mat., Meh.*, Vol. 34, N3, 1970 (in Russian), translated as J. of Appl. Math. and Mech.

[178] KURZHANSKI A.B., Differential Games of Approach Under Constrained Coordinates, *Dok. Akad. Nauk SSSR*, Vol. 192, N3, 1970 (in Russian), translated as Soviet Math. Doklady, Vol. 11, N3, 1970, pp. 658-672.

[179] KURZHANSKI A.B., Differential Games of Observation, *Sov. Math. Doklady*, Vol. 13, N6, 1972, pp. 1556-1560.

[180] KURZHANSKI A.B., On Minmax Control and Estimation Strategies Under Incomplete Information, *Problems of Contr. Inform. Theory*, 4, 1975, pp. 205-218.

[181] KURZHANSKI A.B., *Control and Observation under Conditions of Uncertainty*, Nauka, Moscow, 1977.

[182] KURZHANSKI A.B., Dynamic Decision Making Under Uncertainty, in: *State of the Art in Operations Research*, N.N. Moiseev, ed., Nauka, Moscow, 1979, pp. 197-235 (in Russian).

[183] KURZHANSKI A.B., On the Estimation of Dynamic Control Systems Under Uncertainty Conditions, *Problems of Control and Information Theory*, Part I, Vol. 9, 1980, pp. 395-406; Part II, Vol. 10, 1981, pp. 33-42.

[184] KURZHANSKI A.B., Evolution Equations for Problems of Control and Estimation of Uncertain Systems, in: *Proceedings of the International Congress of Mathematicians*, Warszawa, 1983, pp. 1381-1402.

[185] KURZHANSKI A.B., On the Analytical Description of the Bundle of Viable Trajectories of a Differential System, *Soviet Math. Doklady*, Vol. 33, 1986, pp. 475-478.

[186] KURZHANSKI A.B., Identification: A Theory of Guaranteed Estimates, in: *From Data to Model*, J.C. Willems, ed., Springer-Verlag, 1989.

[187] KURZHANSKI A.B., The Identification Problem: A Theory of Guaranteed Estimates, *Automation and Remote Control*, translated from "Avtomatika i Telemekhanika", Vol. 52, N4, pt.1, 1991, pp. 447-465.

[188] KURZHANSKI A.B., Inverse Problems of Observation and Invertibility for Distributed Systems, *Vestnik Mosk. Univers, Vychislit. Mat.*, N1, 1995 (in Russian).

[189] KURZHANSKI A.B., FILIPPOVA T.F., On the Description of the Set of Viable Trajectories of a Differential Inclusion, *Sov. Math. Doklady*, Vol. 34, 1987.

[190] KURZHANSKI A.B., FILIPPOVA T.F., On the Set-Valued Calculus in Problems of Viability and Control for Dynamic Processes: The Evolution Equation, *Les Annales de l'Institut Henri Poincaré, Analyse Non-linéaire*, 1989, pp. 339-363.

[191] KURZHANSKI A.B., FILIPPOVA T.F., On the Method of Singular Perturbations for Differential Inclusions, *Sov. Math. Doklady*, Vol. 44, 1992, pp. 705-710.

[192] KURZHANSKI A.B., FILIPPOVA T.F., Differential Inclusions with State Constraints: The Singular Perturbation Method, *Trudy Matem. Inst. Ross. Akad. Nauk*, 1995 (in Russian).

[193] KURZHANSKI A.B., FILIPPOVA T.F., On the Theory of Trajectory Tubes: A Mathematical Formalism for Uncertain Dynamics, Viability and Control, in: *Advances in Nonlinear Dynamics and Control: A Report from Russia*, A.B. Kurzhanski, ed., ser. PSCT 17, Birkhäuser, Boston, 1993, pp. 122-188.

[194] KURZHANSKI A.B., NIKONOV O.I., On the Problem of Synthesizing Control Strategies: Evolution Equations and Set-Valued Integration, *Doklady Akad. Nauk SSSR*, 311, 1990, pp. 788-793, Sov. Math. Doklady, Vol. 41, 1990.

[195] KURZHANSKI A., NIKONOV O.I., Evolution Equations for Tubes of Trajectories of Synthesized Control Systems, *Russ. Acad. of Sci. Math. Doklady*, Vol. 48, N3, 1994, pp. 606-611.

[196] KURZHANSKI A., OSIPOV Yu.S., On Optimal Control Under Restricted Coordinates, *Prikl. Mat. Meh.*, Vol. 33, N4, 1969, translated as Applied Math. and Mech.

[197] KURZHANSKI A., OSIPOV Yu.S., On One Problem of Control Under Bounded Coordinates, *Izv. Akad. Nauk SSSR. Teh. Kiber.*, N5, 1970, translated as Engineering Cybernetics

[198] KURZHANSKI A.B., TANAKA M., On a Unified Framework for Deterministic and Stochastic Treatment of Identification Problems, Working Paper WP-89-13, IIASA, Laxenburg, Austria, 1989.

[199] KURZHANSKI A.B., PSCHENICHNYI B.N., POKOTILO V.G., Optimal Inputs for Guaranteed Identification, *Problems of Control and Information Theory,* Vol. 20, N1, 1991, pp. 13-23.

[200] KURZHANSKI A.B., SIVERGINA I.F., On Noninvertible Evolutionary Systems: Guaranteed Estimates and the Regularization Problem, *Sov. Math. Doklady*, Vol. 42, N2, 1991, pp. 451-455.

[201] KURZHANSKI A.B., SUGIMOTO K., VALYI I., Guaranteed State Estimation for Dynamic Systems: Ellipsoidal Techniques, *International Journal of Adaptive Contr. and Sign. Proceedings*, Vol. 8., 1994, pp. 85-101.

[202] KURZHANSKI A.B., VALYI I., Set Valued Solutions to Control Problems and Their Approximation, in: A. Bensoussan, J.L. Lions, eds., *Analysis and Optimization of Systems*, Lecture Notes in Control and Information Systems, Vol. 111, Springer-Verlag, 1988, pp. 775-785.

[203] KURZHANSKI A.B., VALYI I., Ellipsoidal Techniques for Dynamic Systems: the Problems of Control Synthesis, *Dynamics and Control*, 1, 1991, pp. 357-378.

[204] KURZHANSKI A.B., VALYI I., Ellipsoidal Techniques for Dynamic Systems: Control Synthesis for Uncertain Systems, *Dynamics and Control*, 2, 1992, pp. 87-111.

[205] KURZHANSKI A.B., VELIOV V.M., eds., *Set-Valued Analysis and Differential Inclusions,* ser. PCST16, Birkhäuser, Boston, 1993.

[206] KURZHANSKI A.B., VELIOV V.M., eds., *Modeling Techniques for Uncertain Systems*, ser. PSCT18, Birkhäuser, Boston, 1994.

[207] KWAKERNAAK H., SIVAN R., *Linear Optimal Control Systems,* Wiley-Interscience, New York, 1972.

[208] LAKSHMIKANTAM V., LEELA S., Nonlinear Differential Equations in Abstract Spaces, *Ser. in Nonlin. Math.: Theory, Methods, Appl.*, 2, Pergamon Press, 1981.

[209] LAURENT P.J., *Approximation and Optimization*, Hermann, Paris, 1972.

[210] LEDIAYEV Yu.S., Criteria for Viability of Trajectories of Nonautonomous Differential Inclusions and Their Applications, Preprint, *CRM*, 1573, Univ. de Montréal, 1992.

[211] LEDIAYEV Yu. S., MISCHENKO E.F., Extremal Problems in the Theory of Differential Games, *Trud. Mat. Inst. AN SSSR*, Vol. 185, 1988, pp. 147-170, 1988 (in Russian).

[212] LEE E.B., MARKUS L., *Foundations of Optimal Control Theory*, Wiley, New York, 1961.

[213] LEICHTWEISS K., *Konvexe Mengen*, VEB Deutscher Verlag der Wissenschaft, Berlin, 1980.

[214] LEITMANN G., Deterministic Control of Uncertain Systems via a Constructive Use of Lyapunov Stability Theory, *Proceedings of the 14th IFIP Conference on Systems Modeling and Optimization*, Leipzig, July, 1989.

[215] LEITMANN G., One Approach to the Control of Uncertain Dynamical Systems, *Proceedings of the 6th Workshop on Dynamics and Control*, Vienna, 1993.

[216] LEMARECHAL C., ZOWE J., Approximation to a Multi-Valued Mapping: Existence, Uniqueness, Characterization, *Math. Institut, Universität Bayreuth*, 1987, Report N5.

[217] LIAPUNOV A.M., Problème Generale de Stabilité de Mouvement, *Ann. Fac. Toulouse*, Vol. 9, 1907, pp. 207-474.

[218] LIONS J-L., *Controllabilité Exacte, Perturbations et Stabilization des Systèmes Distribués*, Vols. 1,2, Masson, Paris, 1990.

[219] LIONS P-L., SOUGANIDIS P.E., Differential Games, Optimal Control and Directional Derivatives of Viscosity Solutions of Bellman's and Isaac's Equations, *SIAM Journal on Control and Optimization*, Vol. 23, 1995, pp. 566-583.

[220] LOTOV A.V., Generalized Reachable Sets Method in Multiple Criteria Problems, in: *Methodology and Software for Interactive Decision Support*, Lecture Notes in Econ. and Math. Systems, Vol. 337, Springer-Verlag, 1989.

[221] MARCHUK G.I., *Methods of Numerical Mathematics*, Springer-Verlag, 1975.

[222] MARKOV S.M., ANGELOV R., An Interval Method for Systems of ODE, *Interval Mathematics 1985, Proceedings of International Symposium Freiburg, Lecture Notes in Comput. Sci.*, Vol. 212, Springer-Verlag, 1986.

[223] MELKMAN A.A., MICCHELLI C.A., Optimal Estimation of Linear Operators in Hilbert Spaces from Inaccurate Data, *SIAM J. Num. Anal.*, Vol. 16., N1, 1979, pp. 87-105.

[224] MILANESE M., Properties of Least-Squares Estimates in Set-Membership Identification, *Automatica*, Vol. 31, N2, 1995, pp. 327-332.

[225] MILANESE M., NORTON J., PIET-LAHANIER H., WALTER E., eds., *Bounding Approaches to System Identification*, Plenum Press, 1995,

[226] MILANESE M., VICINO A., Optimal Estimation for Dynamic Systems with Set-Membership Uncertainty: An Overview, *Automatica*, Vol. 27, 1991, pp. 997-1009.

[227] MILANESE M., VICINO A., Information-Based Complexity and Nonparametric Worst-Case System Identification, *Journal of Complexity* Vol. 9, 1993, pp. 427-446.

[228] MOHLER R.R., *Nonlinear Systems: Vol. II Application to Bilinear Control*, Prentice Hall, Englewood Cliffs, NJ, 1991.

[229] MOORE R.E., *Methods and Applications of Interval Analysis*, SIAM, Phil., 1979.

[230] MOREAU J-J., Rafle par un Convexe Variable, *Analyse Convexe, Montpellier*, p.I Ex.N15, 1971, p.II Ex.N3, 1972.

[231] NAGPAL K.M., KHARGONEKAR P.P., Filtering and Smoothing in an H_∞ Setting, *IEEE, Trans. on Aut. Contr.*, Vol. 36, N2, 1991, pp. 152-166.

[232] NATANSON, I.P., *A Theory of Functions of the Real Variable*, Nauka, Moscow, 1972 (in Russian)

[233] NEMYTSKI V.V., STEPANOV V.A., *Qualitative Theory of Differential Equations*, Princeton University Press, 1960.

[234] NESTEROV YU.N., NEMIROVSKII A.S., Selfadjustment Functions and Polynomial Algorithms in Convex Programming, *CEMI Akad. Nauk SSSR*, Moscow, 1989 (in Russian).

[235] NEUMAIER A., *Interval Methods for Systems of Equations*, Cambridge University Press, 1990.

[236] NIKOLSKI M.S., On the Approximation of the Attainability Domain of Differential Inclusions, *Vestn. Mosk. Univ., Ser. Vitchisl. Mat. i Kibern.*, 4, 1987, pp. 31-34 (in Russian).

[237] NORTON J.P., Identification and Application of Bounded-Parameter Models, *Automatica*, Vol. 23, N4, 1987, pp. 497-507.

[238] NURMINSKI, E. A., URYASIEV, S.P., The Difference of Convex Sets, *Doklady AN Ukrainian SSR*, ser. A, 1, 1985 (in Russian).

[239] OLECH C., The Characterization of Weak Closures of Certain Sets of Integrable Functions, *SIAM Journal on Control and Optimization*, Vol. 12, 1974.

[240] OSIPOV Yu.S., The Alternative in a Differential-Difference Game, *Sov. Math. Dokl.*, Vol. 12, N2, 1971, pp. 619-624.

[241] OSIPOV Yu.S., KRYAZHIMSKII A.V., *Inverse Problems of Ordinary Differential Equations: Dynamical Solutions*, Gordon and Breach, 1995.

[242] OVSEEVICH, A.I., Extremal Properties of Ellipsoids Approximating Attainability Sets, *Problems of Control and Information Theory*, Vol.12, N1, pp. 1-11, 1983.

[243] OVSEEVICH A.I., RESHETNYAK Yu.N., Approximation of the Intersection of Ellipsoids in Problems of Guaranted Estimation, *Sov. J. Comput. Syst. Sci.*, Vol. 27, N1, 1989.

[244] PANASYUK A.I., PANASYUK V.I., On an Equation Resulting from a Differential Inclusion, *Mathem. Notes*, Vol. 27, N3, 1980, pp. 213-218.

[245] PANASYUK A.I., Dynamics of Attainable Sets of Control Systems, *Diff. Eqns.*, Vol. 24, N12, 1988.

[246] PANASYUK A.I., Equations of Attainable Set Dynamics, pp. 1,2, *JOTA*, Vol. 64, 1990, p.349.

[247] PETROSIAN L.A., *Differential Games of Pursuit*, Leningrad University Press, 1977.

[248] PETROVSKI I.G., *Lectures on Ordinary Differential Equations*, 6th edition, Nauka, Moscow, 1970.

[249] PETTY C.M., Ellipsoids, in: *Convexity and Its Applications*, P.M. Gruber, J.M. Wills, eds., Birkhäuser, 1983.

[250] PISIER G., *The Volume of Convex Bodies and Banach Space Geometry,* Cambridge University Press, 1989.

[251] POLOVINKIN E.S., *The Theory of Multivalued Mappings,* MFTI, Moscow, 1983 (in Russian).

[252] POLOVINKIN E.S., SMIRNOV G.V., Differentiation of Multivalued Mappings and Properties of Solutions of Differential Inclusions, *Sov. Math. Dokl.*, Vol. 33, 1986, pp. 662-666.

[253] POLYAK B.T., SCHERBAKOV P., SCHMULYIAN S., Circular Arithmetic in Robustness Analysis, in: *Modeling Techniques for Uncertain Systems*, Kurzhanski, A.B., Veliov V.M., eds., ser. PSCT18, 1994, pp. 229-243.

[254] POLYAK B.T., TSYPKIN Ya.Z., Robust Identification, *Automatica*, Vol. 16, N1, 1980.

[255] POLYAK B.T., TSYPKIN Ya.Z., Robust Stability Under Complex Parameter Perturbations, *Autom. and Remote Control*, Vol. 52, 1991, pp. 1069-1077.

[256] PONTRYAGIN L.S., On Linear Differential Games, I., II., *Soviet Mathematical Doklady*, V.8, 1967, pp. 769-771 and pp. 910-912.

[257] PONTRYAGIN L.S., Linear Differential Games of Pursuit, *Mat. Sbornik*, Vol. 112 (154):3(7), 1980 (in Russian).

[258] PONTRYAGIN L.D., BOLTYANSKI V.G., GAMKRELIDZE R.V., MISCHENKO E.F., *Mathematical Theory of Optimal Processes*, Interscience, New York, 1962.

[259] PSCHENICHNYI B.N., The Structure of Differential Games, *Sov. Math. Dokl.*, Vol. 10, 1969, pp. 70-72.

[260] PSCHENICHNYI B.N., *Necessary Conditions of Extremum*, Marcel Decker, 1972.

[261] PSCHENICHNYI B.N., *Convex Analysis and Extremum Problems*, Nauka, Moscow, 1980 (in Russian).

[262] PSCHENICHNYI B.N., OSTAPENKO V.V., *Differential Games*, Naukova Dumka, Kiev, 1992 (in Russian).

[263] PSCHENICHNYI B.N., POKOTILO V.G., KRIVONOS I.V., On Optimization of the Observation Process, *Prik. Mat. Meh*, Vol. 54, N3, 1990, translated as Appl. Mat. and Mech.

[264] QUINCAMPOIX M., Differential Inclusions and Target Problems, *SIAM Journal on Control and Optimization*, Vol. 30, 1992.

[265] ROCKAFELLAR R.T., *Convex Analysis*, Princeton University Press, 1970.

[266] ROCKAFELLAR R.T., State Constraints in Convex Problems of Bolza, *SIAM Journal on Control and Optimization*, Vol. 10, 1972, pp. 691-715.

[267] ROCKAFELLAR R.T., The Theory of Subgradients and its Application to Problems of Optimization of Convex and Nonconvex Functions, Helderman Verlag, Berlin, 1981.

[268] ROCKAFELLAR R. T., Linear-Quadratic Programming and Optimal Control, *SIAM Journal on Control and Optimization*, Vol. 25, 1987, pp. 781-814.

[269] ROCKAFELLAR R.T., WETS R.J.B., Generalized Linear-Quadratic Problem of Deterministic and Stochastic Optimal Control in Discrete Time, *SIAM Journal on Control and Optimization*, Vol. 28, 1990, pp. 810-822.

[270] ROXIN E., On the Generalized Dynamical Systems Defined by Contingent Equations, *J. Diff. Eqns*, 1, 1965, pp. 188-205.

[271] SAINT-PIERRE P., Approximation of the Viability Kernel, *Applied Math. and Optim.*, Vol. 29, 1994, pp. 187-209.

[272] SAMARSKII A.A., *The Theory of Difference Schemes*, Nauka, Moscow, 1983 (in Russian).

[273] SAMARSKI A.A., GULIN A.V., *Numerical Methods*, Nauka, Moscow, 1989 (in Russian).

[274] SCHLAEPFER F., SCHWEPPE F., Continuous-Time State Estimation Under Disturbances Bounded by Convex Sets, *IEEE Trans. Autom. Contr.*, AC-17, 1972, pp. 197-206.

[275] SCHNEIDER R., Steiner Points of Convex Bodies, *Isr. J. of Math.*, Vol. 9, 1971, pp. 241-249.

[276] SCHNEIDER R., *Convex Bodies: The Brunn-Minkowski Theory*, Cambridge University Press, 1993.

[277] SCHWEPPE F.C., Recursive State Estimation: Unknown but Bounded Errors and System Inputs, *IEEE, Trans. Aut. Cont.*, AC-13, 1968.

[278] SCHWEPPE F.C., *Uncertain Dynamic Systems*, Prentice Hall, Englewood Cliffs, NJ, 1973.

[279] SEEGER A., Direct and Inverse Addition in Convex Analysis and Applications, *J. Math. Anal. Appl.*, Vol. 148, N2, 1990, pp. 317-349.

[280] SERRA J., *Image Analysis and Mathematical Morphology*, Academic Press, 1982.

[281] SHARY P., On Controlled Solution Set to Interval Algebraic Systems, *Interval Computations*, Vol. 4., N6, 1992, pp. 66-75 (in Russian).

[282] SHOR N.Z., BEREZOVSKI O.A., New Algorithms for Constructing Optimal Circumscribed and Inscribed Ellipsoids, *Optimization Methods and Software*, Vol. 1, N4, 1992, pp. 283-299.

[283] ŠILJAK D.D., *Decentralized Control of Complex Systems*, Academic Press, 1991.

[284] SKOWRONSKI J.M., A Competitive Differential Game of Harvesting Uncertain Resources, in: D.F. Batten, P.F. Lesse, eds., *New Mathematical Advances in Economic Dynamics*, Croom Helm, London, Sydney, 1985, pp. 105-118.

[285] SNYDER J.M., Interval Analysis for Computer Graphics, *ACM Computer Graphics*, Vol. 26, N2, 1992, pp. 121-130.

[286] STOER J., WITZGALL C., *Convexity and Optimization in Finite Dimensions I*, Springer-Verlag, 1970.

[287] SONNEVEND G., STOER J., *Global Ellipsoidal Approximations and Homotopry Methods for Solving Convex Analytical Programs*,

Institut für Angewandte Mathematik und Statistik, Universität Würzburg, Report N40, 1988, also appeared as "The Theory of Analytical Centres" in *Appl. Math. and Opt.*, 1990.

[288] SUBBOTIN A.I., A Generalization of the Basic Equation of Differential Games, *Sov. Math. Dokl.*, Vol. 22, N2, 1980, pp. 358-362.

[289] SUBBOTIN A.I., Existence and Uniqueness Results for Hamilton-Jacobi Equations, *Nonlinear Analysis*, Vol. 16, 1991, pp. 683-689.

[290] SUBBOTIN A.I., *Generalized Solutions of First-Order PDE's: The Dynamic Optimization Perspective*, ser. SC, Birkhäuser, Boston, 1995.

[291] SUBBOTIN A.I., CHENTSOV A.G., *Optimization of Guarantees in Problems of Control*, Nauka, Moscow, 1981 (in Russian).

[292] SUBBOTINA N.N., The Maximum Principle and the Superdifferential of the Value Function, *Problems of Contr. and Inf. Theory*, Vol. 18, N3, 1989, pp. 151-160.

[293] TANAKA M., OSADA A., TANINO T., Ellipsoidal Approximations in Set-Membership State Estimation of Linear Discrete-Time Systems, *Transactions of the Society of Instrument and Control Engineers*, Vol. 27, N.12, 1991, pp. 1374-1381 (in Japanese).

[294] TARASYEV A.M., Approximation Schemes for Construction of the Generalized Solution of the Hamilton-Jacobi (Bellman-Isaacs) Equation, *Prikl. Mat. Meh.*, Vol. 58, 1994, pp. 22-26 (in Russian).

[295] TIKHONOV A.N., On the Dependence of the Solutions of Differential Equations on a Small Parameter, *Matem. Sbornik*, Vol. 22, 1948, pp. 198-204.

[296] TIKHONOV A.N., Systems of Differential Equations Containing a Small Parameter Multiplying the Derivative, *Matem. Sbornik*, Vol. 31, 1952, pp. 575-586.

[297] TITTERINGTON D.M., Estimation of Correlation Coefficients by Ellipsoidal Trimming, *Appl. Stat.*, Vol. 27, N3, 1978, pp. 227-234.

[298] TODD M., On Minimum-Volume Ellipsoids Containing Part of Given Ellipsoid, *Math. of Operat. Res.*, Vol. 7, N2, 1982, pp. 253-261.

[299] TOLSTONOGOV A.A., *Differential Inclusions in Banach Space*, Nauka, Novosibirsk, 1986.

[300] TRAUB J.F., WASILKOWSKI G.W., WOZNIAKOWSKI H., *Information-Based Complexity*, Academic Press, 1988.

[301] TSAI W.K., PARLOS A.G., VERGHESE G.C., Bounding the States of Systems with Unknown-but-bounded Disturbances, *Int. J. of Contr.*, Vol. 52, N4, 1990, pp. 881-915.

[302] USORO P.B., SCHWEPPE F.C., WORMLEY D.N., GOULD L.A., Ellipsoidal Set-Theoretic Control Synthesis, *J. Dyn. Syst. Measur. Cont.*, Vol. 104, 1982, pp. 331-336.

[303] USTIUZHANIN A.M., On the Problem of Matrix Parameter Identification, *Problems of Control and Inf. Theory*, Vol. 15, N4, 1986, pp. 265-274.

[304] VALENTINE F., *Convex Sets*, MacGrawhill, New York, 1964.

[305] VALYI I., Ellipsoidal Approximations in Problems of Control, in: *Modelling and Adaptive Control*, Ch. Byrnes, A.B. Kurzhanski, eds, Lecture Notes in Contr. and Inform. Sci., Vol. 105, Springer-Verlag, 1988.

[306] VALYI I., Ellipsoidal Methods in Time-Optimal Control, in: *Modeling Techniques for Uncertain Systems*, A.B. Kurzhanski, V.M. Veliov, eds., ser. PCST 18, Birkhäuser, Boston, 1994.

[307] VARAIYA P., On the Existence of Solutions to a Differential Game, *SIAM Journal on Control and Optimization*, Vol. 5, N1, 1967, 153-162.

[308] VARAIYA P., LIN J., Existence of Saddle Points in Differential Games, *SIAM Journal on Control and Optimization*, Vol. 7, 1, 1969, pp. 142-157.

[309] VELIOV V.M., Second Order Discrete Approximations to Strongly Convex Differential Inclusions, *Systems and Control Letters*, Vol. 13, 1989, pp. 263-269.

[310] VELIOV V.M., Approximations to Differential Inclusions by Discrete Inclusions, IIASA Working Paper WP-89-17, Laxenburg, Austria, 1989.

[311] VICINO A., MILANESE M., Optimal Inner Bounds of Feasible Parameter Set in Linear Estimation with Bounded Noise, *IEEE, Trans. Aut. Contr.*, Vol. 36, 1991, pp. 759-763.

[312] VINTER R.B., A Characterization of the Reachable Set for Nonlinear Control Systems, *SIAM Journal on Control and Optimization*, Vol. 18, 1980, pp. 599-610.

[313] VINTER R.B., WOLENSKI P., Hamilton-Jacobi Theory for Optimal Control Problems with Data Measurable in Time, *SIAM Journal on Control and Optimization*, Vol. 28, 1990, pp. 1404-1419.

[314] WAZEWSKI T., Systèmes de Commande et Équations au Contingent, *Bull. Acad. Pol. Sci.*, 9, 1961, pp. 151-155.

[315] WIENER N., Extrapolation, Interpolation and Smoothing of Stationary Time Series, MIT Press, Cambridge, MA, 1949.

[316] WILLEMS J.C., Dissipative Dynamical Systems, Part I: General Theory, *Arch. Rat. Mech. Anal.*, Vol. 45, 1972, pp. 321-351.

[317] WILLEMS J.C., Feedback in a Behavorial Setting, in: *Systems, Models and Feedback: Theory and Applications*, Isidori A., Tarn T.J., eds., ser. PCST 12, Birkhäuser, Boston, 1992.

[318] WITSENHAUSEN H.S., Set of Possible States of Linear Systems Given Perturbed Observations, *IEEE Trans. Autom. Cont.*, AC-13, 1968, pp. 556-558.

[319] WOLENSKI P.R., The Exponential Formula for the Reachable Set of a Lipschitz Differential Inclusion, *SIAM Journal on Control and Optimization*, Vol. 28, 1990, pp. 1148-1161.

[320] WONHAM W.M., *Linear Multivariable Control: A Geometric Approach*, Springer-Verlag, 1985.

[321] YOUNG L.C., *Lectures on the Calculus of Variations and Optimal Control Theory*, Saunders, Phil., 1969.

[322] ZADEH L.A., Fuzzy Sets, *Inform. and Control*, Vol. 8, 1965, pp. 338-353.

[323] ZAGUSKIN V.L., On Circumscribed and Inscribed Ellipsoids of the Extremal Volume, *Uspekhi Mat. Nauk.*, Vol. 13, 6(64), 1958 (in Russian), translated as Soviet Math. Surveys.

Index

Systems & Control: Foundations & Applications

Founding Editor
Christopher I. Byrnes
School of Engineering and Applied Science
Washington University
Campus P.O. 1040
One Brookings Drive
St. Louis, MO 63130-4899
U.S.A.

Systems & Control: Foundations & Applications publishes research monographs and advanced graduate texts dealing with areas of current research in all areas of systems and control theory and its applications to a wide variety of scientific disciplines.

We encourage the preparation of manuscripts in TEX, preferably in Plain or AMS TEX—LaTeX is also acceptable—for delivery as camera-ready hard copy which leads to rapid publication, or on a diskette that can interface with laser printers or typesetters.

Proposals should be sent directly to the editor or to: Birkhäuser Boston, 675 Massachusetts Avenue, Cambridge, MA 02139, U.S.A.

Estimation Techniques for Distributed Parameter Systems
H.T. Banks and K. Kunisch

Set-Valued Analysis
Jean-Pierre Aubin and Hélène Frankowska

Weak Convergence Methods and Singularly Perturbed Stochastic Control and Filtering Problems
Harold J. Kushner

Methods of Algebraic Geometry in Control Theory: Part I
Scalar Linear Systems and Affine Algebraic Geometry
Peter Falb

H^∞-Optimal Control and Related Minimax Design Problems
Tamer Başar and Pierre Bernhard

Identification and Stochastic Adaptive Control
Han-Fu Chen and Lei Guo

Viability Theory
Jean-Pierre Aubin

Representation and Control of Infinite Dimensional Systems, Vol. I
A. Bensoussan, G. Da Prato, M. C. Delfour and S. K. Mitter

Representation and Control of Infinite Dimensional Systems, Vol. II
A. Bensoussan, G. Da Prato, M. C. Delfour and S. K. Mitter

Mathematical Control Theory: An Introduction
Jerzy Zabczyk

H_∞-Control for Distributed Parameter Systems: A State-Space Approach
Bert van Keulen

Disease Dynamics
Alexander Asachenkov, Guri Marchuk, Ronald Mohler, Serge Zuev

Theory of Chattering Control with Applications to Astronautics, Robotics, Economics, and Engineering
Michail I. Zelikin and Vladimir F. Borisov

Modeling, Analysis and Control of Dynamic Elastic Multi-Link Structures
J. E. Lagnese, Günter Leugering, E. J. P. G. Schmidt

First Order Representations of Linear Systems
Margreet Kuijper

Hierarchical Decision Making in Stochastic Manufacturing Systems
Suresh P. Sethi and Qing Zhang

Optimal Control Theory for Infinite Dimensional Systems
Xunjing Li and Jiongmin Yong

Generalized Solutions of First-Order PDEs: The Dynamical Optimization Process
Andreĭ I. Subbotin

Finite Horizon H_∞ and Related Control Problems
M. B. Subrahmanyam

Control Under Lack of Information
A. N. Krasovskii and N. N. Krasovskii

H^∞-Optimal Control and Related Minimax Design Problems
A Dynamic Game Approach
Tamer Başar and Pierre Bernhard

Control of Uncertain Sampled-Data Systems
Geir E. Dullerud

Robust Nonlinear Control Design: State-Space and
Lyapunov Techniques
Randy A. Freeman and Petar V. Kokotović

Adaptive Systems: An Introduction
Iven Mareels and Jan Willem Polderman

Sampling in Digital Signal Processing and Control
Arie Feuer and Graham C. Goodwin

Ellipsoidal Calculus for Estimation and Control
Alexander Kurzhanski and István Vályi

MIX
Papier aus verantwortungsvollen Quellen
Paper from responsible sources
FSC® C105338

If you have any concerns about our products,
you can contact us on
ProductSafety@springernature.com

In case Publisher is established outside the EU,
the EU authorized representative is:
Springer Nature Customer Service Center GmbH
Europaplatz 3, 69115 Heidelberg, Germany

Printed by Libri Plureos GmbH
in Hamburg, Germany